Lecture Notes in Computer Science 15622

Founding Editors

Gerhard Goos
Juris Hartmanis

AF167705

The series Lecture Notes in Computer Science (LNCS), including its subseries Lecture Notes in Artificial Intelligence (LNAI) and Lecture Notes in Bioinformatics (LNBI), has established itself as a medium for the publication of new developments in computer science and information technology research, teaching, and education.

LNCS enjoys close cooperation with the computer science R & D community, the series counts many renowned academics among its volume editors and paper authors, and collaborates with prestigious societies. Its mission is to serve this international community by providing an invaluable service, mainly focused on the publication of conference and workshop proceedings and postproceedings. LNCS commenced publication in 1973.

Modesto Castrillón-Santana ·
Carlos M. Travieso-González ·
Oscar Deniz Suarez · David Freire-Obregón ·
Daniel Hernández-Sosa ·
Javier Lorenzo-Navarro · Oliverio J. Santana
Editors

Computer Analysis of Images and Patterns

21st International Conference, CAIP 2025
Las Palmas de Gran Canaria, Spain, September 22–25, 2025
Proceedings, Part II

 Springer

Editors
Modesto Castrillón-Santana ⓘ
University of Las Palmas de Gran Canaria
Las Palmas de Gran Canaria, Spain

Carlos M. Travieso-González ⓘ
University de Las Palmas de Gran Canaria
Las Palmas de Gran Canaria, Spain

Oscar Deniz Suarez ⓘ
University of Castilla-La Mancha
Ciudad Real, Spain

David Freire-Obregón ⓘ
University of Las Palmas de Gran Canaria
Las Palmas de Gran Canaria, Spain

Daniel Hernández-Sosa ⓘ
University of Las Palmas de Gran Canaria
Las Palmas de Gran Canaria, Spain

Javier Lorenzo-Navarro ⓘ
University of Las Palmas de Gran Canaria
Las Palmas de Gran Canaria, Spain

Oliverio J. Santana ⓘ
University of Las Palmas de Gran Canaria
Las Palmas de Gran Canaria, Spain

ISSN 0302-9743   ISSN 1611-3349 (electronic)
Lecture Notes in Computer Science
ISBN 978-3-032-05059-5   ISBN 978-3-032-05060-1 (eBook)
https://doi.org/10.1007/978-3-032-05060-1

Preface

CAIP 2025 marked the 21st edition of the International Conference on Computer Analysis of Images and Patterns—a well-established biennial series dedicated to cutting-edge research in computer vision, image processing, pattern recognition, and closely related fields. Over the years, CAIP has built a strong legacy, with past conferences hosted in locations such as Limassol, Salerno, Ystad, Valletta, York, Seville, Münster, Vienna, and Paris.

This year's scientific program was designed as a single-track event, fostering a cohesive and engaging atmosphere for discussion and exchange. From 109 submitted papers, each reviewed in a rigorous double-blind process by at least two experts, 65 were selected for oral presentation. These were grouped into the following nine thematic sessions:

SESSION 1: Facial and Video Recognition
SESSION 2: Image Segmentation
SESSION 3: Object Detection and Applications
SESSION 4: 3D Vision and Reconstruction
SESSION 5: Biomedical Imaging and Diagnostics
SESSION 6: Model Robustness and Generalization
SESSION 7: Multimodal and Vision-Language Models
SESSION 8: Robotics, Interaction and Intelligent Systems
SESSION 9: Emerging Methods and Vision Applications

In addition to the paper sessions, CAIP 2025 featured the Pedestrian Attributes Recognition (PAR) Contest, focusing on Multi-Task Learning—a timely challenge organized by Antonio Greco (University of Salerno) and Bruno Vento (University of Naples).

We were honored to welcome three distinguished keynote speakers, whose insights bridged academic excellence and societal impact: Anil K. Jain, Jacques Bulchand-Gidumal, and Nadia Bianchi-Berthouze.

We extend our heartfelt thanks to the organizing and technical program committees, and the many reviewers whose careful work ensured the quality and success of this conference. Our sincere appreciation goes to all authors who entrusted us with their research, and to all participants for enriching CAIP 2025 with your presence, ideas, and engagement.

We would like to express our gratitude for the sponsorship of the University of Las Palmas de Gran Canaria, the City Council of Las Palmas de Gran Canaria, the Elder Museum of Science and Technology, and the Gran Canaria Convention Bureau.

A very special thank you goes to AVANTE Canarias, and in particular to Marta Ortega and Raquel Granados, for their exceptional support, attention to detail, and tireless efforts behind the scenes to ensure everything ran smoothly.

September 2025

Modesto Castrillón-Santana
Carlos M. Travieso-González
Oscar Deniz Suarez
David Freire-Obregón
Daniel Hernández-Sosa
Javier Lorenzo-Navarro
Oliverio J. Santana

Organization

Honorary Chair

Nicolai Petkov   University of Groningen, Netherlands

General Chairs

Modesto Castrillón-Santana   University of Las Palmas de Gran Canaria, Spain
Carlos M. Travieso-González   University of Las Palmas de Gran Canaria, Spain

Program Chairs

Oscar Deniz Suarez   University of Castilla-La Mancha, Spain
David Freire-Obregón   University of Las Palmas de Gran Canaria, Spain
Daniel Hernández-Sosa   University of Las Palmas de Gran Canaria, Spain
Javier Lorenzo-Navarro   University of Las Palmas de Gran Canaria, Spain
Oliverio J. Santana   University of Las Palmas de Gran Canaria, Spain

Publicity Committee

Aythami Morales Moreno   Autonomous University of Madrid, Spain
Juan Tapia Farias   Hochschule Darmstadt University of Applied
             Sciences, Germany
Hazım Kemal Ekenel   Istanbul Technical University, Turkey
Enrique Alegre   University of León, Spain
Dorota Kamińska   Łódź University of Technology, Poland
Bruno Vento   University of Napoli, Italy
Sondos Mohamed   University of Cagliari, Italy

Steering Committee

Andreas Lanitis (Co-chair CAIP 2023)
Constantinos S. Pattichis (Co-chair CAIP 2021)
Mario Vento (Chair CAIP 2019)

Michael Felsberg (Chair CAIP 2017)
Nicolai Petkov (Permanent Member)

Program Committee

Ahmad Alsahaf	University Medical Center Groningen, Netherlands
Alain Tremeau	Jean Monnet University, France
Albert Ali Salah	Utrecht University, Netherlands
Alberto Marchisio	New York University Abu Dhabi, UAE
Alessia Saggese	University of Salerno, Italy
André P. Kelm	University of Hamburg, Germany
Andreas Lanitis	Cyprus University of Technology, Cyprus
Antonio Greco	University of Salerno, Italy
Aythami Morales	Universidad Autónoma de Madrid, Spain
Bastian Leibe	RWTH Aachen University, Germany
Bhavin Jawade	University at Buffalo, USA
Carmen Bisogni	Università degli Studi di Salerno, Italy
Christian Rathgeb	Hochschule Darmstadt, Germany
Cosimo Distante	National Research Council of Italy - Institute of Applied Sciences & Intelligent Systems, Italy
Cristina Carmona-Duarte	University of Las Palmas de Gran Canaria, Spain
Cristobal Curio	Reutlingen University, Germany
Daniel Cores	University of Santiago de Compostela, Spain
Daniel Riccio	University of Naples Federico II, Italy
Daniel Santana-Cedrés	University of Las Palmas de Gran Canaria, Spain
Delia A. Mitrea	Technical University of Cluj-Napoca, Romania
Di Huang	Beihang University, China
Dibio Leandro Borges	University of Brasília, Brazil
Elena Lazkano	University of the Basque Country, Spain
Eleni A. Dimitriadou	Cyprus University of Technology & CYENS, Cyprus
Fabio Narducci	University of Salerno, Italy
Fernando Alonso-Fernandez	Halmstad University, Sweden
Fernando C. Monteiro	Polytechnic Institute of Bragança, Portugal
Francesco Longobardi	University of Naples Federico II, Italy
Francisco M. Castro	University of Málaga, Spain
Gennaro Percannella	University of Salerno, Italy
Georg Stemmer	Intel Labs, Germany
George Azzopardi	University of Groningen, Netherlands
Giorgio Fumera	University of Cagliari, Italy

Haiyu Wu	University of Notre Dame, USA
Hakeoung Hannah Lee	University of Texas at Austin, USA
Hatem Abdellatif Rashwan	Rovira i Virgili University, Spain
I-Hsien Ting	National University of Kaohsiung, Taiwan
Inês Domingues	Instituto Superior de Engenharia de Coimbra, Centro Investigação IPO, Portugal
Ioannis Pratikakis	Democritus University of Thrace, Greece
Jan Flusser	UTIA, Czech Academy of Sciences, Czech Republic
Javier Sánchez	University of Las Palmas de Gran Canaria, Spain
Jing-Hao Xue	University College London, UK
Jonay Suárez-Ramírez	University of Las Palmas de Gran Canaria — Qualitas Artificial Intelligence and Science, Spain
Jose Alba-Castro	Vigo University, Spain
José M. Buenaposada	Rey Juan Carlos University, Spain
José María Martínez-Otzeta	University of the Basque Country, Spain
José Ignacio Salas-Cáceres	University of Las Palmas de Gran Canaria, Spain
Juan E. Tapia	Hochschule Darmstadt, Germany
Kalman Palagyi	University of Szeged, Hungary
Kerstin Bunte	University of Groningen, Netherlands
Luigi Di Biasi	University of Salerno, Italy
Luis Baumela	Technical University of Madrid, Spain
Manuel Mucientes	University of Santiago de Compostela, Spain
Manuele Bicego	University of Verona, Italy
Marco Huber	Fraunhofer IGD, Germany
Marco La Cascia	Università degli Studi di Palermo, Italy
Maria De Marsico	Sapienza University of Rome, Italy
Mark S. Nixon	University of Southampton, UK
Martin Kampel	Vienna University of Technology, Austria
Mayer Aladjem	Ben-Gurion University of the Negev, Israel
Michal Haindl	Czech Technical University in Prague, Czech Republic
Mounim A. El Yacoubi	Télécom SudParis, France
Nelson Monzón	University of Las Palmas de Gran Canaria, Spain
Paula Ruiz-Barroso	University of Málaga, Spain
Rachael Jack	University of Glasgow, UK
Rikke Gade	Aalborg University, Denmark
Roberto Casula	University of Cagliari, Italy
Sanaz Nikghadam-Hojjati	NOVA University Lisbon, Portugal
Sicong Chen	Syracuse University, USA
Sondos Mohamed	University of Cagliari, Italy

Contents – Part II

Model Robustness and Generalization

Multimodal and Vision-Language Models

Robotics, Interaction and Intelligent Systems

Emerging Methods and Vision Applications

Contents – Part I

Object Detection and Applications

3D Vision and Reconstruction

Biomedical Imaging and Diagnostics

MDTACNet: MobileNet-DenseNet and Transformer Attention Hybrid Network for Multi-crop and Multi-disease Classification

Anand Kumar Jain[✉] and Neeta Nain

Department of CSE, MNIT, Jaipur, India
{2023rcp9026,nnain.cse}@mnit.ac.in

Abstract. In this study, we propose a novel deep learning framework combining MobileNetV2 and DenseNet121 architectures with a Transformer attention mechanism for multi-crop and multi-disease classification using the CCMT dataset. The MobileNet model, known for its lightweight design and efficient feature extraction, is concatenated with the DenseNet model, which excels in preserving feature continuity and improving gradient flow. To enhance feature refinement and boost performance, a transformer-based attention module is integrated, enabling the model to effectively capture intricate patterns and contextual dependencies across crops and diseases. The CCMT dataset, characterized by its diverse range of crop types and complex disease symptoms, presents significant challenges that are effectively addressed by the proposed hybrid model. The model is trained and evaluated on key performance metrics such as training accuracy 99.30%, test accuracy 92.04%, precision 92%, recall 93%, and F_1-score 93%. Experimental results demonstrate that the proposed MDTACNet model outperforms traditional models, achieving superior classification accuracy and robustness. This research contributes to the advancement of intelligent agricultural systems, providing a reliable solution for early detection and effective management of crop diseases.

Keywords: Multi-Crops · Cashew · Cassava · Maize · Tomato · CCMT dataset · Multi-Disease · MDTACNet

1 Introduction

Agricultural productivity is vital for food security, but crop diseases threaten yields and sustainability. Traditional identification methods are labor-intensive and error-prone, making deep learning a promising alternative. CNNs like MobileNet and DenseNet excel in feature extraction, with MobileNet offering efficiency and DenseNet enhancing information flow. Transformer-based attention further refines feature representation by focusing on critical image [1] regions. This study proposes MDTACNet, a hybrid model combining MobileNet,

M. Castrillón-Santana et al. (Eds.): CAIP 2025, LNCS 15622, pp. 3–13, 2026.
https://doi.org/10.1007/978-3-032-05060-1_1

DenseNet, and Transformer Attention for multi-crop, multi-disease classification using the diverse CCMT dataset. MDTACNet is a hybrid architecture that integrates MobileNet (M) for lightweight feature extraction, DenseNet (D) for deep feature propagation, Transformer Attention (TA) to capture global contextual relationships, and concatenation (C) to combine both models. This combination leverages the strengths of each component, enhancing the model's ability to detect subtle and complex disease patterns across diverse crops. The model enhances accuracy, robustness, and generalization, supporting precision agriculture. The proposed model contributed to improving classification accuracy around 3% to 4%, making multi-crop multi-disease detection possible with good results in all parameters.

2   Literature Review

Table 1. State-of-the-art techniques multi disease detection in crops.

S.No	Study	Year	Plant	Methods	Accuracy (%)
1	Durmus et al. [2]	2017	Tomato	SqueezeNet	94.00
2.	J. Shiji et al. [4]	2017	Tomato	VGG16	88.00
3	Zhang et al. [3]	2018	Maize	GoogLeNet	98.80
4	Zhang et al. [5]	2018	Tomato	AlexNet, GoogLeNet, ResNet	97.28
5	Elhassouny et al. [6]	2019	Tomato	MobileNet	90.30
6.	S. Rallapalliet al. [8]	2021	Fruit	M-net	71.00
7.	R. Deng et al. [9]	2021	Rice	ResNet-197	99.58
8	Umesh et al. [7]	2022	Cassava	ECNN	99.30
9	Timothy et al. [10]	2022	Cashew	ResNet50	97.76
10.	Mizan et al. [11]	2024	Rice	EfficientNet-B3	92.00
11	Jain [12]		2024	Sugarcane	MobileNet-DenseNet Concate	96.00

Recent advancements in deep learning have significantly improved crop disease detection, leveraging convolutional neural networks (CNNs) such as VGG16, ResNet, MobileNet, DenseNet, and Inception. These models have demonstrated remarkable accuracy in classification tasks. However, their high computational complexity presents challenges for real-time deployment in resource-constrained agricultural settings. Durmus et al. [2] conducted a study on the Tomato dataset using the SqueezeNet model, achieving an accuracy of 94%. Another study by J. Shiji [4] employed the VGG16 model on the same dataset, reporting an accuracy of 88%. Zhang et al. [5] further investigated tomato disease classification using AlexNet, GoogLeNet, and ResNet, obtaining a high accuracy of 97.28%. Additionally, Zhang's [3] research on maize crop disease classification demonstrated an impressive 98.80% accuracy using the GoogLeNet model. Elhassouny [6] utilized the MobileNet architecture, achieving 90.30% accuracy in tomato disease detection by balancing speed and precision. Rallapalli [8] proposed a lightweight M-Net model with fewer layers than AlexNet, reaching 71% accuracy on a rice disease dataset. Similarly, R. Deng's study [9] on rice disease classification employed the ResNet-197 model, achieving a near-perfect accuracy of 99.58%. Studies on other crops also report promising results. Umesh et al. [7] achieved 99.30% accuracy on cassava disease classification using an Enhanced Convolutional Neural Network (ECNN). Timothy's research [10] on cashew disease classification employed the ResNet50 model, obtaining 97.76%accuracy. Mizan [11] investigated rice disease detection using the EfficientNet-B3 model. However, the reported accuracy was 92.00%. Jain [12] study on sugarcane disease detection utilized a MobileNet-DenseNet concate model, achieving an accuracy of 96%. Table 1 summarizes the state-of-the-art techniques. This study introduces MDTACNet, a novel hybrid deep learning model integrating MobileNet, DenseNet, and transformer-based attention mechanisms for multi-crop, multi-disease classification. While MobileNet ensures efficient feature extraction, DenseNet enhances learning through dense connections, and transformers refine spatial features. Addressing the gap in combining lightweight CNNs with attention mechanisms, MDTACNet aims to optimize computational efficiency while achieving high classification accuracy, making it ideal for precision agriculture.

3 Proposed Work

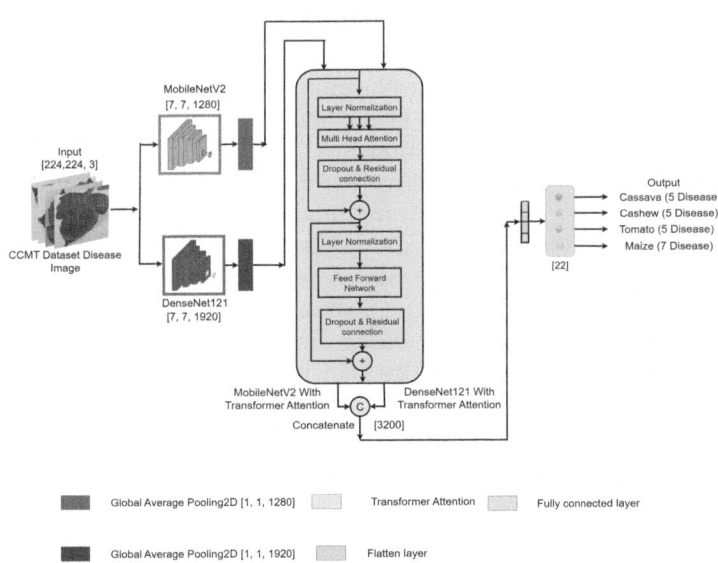

Fig. 1. Proposed MDTACNet model architecture diagram.

The proposed MDTACNet model, as shown in Fig. 1 integrates MobileNetV2 and DenseNet121 through a concatenation-based approach, leveraging their strengths for enhanced feature extraction. An input image (224 × 224 × 3) is processed through both models. Each model's output passes through a separate GlobalAveragePooling2D (MobileNetV2: 1280, DenseNet121: 1920), and transformer attention mechanism, which dynamically refines feature representation by focusing on key image regions. The outputs of both models are concatenated and given a weight (3200). A final fully connected layers classify the extracted features using weights (3200), optimizing both efficiency and accuracy for image classification tasks.

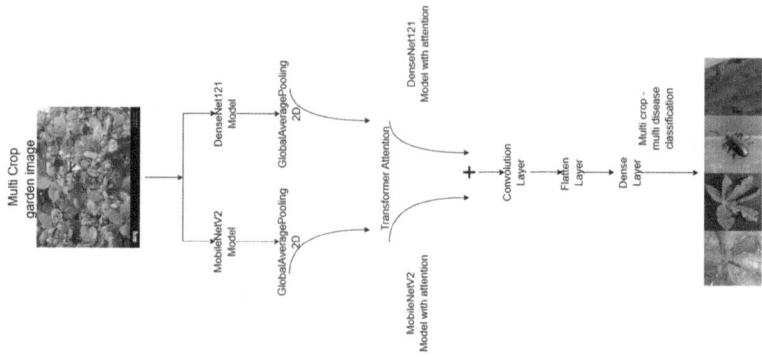

Fig. 2. Proposed MDTACNet model architecture flow diagram.

MobileNetV2 extracts features efficiently using depthwise separable convolutions and inverted residual blocks, beginning with a 3 × 3 convolution and utilizing expansion, depthwise, and projection layers for minimal computation. DenseNet121 employs dense connections and transition layers, using 1 × 1 and 3 × 3 convolutions to maximize feature reuse while maintaining efficiency. The transformer mechanism with 128 feature dimensions and 4 attention heads captures long-range dependencies through self-attention, enabling effective parallel processing. It enhances feature extraction by focusing on important regions, improving classification accuracy, and making the model robust to variations like lighting, occlusion, and complex backgrounds. Before concatenating the outputs of MobileNetV2 and DenseNet121, they pass through transformer attention as shown in Fig. 2. The process can be mathematically described based on the operations performed. Assuming that both networks generate feature maps after processing an input image, the concatenation and transformer attention can be represented as follows:

1. Input image:
$$X \in \mathbb{R}^{224 \times 224 \times 3}$$

2. Feature Extraction: MobileNetV2 and DenseNet121 feature vectors:
$$F_M = \text{GAP}(\text{MobileNetV2}(X)) \quad \in \mathbb{R}^d$$
$$F_D = \text{GAP}(\text{DenseNet121}(X)) \quad \in \mathbb{R}^d$$

3. Transformer Block:
   Project input to Q, K, V:

$$Q = FW_Q, \quad K = FW_K, \quad V = FW_V$$

   Attention output:

$$A = \text{softmax}\left(\frac{QK^T}{\sqrt{d_k}}\right)V$$

   Feed-forward output:

$$\text{FFN}(X) = \text{ReLU}(XW_1 + b_1)W_2 + b_2$$

   Transformer output:

$$T = \text{Add}(\text{Dropout}(\text{FFN}(\text{LayerNorm}(A))), A)$$

   where T: Final transformer block output.
   d: The feature dimension 128.
   Query Q: A projection of the input that represents what we are looking for.
   Key K: A projection of the input that represents what we have.
   Value V: A projection of the input that contains the information to be passed
   along if its corresponding key matches the query.
4. Concatenation and Classification: T_M and T_D are output of both
   mobileNetV2 and DenseNet121 after pass through transformer attention.
   Concatenated vector:

$$F = \text{Concat}(T_M, T_D) \in \mathbb{R}^{2d}$$

   Classification:

$$\hat{y} = \text{softmax}(FW + b)$$

5. Loss:

$$\mathcal{L} = -\log(\hat{y}_{y_{\text{true}}})$$

   In the transformer-attention mechanism, there were 4 heads and a dimension
   of 128.

4   Result and Analysis

This study utilizes the CCMT disease dataset [13], comprising $38,489$ labeled
images of cassava, cashew, maize, and tomato diseases. The crop wise distribution is shown in Table 2. The dataset poses real-world challenges due to
crop diversity, environmental variation, and differing disease symptoms. Figure 3
shows a few samples from the dataset. To address class imbalance and improve
generalization, data augmentation techniques such as rotation, flipping, zoom-
ing, and brightness adjustment are applied. Images are organized into train and
test splits to support supervised learning. The dataset's complexity and variability make it a valuable benchmark for evaluating the MDTACNet model's
robustness in multi-crop plant disease detection.

Fig. 3. CCMT disease dataset [13] samples.

Table 2. Crop wise class and image distribution of CCMT dataset [13].

S.No	Crops	Class	No. of images
1.	Cashew	5	7900
2.	Cassava	5	11139
3.	Maize	7	9380
4.	Tomato	5	10110

4.1 Experiment Results

Feature maps and Grad-CAM heatmaps play a crucial role in improving CNN interpretability and performance for plant disease classification. Feature maps help determine whether the model captures relevant disease patterns, while heatmaps confirm that the network focuses on critical regions of the input image. Inconsistent activation may indicate dataset issues, whereas poor heatmap localization can highlight model limitations. Advanced techniques like multi-scale feature extraction, attention mechanisms, and Transformer-based models enhance feature localization and reduce false positives. As shown in Fig. 4, activation maps reveal the model's focus areas are very precise, aiding in the evaluation of disease-specific feature learning. Whole datasets were divided into 80% training set, 20% test set, and all the experiments were conducted on this ratio. GPU100 is used for model training as a hardware device. Experiment were conducted with a batch size of 32, Epochs 50, momentum 0.9, learning rate 0.0001, and TensorFlow libraries were used.

The CCMT leaf disease precision Table 3 shows good precision in most of the classes, while a few classes show average precision. Our proposed model when compared with the state-of-the-art methods like Babu et al. [14] with accuracy 90.85%, Wang et al. [15] with accuracy 95.35%, Karthik et al. [16] with accuracy 95.68%, Palaniappan et al. [18] with accuracy 81.7%, Sakira et al. [17] with accuracy 98.09%, Jain [19] with accuracy 99.40%, and Jain [20] with

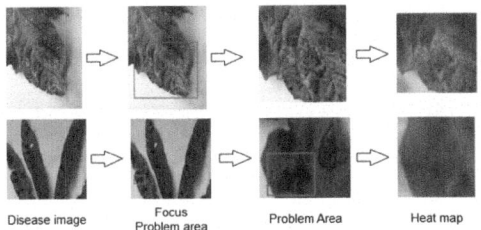

Disease image   Focus Problem area   Problem Area   Heat map

Fig. 4. Disease visualization activation maps.

Table 3. Model precision, recall, and F_1-score comparison.

Diseases	Precision (%)	Recall (%)	F_1-Score (%)	# Test images	# Training images
Cassava Healthy	97	99	98	516	1095
Maize Healthy	65	90	75	96	374
Tomato Healthy	92	98	95	168	668
Anthracnose	64	79	70	264	1050
Bacterial Blight	87	93	90	576	3830
Brown Spot	87	88	87	288	1114
Fall Armyworm	98	94	96	252	492
Grasshoper	94	91	92	384	1243
Green Mite	97	86	91	420	1602
Gumosis	98	100	99	180	538
Cashew Healthy	85	86	85	240	1245
Leaf Beetle	87	98	92	432	1437
Maize Leaf Blight	80	94	86	432	2076
Leaf Curl	86	75	80	240	778
Leaf Miner	82	82	82	288	1114
Leaf Spot	74	31	43	312	1260
Mosaic	94	93	93	276	1098
Red Rust	100	85	92	480	1823
Septoria Leaf Spot	67	86	75	528	3913
Streak Virus	89	94	92	384	1343
Tomato Leaf Blight	62	31	41	480	1848
Verticulium Wilt	54	81	65	264	1048

accuracy 98.20%. In this comparison, our proposed model MDTACNet performs better compared to other models, with 92.04% test accuracy and 99.30% training accuracy as shown in Table 4.

4.2   Ablation Study

This section examines how individual components influence the overall performance of the proposed model. The goal is to identify the optimal combination of

Table 4. Model accuracy comparison on CCMT dataset [13] with the state-of-the-art.

S.No	Model	Datasets	Accuracy (%)
1.	Babu et al. [14]	CCMT(Cashew, cassava, Maize)	90.85
2.	Wang et al. [15]	CCMT(All)	95.35
3.	Karthik et al. [16]	CCMT(All)	95.68
4	Sakira et al. [17]	CCMT(Cassava)	98.09
5.	Palaniappan et al. [18]	CCMT(Cashew)	81.70
6.	Jain [19]	CCMT(Cassava)	99.40
7.	Jain [20]	CCMT(Cashew)	98.20
8.	Proposed Model	CCMT(All)	99.30
	MDTACNet		

components that maximizes effectiveness. Insights from this analysis provide a deeper understanding of the model's behavior and guide informed enhancements. The evaluation of the proposed network is based on key performance metrics, including accuracy, precision, recall, and F_1-score (Table 5).

Table 5. Model accuracy comparison on CCMT dataset [13] with the state-of-the-art.

S.No	Model	Training ACC. (%)	Test ACC. (%)	# Parameters (Millions)	Epoch time (Minutes)	Time per image (seconds)
1.	MobileNetV2	89.20	71.51	2.4	0.45	0.05
2.	DenseNet121	90.52	73.87	7.1	1.10	0.09
3.	MobileNetV2 (+TA)	98.43	87.36	28.8	1.10	0.09
4.	DenseNet121 (+TA)	98.58	89.22	24.1	2.20	0.17
5	Model (-TA)	98.14	84.88	27.8	3.11	0.24
6.	Proposed model (+TA)	99.30	92.04	52.9	4.15	0.26

The accuracy and loss graphs of the individual MobileNetV2 and DenseNet121 models with transformer attention show that their maximum accuracy reached only 98.43% and 98.58% respectively, as shown in (a) and (b) parts of Fig. 5, with a slight increase in loss. However, after combining both models, the concatenated model without attention transformer achieve 98.14% and with transformer attention proposed model achieved a higher accuracy of (99.30%) and significantly reduced loss as shown in (C) and (d) parts of Fig. 5, the accuracy steadily improves from the initial epochs, peaking at 99.30%, indicating strong feature learning. The loss graph declines smoothly and stabilizes after around 10 epochs, reflecting good convergence. Minor spikes in the loss suggest occasional layer mismatches during optimization. To address this, L_2 regularization with $\lambda = 0.1$ is applied to the fully connected layer, improving stability and reducing over-fitting.

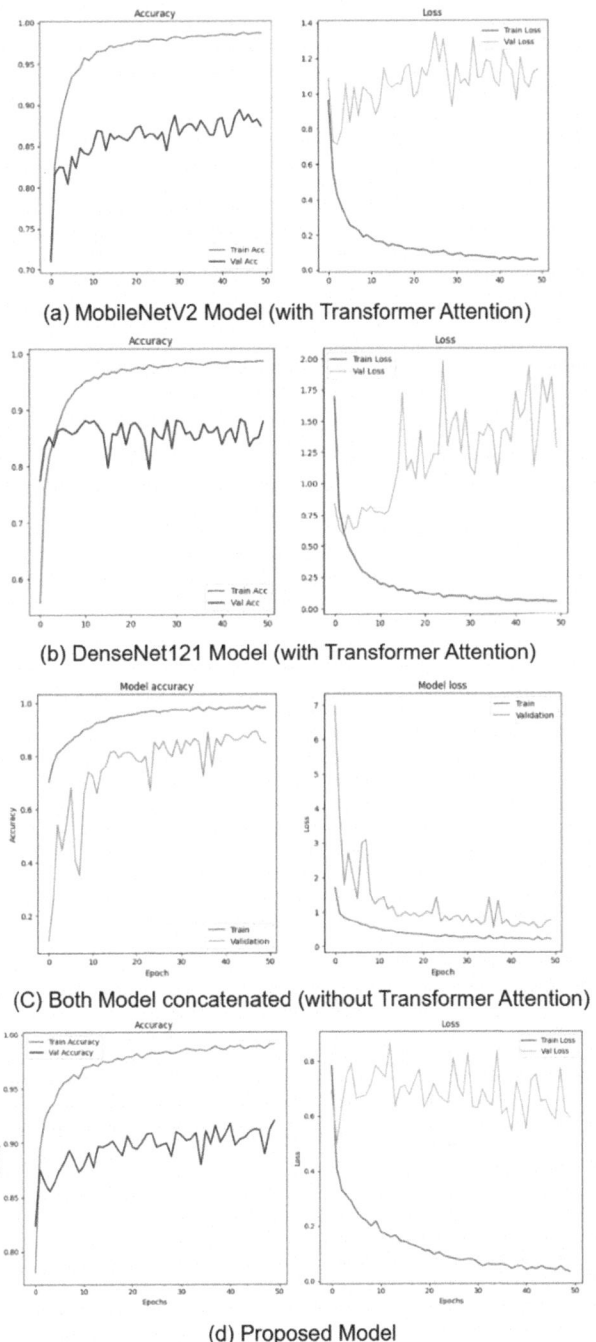

(a) MobileNetV2 Model (with Transformer Attention)

(b) DenseNet121 Model (with Transformer Attention)

(C) Both Model concatenated (without Transformer Attention)

(d) Proposed Model

Fig. 5. Accuracy and loss graph comparison on constituent components of the proposed model.

5   Conclusion

In this study, we proposed a novel deep learning framework, MDTACNet, integrating MobileNetV2, DenseNet121, and Transformer attention mechanisms for multi-crop and multi-disease classification using the CCMT dataset with parameters reduced by various techniques such as replacing dense layers. The combination of MobileNet's lightweight feature extraction, DenseNet's efficient feature propagation, and Transformer attention enabled the model to capture complex patterns and contextual dependencies. Our results showed training accuracy 99.30%, test accuracy 92.04%, precision 92%, recall 93%, and F_1-score 93%. We conclude that MDTACNet outperformed traditional CNN-based architectures, effectively handling diverse visual characteristics of multiple crops like cassava, cashew, maize, and tomato in challenging conditions like varying lighting and occlusions. Transformer attention improved the model's focus on relevant image regions, enhancing generalization and robustness. Data augmentation boosted performance, ensuring adaptability for real-world agricultural applications. As future work, we propose to optimize the number of parameters used.

References

1. Deb, D., Nain, N., Jain, A.K.: Longitudinal study of child face recognition. In: 2018 International Conference on Biometrics (ICB), pp. 225–232 (2018)
2. Durmuş, H., Güneş, E.O., Kırcı, M.: Disease detection on the leaves of the tomato plants by using deep learning. In: 2017 6th International Conference on Agro-Geoinformatics, pp. 1–5 (2017)
3. Zhang, X., Qiao, Y., Meng, F., Fan, C., Zhang, M.: Identification of maize leaf diseases using improved deep convolutional neural networks. IEEE Access **6**, 30370–30377 (2018)
4. Shijie, J., Peiyi, J., Siping, H., et al.: Automatic detection of tomato diseases and pests based on leaf images. In: 2017 Chinese Automation Congress (CAC), pp. 2537–2510. IEEE (2017)
5. Zhang, K., Qiufeng, W., Liu, A., Meng, X.: Can deep learning identify tomato leaf disease? Adv. Multimedia **2018**(1), 6710865 (2018)
6. Elhassouny, A., Smarandache, F.: Smart mobile application to recognize tomato leaf diseases using convolutional neural networks. In: 2019 International Conference of Computer Science and Renewable Energies (ICCSRE), pp. 1–4 (2019)
7. Lilhore, U.K., et al.: Enhanced convolutional neural network model for cassava leaf disease identification and classification. Mathematics **10**(4), 580 (2022)
8. Rallapalli, S., Saleem Durai, M.A.: A contemporary approach for disease identification in rice leaf. Int. J. Syst. Assur. Eng. Manag. 1–11 (2021)
9. Deng, R., et al.: Automatic diagnosis of rice diseases using deep learning. Front. Plant Sci. **12**, 701038 (2021)
10. Timothy, M., John, O., Aibinu, A., Adebisi, B.: Detection and classification system for cashew plant diseases using convolutional neural network. In: Proceedings of the 5th International Conference on Future Networks and Distributed Systems, pp. 225–232 (2021)

11. Mizan, M.A.I., Ahmed, M., Ali, M.H.: Bangladeshi crop disease detection using convolutional neural network. In: 2024 6th International Conference on Electrical Engineering and Information & Communication Technology (ICEEICT), pp. 1304–1309 (2024)
12. Jain, A.K., Sharma, S., Nain, N.: Multi-disease detection and classification of sugarcane leaves using transfer learning. In: 2024 11th International Conference on Soft Computing & Machine Intelligence (ISCMI), pp. 308–312. IEEE (2024)
13. Mensah, P., et al.: CCMT: dataset for crop pest and disease detection. Data Brief **49**, 109306 (2023)
14. Ramesh Babu, P., Srikrishna, A., Gera, V.R.: Diagnosis of tomato leaf disease using OTSU multi-threshold image segmentation-based chimp optimization algorithm and LeNet-5 classifier. J. Plant Dis. Prot. 1–16 (2024)
15. Wang, Y., et al.: Classification of plant leaf disease recognition based on self-supervised learning. Agronomy **14**(3), 500 (2024)
16. Karthik, R., Ajay, A., Bisht, A.S., Illakiya, T., Suganthi, K.: A deep learning approach for crop disease and pest classification using swin transformer and dual-attention multi-scale fusion network. IEEE Access (2024)
17. Sakira, T., Fajar, M., Kaswar, A., Gunawan, S.: Application of the transfer learning method to detect diseases in cassava. In: AIP Conference Proceedings, vol. 3140 (2024)
18. Palaniappan, S., Pazhamalai, K.: ELDA: enhanced linear discriminant analysis for cashew crop disease detection using precision agriculture. J. Electron. Imaging **33**(2), 023030 (2024)
19. Jain, A.K., Nain, N.: Detection and classification of cassava diseases using concatenate model. In: 13th International Conference on Soft Computing for Problem Solving - SocProS 2025, pp. 1–6. Springer, IIT Roorkee (2025)
20. Jain, A.K., Nain, N.: Cashew diseases classification using concatenate of models. In: European Conference on Computer Vision, Institute for Scientific and Engineering Research, Paris, France, pp. 1–6 (2025)

Ensuring the Origin of Cytological Whole Slide Images Through Preparation and Scanner Detection

Paul Barthe[1,2]([📧]) [ID], Romain Brixtel[1] [ID], Mathieu Fontaine[1] [ID],
Arnaud Renouf[1] [ID], Sébastien Bougleux[2] [ID], and Olivier Lézoray[2] [ID]

[1] Datexim, Caen, France
{paul.barthe,romain.brixtel,mathieu.fontaine,arnaud.renouf}@datexim.ai
[2] Normandie Univ, UNICAEN, ENSICAEN, CNRS, GREYC, Caen, France
{sebastien.bougleux,olivier.lezoray}@unicaen.fr

Abstract. Domain shift is one of the main obstacles to the safe deployment of machine learning (ML) models. In cytology, whole slide images (WSIs) can have various characteristics due to the diversity of preparation protocols and scanners. To prevent domain shift, we propose a generic model that identifies the origin of a WSI, by detecting the preparation and the scanner used to produce it. Depending on the task, preparation or scanner detection, different WSI representations are suitable. We introduce a two-branch architecture to handle these representations. For scanner detection, we propose a method inspired by forensic research on device identification, and demonstrate its effectiveness with WSIs.

Keywords: Automated quality control · Whole slide image analysis · Cytology · Domain shift detection · Deep learning

1 Introduction

The digitization revolution allows digital pathology (DP) to benefit from computer-aided diagnosis systems. The digital representation of glass slides, called whole slide images (WSIs), is nowadays used conjointly with machine learning (ML), and more particularly with deep neural networks (DNNs), to enhance diagnostic accuracy and efficiency of pathology experts. For cytology, a branch of DP focusing on the analysis of cells in biological samples, a biofluid sample is turned into a WSI through two main stages: sample preparation and digitization. For preparation, sample cells are collected, distributed on a glass slide, and stained. Then, a scanner is used to digitize it to a WSI, which can be processed by ML systems. However, the performance of such systems is hindered by domain shift [14], especially when processing WSIs, as variations in features can be due to various factors. In cytology, feature variations can be the result

---

P. Barthe—Work partly supported by the ANRT with Cifre number 2023/0122.

M. Castrillón-Santana et al. (Eds.): CAIP 2025, LNCS 15622, pp. 14–24, 2026.
https://doi.org/10.1007/978-3-032-05060-1_2

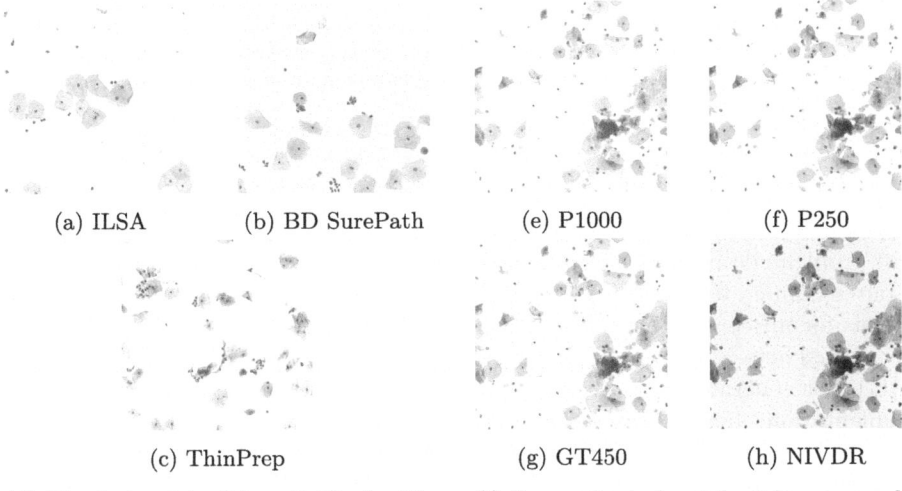

(a) ILSA   (b) BD SurePath   (e) P1000   (f) P250

(c) ThinPrep   (g) GT450   (h) NIVDR

(d) Physical patch slides, digitized with the same scanner, and prepared with different preparation protocols.

(i) Same physical patch slide, prepared with the same protocol, and digitized with different scanners.

Fig. 1. Variations in slide characteristics.

of sample preparation or digitization. Different automated system manufacturers use unique settings and methods, resulting in different staining characteristics, see Fig. 1(d). Additionally, each scanner has its own hardware and software components, leading to significant variations, see Fig. 1(i). While human vision is robust to such feature deviations, ML and DNN models are sensitive to them [6], making it essential to verify the origin of WSIs for optimal performance and safe deployment. Although a WSI's origin can be identified through the associated meta-data, these reflect the method rather than the outcome, making it essential to inspect WSI contents to prevent domain shift. However, manual review of WSI is time-consuming, and prone to interpretation, highlighting the need for automated and reproducible approaches to evaluate WSI quality [3].

We propose a two-branch architecture for detecting WSI origin via preparation and scanner, with key contributions summarized as follows:

- Proposing a generic two-branch architecture for detecting WSI origin, applicable to various DP fields, including histology and cytology.
- Showing the need for distinct WSI representations, depending on the branch.
- Demonstrating the effectiveness of WSI scanner detection through an image residual noise approach.

In Sect. 2, we explore existing research on domain shift and device identification. Section 3 details the method pursued to create our two-branch detection network. Finally, we present experiments and results in Sect. 4, and discuss them in Sect. 5 before concluding.

2   Related Works

This section introduces strategies to manage domain shift in histology and cytology: domain adaptation, generalization, and quantification. Cytology research remains limited due to sample thickness and staining variability. We also examine work on device identification.

Domain Shift. Two main approaches are usually considered to tackle the domain shift problem: domain adaptation and generalization. Domain adaptation adjusts a source domain to a target, while domain generalization builds domain-agnostic models for unseen targets without explicit adaptation. For domain adaptation, adversarial networks are used to remove domain-specific knowledge [8] or transfer it [17]. For generalization, the domain-agnostic model can be created using data augmentation, data normalization or a combination of both [16].

Domain shift is often detected by catching out-of-distribution (OOD) samples, using model uncertainty analysis or data description methods that characterize the training set for comparison. While such OOD methods are explored in various DP fields [10], they remain, to the best of our knowledge, poorly explored in cytology. To quantify domain shift, measures of discrepancy between domain distributions are commonly considered, e.g., in histology [15]. As multiple-instance learning (MIL) is common with WSI, the Fréchet Domain Distance [13] has been introduced to quantify domain shift in MIL. For cytology, a nuclei-based measure was proposed in [1] to quantify similarity between sets of WSIs.

Device Identification. The standard approach for device identification in digital forensics relies on the extraction of the photo-response non-uniformity (PRNU) noise caused by pixel non-uniformity [12]. The idea is to extract noise residuals that approximate a PRNU pattern, by image filtering. First works determined a reference noise pattern using wavelet-based denoising [12], later enhanced with alternative filters or correlation scores [7]. More recently, deep learning techniques have been considered to process residuals and develop learned denoising filters [19], using constrained convolutional kernels to enforce high-pass filtering and extract low-level features [2]. This has been applied to magnetic resonance imaging scanner identification [5], but, to our knowledge, not to WSIs. This suggests a novel avenue for research, where similar methods could be adapted to verify WSI scanner compatibility and configuration, thereby enhancing ML model safety.

3   Method

Our method for WSIs origin detection adopts a two-branch architecture to isolate scanner and preparation specific features, enabling independent pathway optimization, see Fig. 2. In this section, we first outline MIL-based formulation, then describe the training process, and finally detail the scanner identification method, which emerges as the distinguishing factor between the two branches.

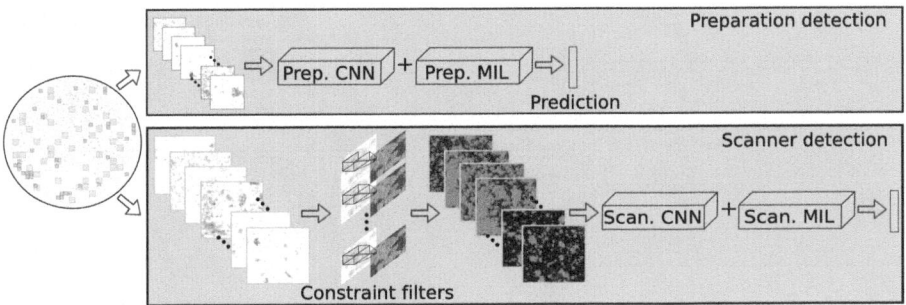

Fig. 2. Overview of the proposed two-branch network. The WSI is represented by patches with branch-specific parameters. For scanner, patches are transformed into image residuals. Each branch makes its prediction via a MIL approach.

3.1 MIL Formulation

Standard supervised learning relies on instance-level annotations, which is challenging for WSIs, where only slide-level labels are commonly available and each WSI is gigapixel-scale. MIL models manage these limitations by receiving a set of labeled sets, denoted X, where each WSI i is represented by a set $X_i = \{x_{i,1}, \ldots, x_{i,K}\}$, containing K instances. Representing a WSI by its instances enables the use of slide-level annotations only and effectively addresses the challenges of large image size. These instances are extracted from the WSI, by dividing it into non-overlapping squared patches of fixed area (in px^2) at a given magnification. Different area sizes and magnifications are explored in Sect. 4.3, as each of these combinations does not capture the same content. We do not consider background patches, as we are interested in the objects of study for cytological ML models that carry cytological information for diagnostic tasks, i.e., cells. For this, we ignore patches with an entropy value below a threshold of 3, which was manually established by visual investigation.

Based on the set representation X_i, MIL models then learn to predict a label $y_i \in \{1, \ldots, N\}$, where N represents the number of possible classes. There exist two main approaches to address this problem:

- Instance-level approach: an instance-level classifier assigns a score to each instance. These scores are subsequently pooled.
- Embedding-level approach: a model first maps instances to a low-dimensional space. Then, a pooling function is used to obtain a WSI representation, which is processed by a WSI-level classifier.

We used the second approach, as it creates a joint representation of the WSI and avoids introducing additional bias into the WSI-level classifier [18].

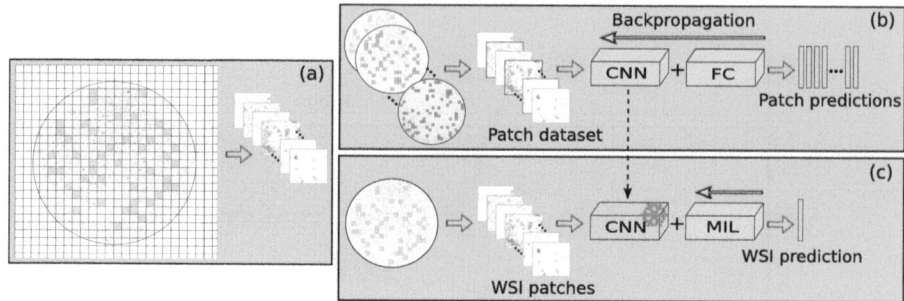

Fig. 3. Training process overview. (a) Patches are extracted from WSIs and are labeled according to the WSI label. (b) Patches are then used to fine-tune a DNN, composed of a CNN and a fully connected layer (FC). (c) Once trained, the CNN is frozen and used to extract instance features for the MIL training.

3.2 Training Process

For each branch, a multiclass classification paradigm is used for training. For the sake of brevity, as our training procedure is similar for both branches, we focus here only on the training process of the preparation branch.

Although many MIL studies used pre-trained networks to extract instance features, we decided to fine-tune an existing one, as we consider that cytological images have characteristics that differ significantly from the common images used in pre-trained networks. For this, we fine-tuned a convolutional neural network (CNN) based DNN, via an instance multiclass classification task, without freezing any layers. As we only have annotation labels for WSIs, we created the instance dataset by giving to each instance of a WSI i, i.e., $x_{i,k}$ $\forall k \in K$, the pseudo-label y_i, which corresponds to the preparation label of i. Training can therefore be processed thanks to these pseudo-labels, as illustrated in Fig. 3.

For WSI classification, the CNN part of the previously trained model is frozen and used to extract features for the K instances of a WSI i. From these K instances, the clustering-constrained attention MIL [11] (CLAM) model predicts N sets of F attention scores, via an attention network. This corresponds to the N classes of the multiclass classification problem and to the F features produced by the CNN. This $N \times F$ matrix is then processed to predict y_i. Note that such an architecture allows the model to manage variable values of K.

3.3 Scanner Identification from Residuals

Inspired by works on device camera identification, we identify the scanner using image residuals rather than raw pixel values, as done for preparation detection. To extract image residuals, we consider C convolution filters of k kernels of size $s \times s$, where $C = 3$ is the number of image channels. We denote by $w_c^k \in \mathbb{R}^{s \times s}$ the kth kernel dedicated to channel c.

The set of residuals from a patch P is then defined as:

$$P^r := \left\{ \sum_k w_c^k \otimes P_c, \ c = 1, \ldots, C \right\}, \tag{1}$$

where $\otimes$ denotes 2D convolution. Low-level features are extracted using the constraints defined in [2]:

$$\begin{cases} w_c^k(0,0) = -1, \\ \sum_{(m,n) \neq (0,0)} w_c^k(m,n) = 1 \end{cases} \tag{2}$$

with $(0,0)$ the central value of the kernel. As the values are not individually constrained, very large positive and negative values can appear in some learned filters [2], which therefore dominate other filters. To avoid this, corrected functions are used [5] to limit the magnitude of all filter values to $[0,1]$. However, once applied, the kernels no longer respect the constraints defined by Eq. 2. To ensure that the kernels satisfy the constraints, we define a correction function in two steps:

1. Normalize kernel values as:

$$\tilde{w}_c^k(m,n) = \frac{w_c^k(m,n)}{||w_c^k||_\infty} \tag{3}$$

We define this normalization as L_∞.

2. Assign -1 to $\tilde{w}_c^k(0,0)$ and adjust other kernel values to sum to 1:

$$\tilde{w}_c^k(m,n) = \tilde{w}_c^k(m,n) - \frac{\sum_{(m,n) \neq (0,0)} \tilde{w}_c^k(m,n) - 1}{s \times s}, \ \forall (m,n) \neq (0,0) \tag{4}$$

We define this two-step normalization as L_∞^+.

4 Experiments and Results

4.1 Materials

We possess a private dataset composed of 3 473 WSIs from 7 different WSI preparation pipelines (PP), denoted as PP_i, and from 29 different sources, see Table 1(a). Two sources are considered distinct if they originate from different laboratories, or from the same laboratory but with different preparation pipelines or a time gap of over a year, as temporal variations (e.g., scanner recalibration or dye degradation) can cause significant shifts. Slides were prepared with one of the following liquid-based cytology technologies: ILSA (iLsa Diagnostics), Thin-Prep (Hologic), and BD SurePath (BD Diagnostics). Slides were then digitized with one of the following scanners: NanoZoomer S360 IVDR and NanoZoomer S360 RUO (both from Hamamatsu, 0.23 micron/pixel under 40× magnification), Aperio GT 450 (Leica Biosystem, 0.263 micron/pixel under 40× magnification), Pannoramic 1000 or Pannoramic 250 (both from 3DHISTECH, with 0.243 micron/pixel under 40× magnification).

We composed the training set D_{train} by taking, for each PP_i, all WSIs from the source with the highest number of WSIs, resulting in $|D_{\text{train}}| = 2\,073$ WSIs. The validation set D_{val} comprises the remaining 1 400 WSIs, see Table 1(b-c).

Table 1. Description and composition of the datasets.

(a) Dataset description

PP	Prep.	Scan.	WSIs	Sources
PP$_1$	BD	P1000	595	2
PP$_2$	ThinPrep	P1000	850	12
PP$_3$	ILSA	P1000	282	2
PP$_4$	ILSA	P250	550	6
PP$_5$	ILSA	NIVDR	350	3
PP$_6$	ILSA	NRUO	400	2
PP$_7$	ILSA	GT450	446	2

(b) Prep. datasets

Name	D_{train}	D_{val}
BD	395	200
ThinPrep	300	550
ILSA	1378	650

(c) Scan. datasets

Name	D_{train}	D_{val}
P1000	927	800
P250	300	250
NIVDR	300	50
NRUO	300	100
GT450	246	200

4.2 Implementation Details

During instance-level classifier training, parameters are optimized using the Adam optimizer, with a constant learning rate of 10^{-4}. Cross-entropy loss supervised the model during 3 epochs. The AutoAugment policy learned on ImageNet is used for data augmentation (DA). Batch size is adapted to the patch area, as explained in Sect. 4.3. For the preparation, we fine-tuned a ResNet18 model trained on 400K histopathological WSIs [4] on our datatset and trained a ResNet18 model from scratch for the scanner branch. We used a ResNet18 for its strong performance observed in our preliminary experiments and its lightweight architecture, making it well-suited for limited data and computational resources.

During MIL classifier training, parameters are optimized as above but with a constant learning rate of 10^{-3}, during 5 epochs, with a batch size of 1 WSI, and without DA. Each WSI is represented by all its instances.

To manage the imbalanced dataset setting, the sampling probability of each sample $X_i \in X$ of label l, is equal to $p_i = |X|/|X^l|$, where X^l is the set of samples with label l. In all experiments, each patch $P \in \mathbb{R}^{C \times H \times W}$ with C channels and $H \times W$ pixels is scaled as $P'_{c,h,w} = (P_{c,h,w} - \min_c P_{c,h,w})/(\max_c P_{c,h,w} - \min_c P_{c,h,w})$. For the preparation branch, each scaled patch $P'_{c,h,w}$ is then normalized to $\tilde{P}_{c,h,w} = (P'_{c,h,w} - \mu)/\sigma$, where μ and σ are fixed to 0.5, to be consistent with the normalization used to train the fine-tuned model [4]. All experiments were run on 2 NVIDIA GeForce RTX 3090. Each branch has 11.7M parameters, 11.2M for the CNN (parameter count is negligible for constrained filters), and 500K for the MIL model.

4.3 Patch Parameters

Table 2. Accuracy (%) depending on patch area and magnification (magn.). Left table concerns preparation detection while right concerns scanner detection.

		Area (px^2)			
		2048	1024	512	256
	5x	95.5	**98.8**	98.1	96.9
Magn.	10x	97.4	97.3	98.0	90.0
	20x	96.2	93.9	95.9	88.3

		Area (px^2)			
		2048	1024	512	256
	5x	78.4	88.1	85.9	79.8
Magn.	10x	91.9	87.6	**95.9**	79.9
	20x	92.0	83.6	95.5	80.5

Table 3. Accuracy (%) depending on constrained filter parameters. For (a), $k = 1$. For (b), we used $s = 9$ and L_∞^+.

(a) Different normalization and kernel sizes (s).

Norm.	$s = 5$	$s = 7$	$s = 9$
L_∞	91.7	92.7	96.0
L_∞^+	93.1	96.0	**96.5**

(b) Different number of filters (k).

	$k = 1$	$k = 2$	$k = 3$
Accuracy	**96.5**	95.4	93.8

In this section, we describe the grid search conducted to empirically determine the optimal patch area and magnification. As patch parameters affect the number of patches extracted from a WSI, each parameter combination *comb* produced its own datasets D_{train}^{comb} and D_{val}^{comb}. After filtering low-entropy patches, some WSIs may not have any valid instances for representation. For training, we excluded these WSIs from D_{train}^{comb}. For validation, to ensure comparability of results, each model is evaluated on the same set of WSIs, which is the intersection of all D_{val}^{comb}, resulting in 1 375 WSIs. Additionally, for CNN training, since the number of instances produced by the WSIs of each D_{train}^{comb} fluctuates significantly, we limited the number of instances processed per epoch to 20K. This ensures that the experience remains unbiased by these variations. For batch size, we adapted it to the patch area, and used sizes of 8, 32, 64 and 128, for areas of 2 048 px^2, 1 024 px^2, 512 px^2 and 256 px^2, respectively. We do not used constrained filters and report the accuracy scores of each branch in Table 2. Based on these results, we have determined that optimal patch parameters are 1 024 px^2 and 5× for preparation detection, and 512 px^2 and 10× for scanner detection.

4.4 Constrained Filters

To explore the impact of constrained filters on scanner device identification, we compared the results obtained with and without them. Moreover, we tested different kernel sizes s, as well as the impact of L_∞^+ normalization, see Eq. 4. Optimal patch parameters are used, see Sect. 4.3. To be consistent with previous experiments, we limited the number of instances processed per epoch during CNN training to 20K. The results are reported in Table 3(a).

Table 4. Accuracy (%) and weighted-averaged F1 scores obtained with proposed model. We train and infer it 10 times and write results as mean±std.

	Preparation		Scanner		Global	
	Accuracy	F1	Accuracy	F1	Accuracy	F1
Vote	96.9±1.5	96.8±1.5	92.8±4.5	93.0±4.2	90.6±4.3	95.3±2.1
MIL	**97.7±0.9**	**97.8±0.9**	**95.6±3.0**	**95.8±2.8**	**93.8±2.6**	**96.9±1.3**

We then fixed the normalization to L_∞^+ and s to 9 and compared the results obtained with different numbers of kernels per filter. We found that 1 kernel appears to be the most effective for our configuration, as shown in Table 3(b).

4.5   Results

We ensure stable results by training our two-branch network on 10 different seeds. We show the positive contribution of MIL by comparing it to an instance-level vote classification, which assigns the final WSI label to the label most frequently predicted by the instance-level model, described in Fig. 3(b). Unlike previous experiments, all instances are processed at each epoch. For the preparation branch, WSIs are represented with $1\,024$ px^2 patches at $5\times$ magnification, and without constrained filters. For scanner detection, WSIs are represented with 512 px^2 patches obtained at $10\times$ magnification, and with constrained filters composed of $k = 1$ kernel of size $s = 9$. We trained our two-branch model with these parameters, and report the accuracy and weighted-averaged F1 scores for each branch and for the global classification task, see Table 4. For the global classification task, the global prediction vector p_i^g, corresponding to the WSI i, is defined by concatenating the prediction vectors from each branch, resulting in an 8-dimensional two-hot vector. Similarly, the global ground truth y_i^g is obtained by concatenating the preparation and scanner ground truth vectors of the WSI i. Evaluation metrics are computed between p_i^g and y_i^g.

5   Discussion

WSI Representation Matter. We have shown that, depending on the classification task, the way a WSI is represented in MIL has a significant impact on the results. For preparation detection, our method seems resistant to changes in patch parameters. However, the higher the magnification, the weaker the results. As a change in the preparation protocol could impact cells disposition and foreign object in the background, we suppose that it is better to have global rather than local information. For scanner detection, having a higher magnification, and therefore higher image resolution, seems to be better. As a change in the scanner does not change WSI content or structure, relying on higher resolution patches is therefore more important than having global information.

Table 5. Confusion matrices

(a) Preparation detection

		Predicted		
		ILSA	ThinPrep	BD
Actual	ILSA	0.987	0.0078	0.0048
	ThinPrep	0.04	0.956	0.004
	BD	0.008	0	0.992

(b) Scanner detection

		Predicted				
		P1000	P250	NIVDR	NRUO	GT450
Actual	P1000	0.944	0.056	0	0	0
	P250	0.058	0.933	0	0.003	0.006
	NIVDR	0.006	0	0.991	0.001	0.002
	NRUO	0.001	0	0.01	0.988	0.001
	GT450	0	0	0	0	1

Learning from Patch Residuals. The use of constrained filters has led to improved results for the scanner detection task. As shown in Sect. 4.4, kernel sizes and numbers are important parameters in the design of such a system. Moreover, we have shown that the proposed L_∞^+ normalization, which forces the network more strictly to satisfy the constraints, appears to be beneficial. We observe in Table 4 that scanner scores are less stable than preparation scores. Further research includes the analysis of this difference, to determine whether this is due to the difficulty of the task or to the use of constrained filters.

Usually, device identification results are split into manufacturer and model identification. The scanner accuracy score of 95.6% provided in the present work focuses on model identification. At the manufacturer level, we achieved 99.8% average accuracy with a standard deviation of $\pm 0.1\%$ over the 10 seeds. The accuracy gap is illustrated in Table 5(b), which was generated by summing the 10 confusion matrices obtained in Sect. 4.5 and then normalizing it so that each row sums to 1. We observe that errors occur almost always between Pannoramic 1000 and Pannoramic 250, which are scanners of the same manufacturer.

Learning from a Single Domain. An approach to managing domain shifts is to develop a model robust to such shifts. When data from several domains are available, robustness can be achieved by injecting these different domains into the training process [9]. However, incorporating all domains, in our case all the sources, into the training set does not allow us to test the model generalization on new domains. It is to counteract this and be able to test model generalization on unseen sources, that the training set D_{train} is composed of one source. We have shown that the proposed method achieved a good generalization despite this splitting strategy, and ensures its generalization capacity on a much larger number of tested sources than if more than one source were injected in D_{train}.

6   Conclusion

To ensure safe deployment of ML models, we proposed a two-branch network designed to detect both the preparation protocol and the scanner used to produce a cytological WSI. The proposed architecture is generic enough to be applied with WSI of any DP fields. Through additional experiments, we highlight the significance of the two branches, demonstrating that they do not require identical patch area and magnification to perform effectively. Moreover, we introduce the use of image residuals for scanner detection, showing that this approach works effectively for cytological WSIs. This suggests a novel avenue for research, where PRNU-based strategies or similar noise residual methods could be adapted to check WSI origin, potentially ensuring safer ML model deployment.

References

1. Barthe, P., et al.: Assessing the quality of whole slide images in cytology from nuclei features. J. Pathol. Inf. **17**, 100420 (2025)
2. Bayar, B., Stamm, M.C.: Constrained convolutional neural networks: a new approach towards general purpose image manipulation detection. IEEE Trans. Inf. Forensics Secur. **13**(11), 2691–2706 (2018)
3. Brixtel, R., et al.: Whole slide image quality in digital pathology: review and perspectives. IEEE Access **10**, 131005–131035 (2022)
4. Ciga, O., et al.: Self supervised contrastive learning for digital histopathology. Mach. Learn. Appl. **7**, 100198 (2021)
5. Fang, S., et al.: A deep learning approach to MRI scanner manufacturer and model identification. J. Electron. Imaging **32**(4), 217-1–217-1 (2020)
6. Geirhos, R., et al.: Generalisation in humans and deep neural networks. In: NeurIPS, pp. 7549–75, 2020
7. Kang, X., et al.: Enhancing source camera identification performance with a camera reference phase sensor pattern noise. IEEE Trans. Inf. Forensics Secur. **7**(2), 393–402 (2021)
8. Lafarge, M.W., et al.: Domain-adversarial neural networks to address the appearance variability of histopathology images. In: DLMIA ML-CDS, 2017
9. Li, H., et al.: Domain generalization with adversarial feature learning. In: CVPR, pp. 5400–5409, 2018
10. Linmans, J., Raya, G., Van Der Laak, J., Litjens, G.: Diffusion models for out-of-distribution detection in digital pathology. Med. Image Anal. **93**, 103088 (2024)
11. Lu, M.Y., et al.: Data efficient and weakly supervised computational pathology on whole slide images. Nat. Biomed. Eng. **5**(6), 555–570 (2021)
12. Luka, J., et al.: Digital camera identification from sensor pattern noise. IEEE Trans. Inf. Forensics Secur. **1**(2), 205–214 (2006)
13. Pocevičiūtė, M., et al.: Detecting domain shift in multiple instance learning for digital pathology using fréchet domain distance. In: MICCAI, pp. 157–167, 2023
14. Quiñonero-Candela, J., et al.: Dataset Shift in Machine Learning. MIT Press, Cambridge (2008)
15. Stacke, K., Eilertsen, G., Unger, J., Lundstrom, C.: Measuring domain shift for deep learning in histopathology. IEEE J. Biomed. Health Inf. **25**(2), 325–336 (2021)
16. Tellez, D., et al.: Quantifying the effects of data augmentation and stain color normalization in convolutional neural networks for computational pathology. Med. Image Anal. **58**, 101544 (2019)
17. Wang, R., et al.: CytoGAN: unpaired staining transfer by structure preservation for cytopathology image analysis. Comput. Biol. Med. **180**, 108942 (2024)
18. Wang, X., et al.: Revisiting multiple instance neural networks. Pattern Recognit. **74**, 15–24 (2018)
19. Yang, P., et al.: Source camera identification based on content-adaptive fusion network. Pattern Recogn. Lett. 195–204 (2019)

Unsupervised Mineral Segmentation with Graph Neural Networks and Multi-modal SEM Data

Samuel Repka[1,2]([✉]), Tuomas Eerola[1], David Motl[3], Jakub Výravský[3], and Pavel Zemčík[1,2]

[1] LUT University, Yliopistonkatu 34, 53850 Lappeenranta, Finland
{samuel.repka,tuomas.eerola,pavel.zemcik}@lut.fi,
{irepka,zemcik}@fit.vutbr.cz
[2] Brno University of Technology, Božetěchova 1/2, 612 00 Brno-Královo Pole, Czechia
[3] TESCAN GROUP a.s., Libušina třída 21, 623 00 Brno, Czechia
{david.motl,jakub.vyravsky}@tescan.com

Abstract. We propose a novel method for multi-modal mineral segmentation that utilises backscattered electron (BSE) images and sparse Energy-Dispersive X-ray spectroscopy (EDS) measurements from Scanning Electron Microscope (SEM). The method uses Graph Neural Networks for simultaneous data fusion and segmentation. The segmentation is unsupervised, allowing for the separation of mineral phases even if they were not included in the training dataset. The segments are created from graph structure, where each BSE pixel is connected to a set of EDS nodes that correspond to pointwise spectral measurements. This connection (edge in the graph) is perceived as a choice, allowing the network to select an EDS measurement to which the BSE pixel most likely belongs. Each pixel is assigned to an EDS measurement, effectively creating segments; inside of each is exactly one EDS measurement. This allows for unsupervised segmentation applicable to any mineral phase. In our experiments with challenging mineral datasets, we show that the proposed method outperforms state-of-the-art segmentation accuracy while scaling more efficiently with sample size.

Keywords: Graph neural networks · Data fusion · Mineral segmentation

1 Introduction

Scanning Electron Microscope (SEM) is a powerful tool for analysing material microstructure and composition, able to capture various data with diverse properties. Backscattered Electrons (BSE) imaging provides phase contrast by exploiting differences in atomic number, while Energy-Dispersive X-Ray Spectroscopy (EDS) offers detailed elemental composition by detecting characteristic

X-rays. By fusing these modalities, researchers can combine textural and compositional information, enabling more comprehensive material analysis. Accurate mineral segmentation (Fig. 1) is vital for applications like optimising mineral processing, understanding geological processes, tracking contaminants in soils, and linking material properties to phase composition in materials science.

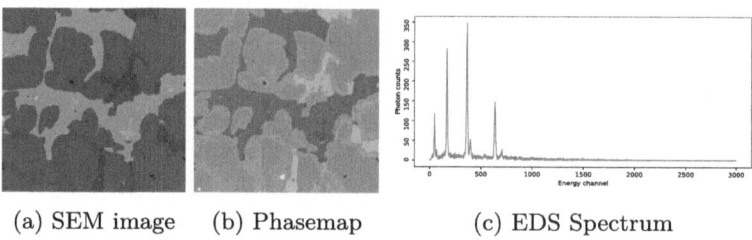

(a) SEM image   (b) Phasemap   (c) EDS Spectrum

Fig. 1. Example of a SEM image, corresponding phase segmentation and a single EDS spectrum.

Traditional phase segmentation approaches rely on either BSE data or dense EDS. The BSE is fast but struggles with minerals of similar atomic numbers. EDS provides precise compositional data for mineral differentiation [4] but its acquisition is significantly slower than BSE, taking over 1 ms per spectrum versus 1 μs per BSE pixel. Therefore, methods relying on dense EDS are accurate but very slow. As an alternative approach, adaptive sampling focuses EDS measurements on key regions but does complicate processing [7, 12]. The main challenge lies in the fundamentally different structures of the modalities. BSE data is grid-like (images), whereas pointwise EDS data has a non-grid-like structure. Constructing a fused representation of these modalities is not straightforward, which limits the fusion approaches that can be utilised.

In this paper, we propose a novel method for SEM image segmentation that fuses high-resolution BSE images with sparse EDS measurements. Building on the Graph Neural Network (GNN)-based supervised segmentation approach proposed in [13], our method extends it to unsupervised segmentation capable of separating minerals not present in the training set. The BSE and EDS data are fused into a unified graph structure, which is processed using a Graph Attention Network (GATv2) [15]. This approach naturally accommodates the fusion of structured image data with unstructured point measurements, eliminating the need for complex preprocessing steps. By learning the relationships between features in BSE images and compositional signatures in EDS data, our method effectively propagates sparse elemental information across regions with similar characteristics, producing accurate segmentation with significantly reduced acquisition times. The final BSE image segmentation is obtained by considering EDS connections between BSE and EDS as possible classes for BSE pixels. This allows for the classification of each BSE pixel into the mineral phases present in the image, regardless of whether that mineral was included in the training set, making the segmentation essentially unsupervised. An overview of the method is shown in Fig. 2.

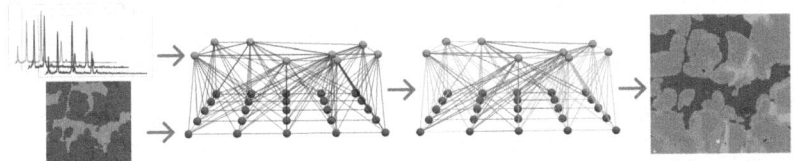

Fig. 2. An overview of the method. EDS and BSE data are put into the green and red nodes, respectively. The graph is processed, and edges are classified as either connected or disconnected. The analysis of edges produces full segmentation. (Color figure online)

The main contributions of this paper are: (a) a generalisation of the GNN-based multimodal segmentation approach from [13] to unsupervised segmentation, (b) a novel method for SEM image segmentation using this approach, and (c) an evaluation of the methods on mineral phase segmentation.

2   Related Work

2.1   Multimodal Segmentation

Data fusion techniques are widely used in image segmentation, with neural networks representing the dominant approach. These methods can be categorised into three main types: (1) early, (2) late, and (3) hybrid fusion approaches [17].

Early fusion approaches integrate data either in raw form or at the feature level [17]. An example of raw-form fusion was presented by Couprie et al. [2], or by Jaritz et al. [6]. Late fusion approaches integrate feature maps from different modalities at the decision level. This requires an initial separate processing of the data without information sharing between feature extractors. A representative example from this category is the work of Gupta et al. [5]. Hybrid fusion methodologies aim to address the limitations inherent in both early and late fusion approaches. In neural network implementations, skip connections typically bridge encoders and decoders to enhance segmentation performance, for example, as in the work proposed by Lee et al. [9].

2.2   Graph Neural Networks

Graphs are versatile data structures suitable for the representation of both structured and unstructured data. To efficiently utilise graph representations in computation, various processing methodologies have been developed, such as GNNs.

GNNs apply differentiable models to all graph components (nodes, edges, and global embeddings), transforming them into a new graph. Data integration between neighbouring nodes is facilitated through *pooling* [14]. For items to be pooled, data are first gathered and then aggregated using functions such as sum or mean. A *message passing layer* combines information from different graph parts. Given the high flexibility and applicability of graphs, the architectures of GNNs are remarkably varied. Variants include Graph Convolutional Network [8], Graph Attention Network [15], and Graph Isomorphism Network [16].

2.3   Mineral Segmentation for SEM Images

Various techniques of image segmentation for automated mineralogical analysis have been developed, such as watershed-based methodologies [12], Mineral Liberation Analyzer [3], and Quantitative Evaluation of Minerals by Scanning Electron Microscopy [10]. Numerous commercial solutions are available, including TESCAN Integrated Mineral Analyzer (TIMA), Zeiss Mineralogic or Maps Min by ThermoFisher. However, the underlying algorithms are frequently proprietary and not publicly disclosed.

Juránek et al. [7] introduced an alternative methodology for mineral segmentation from SEM images. Their approach utilises a joint graph representation derived from the BSE and reduced EDS spectra. Initially, the input EDS spectra are transformed via a Convolutional Neural Network (CNN) to generate compact descriptors for individual spectra. Using this representation, a graph is constructed through Voronoi analysis, with vertices assigned spectrum descriptors and BSE values. The edges receive two values, δ^b and δ^e, representing differences between BSE values and EDS descriptors, respectively. Edges with δ values exceeding thresholds are removed. The resulting graph components serve as initial points for Markov Random Field segmentation, which generates the final segments. The segments lack phase assignments, they just represent sufficiently similar sample regions. The algorithm presents several limitations, the most significant of which is the computational complexity.

Recently, Repka et al. [13] proposed a graph-based approach using GNNs for simultaneous BSE-EDS fusion and segmentation. Although promising, this method is limited to classifying minerals present in the training dataset and cannot generalise to unknown classes. The proposed method aims to address the limitations of the previous approach, providing a more general solution for mineral segmentation in SEM images.

3   Proposed Solution

The method builds on the previous work [13], incorporating many similarities. The aim is to expand the method generalisation, as the previous method was constrained to the minerals from the training dataset. The proposed segmentation method consists of four steps: (1) EDS preprocessing, (2) graph construction, (3) GNN inference, and (4) segmentation. The process is illustrated in Fig. 2.

3.1   EDS Preprocessing

The EDS data contain a large number of channels, 3000 in the case of datasets used in this work. If the data were to be processed raw, it would reflect on the computational time and memory requirements of the graph processing. As reduction of the EDS spectra has been used successfully in the past [7], a sensible choice was to use it in this method as well. As a reduction method, Principal Component Analysis (PCA) is used because of its effectiveness, simplicity, and denoising properties. PCA reduces the data to 64 channels. The PCA was trained on a dataset of simulated high-quality spectra, which were randomly subsampled to simulate noisy and lower-quality spectra. The dataset contained a broad enough range of minerals to be considered representative, but it contained only EDS spectra. Because of this, PCA had to be trained before the GNN.

3.2   Graph Construction

The graph consists of nodes and edges. Edges connect exactly two nodes, and both nodes and edges are assigned an attribute vector. Even though the graph is homogeneous and undirected in terms of graph construction, it can be said that it has two types of nodes: those corresponding to BSE pixels and those corresponding to EDS measurements. The construction works with the assumption that the two modalities are aligned, which means that an xy position on the BSE image corresponds to the same place as the same xy coordinate on the EDS map.

Following from the fact that two node types exist, three types of edges can be identified: edges connecting two BSE nodes, edges connecting two EDS nodes, and edges connecting BSE nodes and EDS nodes. The edges of each of these types need to be constructed differently. The easiest to construct are the edges of the first type - BSE to BSE edges. In this case, the nodes are already aligned in a regular grid, and the edges are created between adjacent nodes in a 4-neighbourhood, as illustrated in Fig. 3a. Constructing the second type of edges, EDS to EDS, is more challenging because the nodes cannot be assumed to be structured in a grid-like manner, making it unclear which nodes should be connected. A Delaunay triangulation (DT) [1] is used to select the edges. DT is a method of creating a triangulation of a set of points where no point is inside the circumcircle of any triangle. A graph constructed with this method is shown in Fig. 3b. The third and last edge type contains intermodal connections. For each BSE node, k edges are created and connected to k nearest EDS nodes. The reasoning behind this is such that those edges will serve as possible choices for the BSE nodes when assigning to EDS nodes. In this work, k was set to 4. An illustration of the complete graph is shown in Fig. 3c.

The nodes of the graph represent either pixels of the BSE image or EDS measurements. Each node holds an attribute vector 65 values long: the first value represents a BSE measurement (pixel value), and the remaining 64 values contain reduced EDS data. If a node represents a BSE, only the first value is set, and the remaining 64 values are initialised with zeroes. On the other hand,

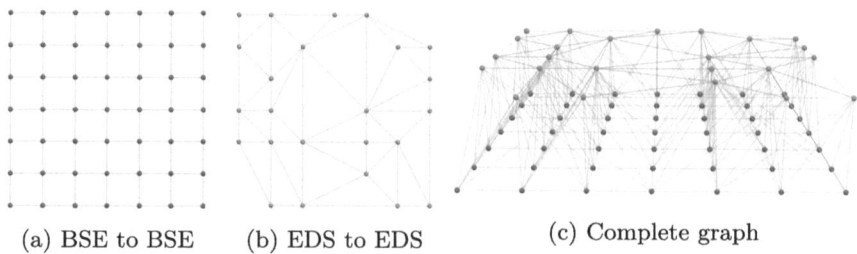

(a) BSE to BSE   (b) EDS to EDS   (c) Complete graph

Fig. 3. Graph construction.

if the node represents an EDS measurement, the first value is initialised to zero, and the rest are filled with EDS data.

Similar to the nodes, edges also have an attribute assigned to them. The attribute is just a single value, which is initialised to the length of the edge. As it is assumed that both modalities are aligned, they are just overlaid over each other, and the Euclidean distance is computed from this representation.

3.3   Network Architecture

The proposed network uses custom modules composed of GATv2 layers and two linear layers. The task is to classify edges, however, the GATv2 layer does not output edge embeddings. Therefore, additional linear layers are used to process the edge information. When running the forward pass, the graph is first run through the GATv2 layer. An activation function is then applied to the output (Mish [11] in this case), creating new node embeddings $\mathbf{x}$. A new feature vector h_i is then created as $h_i = [x_{e_{i1}}||g_i||x_{e_{i2}}]$, where g_i is embedding of i-th edge, e_{i1} and e_{i2} represent node embeddings of the node in which the i-th edge starts and ends, respectively, and $||$ denotes concatenation. This feature vector h_i is then run through two linear layers, each of which is followed by a Mish activation function. An illustration of the architecture can be seen in Fig. 4.

Several such modules are chained, three in the case of the network used here. Lastly, the outputs are again concatenated, and the result is passed through a linear layer, producing a single value for each edge. A logistic sigmoid function is applied to this output, which is then interpreted as a confidence value indicating whether the edge should be connected or not.

Loss Function. The loss function used for training was designed for optimising both node embeddings and edge predictions. The ground truth for node embeddings is a full node embedding (as described in Sect. 3.2), that is, the node attributes that should be filled in the nodes if all data were known. The ground truth for edge predictions were the connected/disconnected states, based on the phases of the nodes the edges connect to. The loss function is formulated as a weighted combination of mean squared error (MSE) for node predictions and binary cross-entropy (BCE) for edge predictions. Given predicted node values $\hat{N}$

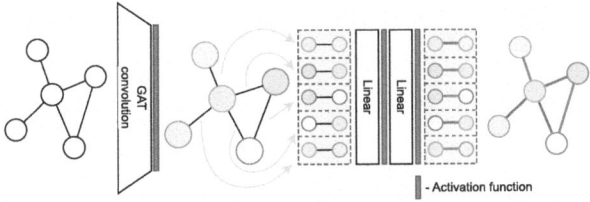

Fig. 4. Architecture of one module of the neural network. First, GATv2 convolution generates new node embeddings. Two linear layers then process edges together with embeddings of the nodes they connect, creating new edge attributes.

and ground truth node values N, as well as edge predictions $\hat{E}$ and ground-truth edge labels E, the loss function is defined as

$$\mathcal{L} = \alpha \cdot \text{MSE}(\hat{N}, N) + (1 - \alpha) \cdot \text{BCE}(\hat{E}, E), \tag{1}$$

where α is a tunable hyperparameter ($0 \leq \alpha \leq 1$) that controls the trade-off between node and edge losses. It was set to 0.2, as predicted node embeddings were not used in the final segmentation.

3.4   Segmentation

After the graph has been processed, it is expected that the edges are classified to be connected or disconnected. A single EDS measurement is, in theory, capable of classifying a mineral to its phase. The problem is that EDS is not available for each BSE measurement. However, it can be assumed that some EDS measurements in spatial proximity should describe the mineral phase of the BSE pixel. Thus, instead of segmenting the image, the task can be reformulated to the assignment of the BSE measurements to the EDS measurements in close proximity. The graph structure allows that, as intermodal edges were created between the closest BSE and EDS nodes. BSE nodes assigned to the same EDS node then constitute a single segment and each segment has exactly one EDS measurement assigned to it. After EDS is classified to a mineral phase, the whole segment is assigned to it, too. An example of segmentation can be seen in Fig. 5.

4   Experiments

4.1   Data

Eight different datasets were used for training, validation, and testing. All datasets contain full resolution EDS and BSE data and have varying sample sizes and rock or ore content. The datasets contain square samples of various rocks and ores, such as: (1) 1596 greisens samples with the side length of 150 px, (2) 7 tin mineralisation samples (1000 px), (3) 205 carbonatite rare earth elements ore samples (300 px), (4) 8 Fe-Ti rich mafic magmatic rock samples

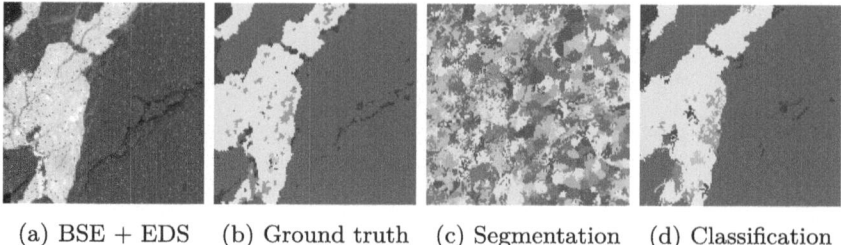

(a) BSE + EDS   (b) Ground truth   (c) Segmentation   (d) Classification

Fig. 5. An example of the segmentation.

(600 px), (5) 164 metamorphic shist with garnet and Fe-Cu sulphides samples
(187 px), (6) 657 pegmatite Li ore samples (200 px), (7) 146 Nb-Ta mineral
samples (1000 px), (8) 14 and heavy minerals concentrate from pegmatite samples (500–1000 px). Example BSE images from the dataset can be seen in Fig. 6.
The quality of EDS spectra varies, as counts of X-ray photons can be anywhere
between 1000 and 30000. Additionally, the samples were analysed using Tescan
TIMA software to generate a mineral phase map. This map segments the BSE
image into different minerals, serving as the ground truth for the segmentation
task. A mineralogy expert prepared these maps using a hand-crafted classification schema to ensure their reliability. Datasets 1, 6, and 7 were used for training,
Datasets 3, 4 and 8 for validation, Datasets 2 and 5 were used for testing.

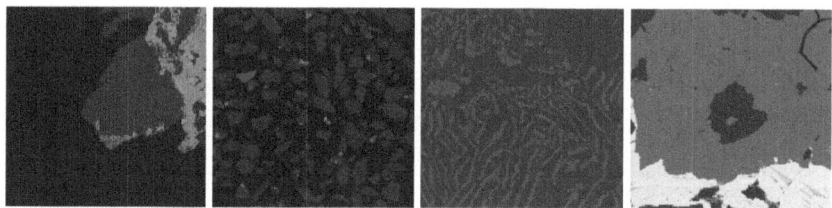

Fig. 6. Example BSE images from datasets.

4.2   Description of Experiments

The data were in full resolution, meaning that the EDS data was available for all
BSE pixels. To simulate the sparse, unstructured EDS data, the full-resolution
data was sampled both randomly, where a percentage of measurements was
retained, and in a sparser grid. The grid sampling was used because the network
should be able to utilise sparse but grid-like data as well.

To prevent overfitting on the limited mineral dataset and improve generalisation, spectral shuffling data augmentation was applied. For each sample, the

reduced EDS spectra were randomly shuffled channel-wise while maintaining channel order across all measurements, simulating a broader variety of minerals.

The method was implemented using PyTorch and PyTorch Geometric. The model used the Adam optimiser with a learning rate of 0.005, a number of hidden channels of node attributes 65, and of edge attributes 4. It was trained on samples of size 150×150 batched by 20 samples for 50 epochs.

4.3 Results

The generated segments are classified based on their EDS measurements. For evaluation purposes, it is assumed that the classification is accurate, meaning that each EDS measurement is assigned the correct mineral phase. This process results in a fully segmented and classified image, enabling a direct comparison of the segmentation performance with the ground truth data.

The proposed method was compared to Graph-based Deep Learning Segmentation (GDLS) [7]. Note that the method could not be compared to [13], as that method produces direct classification and is constrained to classes present in the training dataset. Like the proposed method, GDLS produces unclassified segments but also includes a merging mechanism controlled by user-set thresholds. To avoid potential errors during evaluation, they were set to 0, preventing merging. Segments were classified as before, assuming perfect classification and assigning classes from the ground truth. Example outputs are shown in Fig. 7.

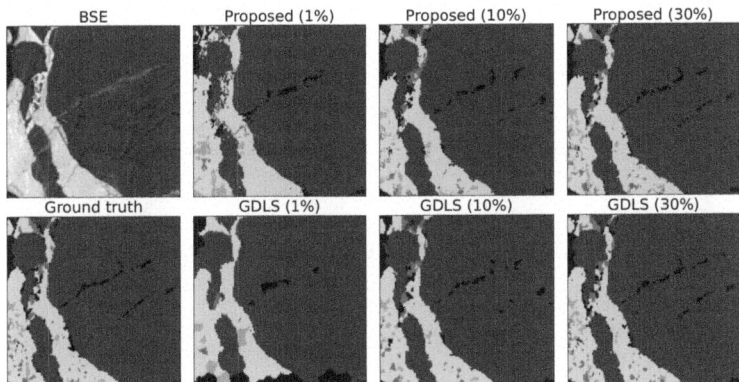

Fig. 7. Example outputs of the proposed method and GDLS. The left column shows the input BSE image and ground truth. The top and bottom rows show outputs of the proposed method and GDLS, respectively.

The evaluation of the segmentation results for both methods is shown in Fig. 8. As can be seen, increasing the number of EDS measurements improves the segmentation for both methods. This is expected, as the additional EDS data provide useful information for the segmentation. Furthermore, the metrics show that the proposed method outperforms GDLS. This is particularly evident with sparse EDS measurements, for which the proposed method achieves the F1-score of 0.81 with just 1% of pixels accompanied by EDS data, compared to the F1-score of 0.77 for GDLS. As the fraction of EDS data increases, the difference gradually diminishes, as the measured F1-scores with 20% of EDS data are 0.918 and 0.917 for the proposed method and GDLS, respectively.

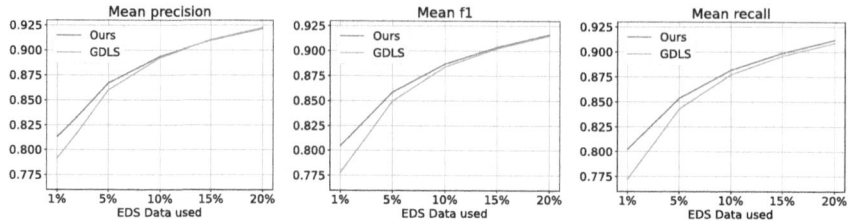

Fig. 8. Evaluation results with comparison of the proposed method to GDLS.

Experiments on time requirements were conducted on a server with an Nvidia A100 80 GB and AMD EPYC 7543 processor. The GDLS implementation provided with [7] was run on Python 3.7.12 due to compatibility with the provided network, while the proposed method used Python 3.12.3. Tests analysed the impact of EDS data density and sample size, with each experiment repeated 20 times. Figure 9 shows the mean and standard deviation of results. The proposed method was consistently faster and scaled better, with an increasing performance margin as sample size or data fraction grew.

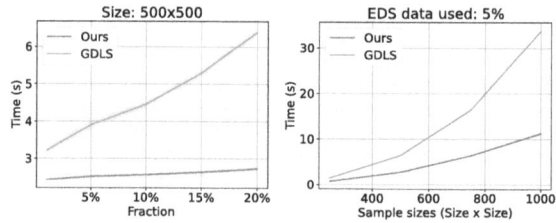

Fig. 9. Processing times vs. sample size and EDS density for both methods.

The results demonstrate that the proposed method efficiently fuses BSE and EDS data, generalises well to unknown minerals, and provides accurate results. Compared to GDLS, it achieves higher accuracy and efficiency while scaling better with increasing data, making it suitable for practical applications.

5   Conclusion

In this work, we proposed a novel method for multi-modal segmentation of minerals in SEM images using GNN. By combining dense BSE and sparse EDS data into a single graph, segmentation is formulated as an edge classification problem, enabling class-agnostic segmentation beyond the training set. The method does not assume specific spatial structures for EDS data, supporting flexible sampling strategies. Compared to GDLS, our approach achieves higher accuracy with limited EDS data and scales better with larger datasets. This opens possibilities for faster SEM data acquisition with minimal accuracy loss. Future work could explore segment merging for low-quality spectra.

Acknowledgments. This work was supported by the Finnish Ministry of Education and Culture's Pilot for Doctoral Programmes (Pilot project Mathematics of Sensing, Imaging and Modelling).

References

1. de Berg, M., Cheong, O., van Kreveld, M., Overmars, M.: Computational Geometry: Algorithms and Applications. Springer, Berlin, Heidelberg (2008). https://doi.org/10.1007/978-3-540-77974-2
2. Couprie, C., Farabet, C., Najman, L., LeCun, Y.: Indoor Semantic Segmentation using depth information. arXiv:1301.3572 (2013)
3. Fandrich, R., Gu, Y., Burrows, D., Moeller, K.: Modern SEM-based mineral liberation analysis. Int. J. Miner. Process. **84**(1), 310–320 (2007)
4. Girão, A.V., Caputo, G., Ferro, M.C.: Application of scanning electron microscopy–energy dispersive X-Ray spectroscopy (SEM-EDS). In: Characterization and Analysis of Microplastics, vol. 75. Elsevier (2017)
5. Gupta, S., Girshick, R., Arbeláez, P., Malik, J.: Learning rich features from RGB-D images for object detection and segmentation. In: Fleet, D., Pajdla, T., Schiele, B., Tuytelaars, T. (eds.) ECCV 2014. LNCS, vol. 8695, pp. 345–360. Springer, Cham (2014). https://doi.org/10.1007/978-3-319-10584-0_23
6. Jaritz, M., Charette, R.D., Wirbel, E., Perrotton, X., Nashashibi, F.: Sparse and dense data with CNNs: depth completion and semantic segmentation. In: International Conference on 3D Vision, pp. 52–60 (2018)
7. Juránek, R., Výravský, J., Kolář, M., Motl, D., Zemčík, P.: Graph-based deep learning segmentation of EDS spectral images for automated mineral phase analysis. Comput. Geosci. **165**, 105109 (2022)
8. Kipf, T.N., Welling, M.: Semi-supervised classification with graph convolutional networks. In: International Conference on Learning Representations (ICLR) (2017)
9. Lee, S., Park, S.J., Hong, K.S.: RDFNet: rgb-d multi-level residual feature fusion for indoor semantic segmentation. In: IEEE International Conference on Computer Vision, pp. 4990–4999 (2017)
10. Miller, P., Zuiderwyk, M., Reid, A.: SEM image analysis in the determination of modal assays, mineral associations and mineral liberation. In: WorldWind IndustriaL Applied Mineral Processing Technology (1983)
11. Misra, D.: Mish: a self regularized non-monotonic neural activation function. arXiv:1908.08681 (2019)

12. Motl, D., Filip, V.: Method and apparatus for material analysis by a focused electron beam using characteristic X-rays and back-scattered electrons (2013). uS Patent App. 13/398,114
13. Repka, S., Reich, B., Zolotarev, F., Eerola, T., Zemčík, P.: Mineral segmentation using electron microscope images and spectral sampling through multimodal graph neural networks. Pattern Recogn. Lett. (2025)
14. Sanchez-Lengeling, B., Reif, E., Pearce, A., Wiltschko, A.B.: A Gentle Introduction to Graph Neural Networks. Distill (2021)
15. Veličković, P., et al.: Graph attention networks. arXiv:1710.10903 (2018)
16. Xu, K., Hu, W., Leskovec, J., Jegelka, S.: How powerful are graph neural networks? In: International Conference on Learning Representations (ICLR) (2019)
17. Zhang, Y., Sidibé, D., Morel, O., Mériaudeau, F.: Deep multimodal fusion for semantic image segmentation: a survey. Image Vis. Comput. **105**, 104042 (2021)

Predicting Sheep Body Condition Scores via Explainable Deep Learning Model

Nourelhouda Hammouda[1,2]([✉]) [iD], Mariem Mahfoudh[1,3] [iD], Rima Grati[4] [iD],
and Khouloud Boukadi[1,5] [iD]

[1] MIRACL Laboratory, University of Sfax, Sfax, Tunisia
mariem.mahfoudh@isims.usf.tn
[2] National School of Electronics and Telecommunications, Sfax, Tunisia
nourelhouda.hammouda@enetcom.u-sfax.tn
[3] Higher Institute of Computer Science and Multimedia, Sfax, Tunisia
khouloud.boukadi@fsegs.usf.tn
[4] Zayed University, Abu Dhabi, UAE
Rima.Grati@zu.ac.ae
[5] Faculty of Economics and Management, Sfax, Tunisia

Abstract. Body Condition Scoring (BCS) is essential for evaluating livestock health, nutrition, and reproduction. For sheep, especially those with long wool, manual scoring requires a level of expertise that many farmers do not possess. Despite its importance, to date, no automated approach has been proposed in the existing literature. In this context, we propose an innovative approach that automatically predicts the BCS of sheep from images, followed by an explainable AI (XAI). The pipeline follows a three-step process: (1) data collection, which was conducted under real-world farming conditions; (2) sheep detection using the Florence-2 object detection model and BCS classification via a Vision Transformer (ViT) trained on 5,848 images (11,000 images after the augmentation), and (3) visual explanation using multiple XAI methods. Our method achieved 72% accuracy using exact BCS class matches and 95% when allowing adjacent classes. This tolerance reflects the variability typically observed among expert scores.

Keywords: Computer vision · Body Condition Scoring · XAI · Deep learning · Sheep

1 Introduction

The Body Condition Score (BCS), as defined by [25], is a systematic approach that evaluates muscle and fat development through palpation rather than visual inspection alone (see Figure 1). The BCS scale ranges from 1 (emaciated) to 5 (obese), with 3 considered ideal for a healthy fat–muscle balance [14]. Scores below 2.5 or above 4.0 may indicate nutritional or health issues, and each unit change in BCS reflects a significant shift in body fat reserves. This classification

M. Castrillón-Santana et al. (Eds.): CAIP 2025, LNCS 15622, pp. 37–48, 2026.
https://doi.org/10.1007/978-3-032-05060-1_4

is fundamental for interpreting BCS implications on sheep physiology, reproduction, and productivity. A BCS at mating (2.5–3.5) improves fertility, litter size, and lambing timing, while a low BCS (<1.5) leads to poor reproductive outcomes [24]. BCS, influenced by genetic and environmental factors, remains a crucial tool for reproductive success, metabolic health monitoring, behavior analysis, and overall herd management [8,14].

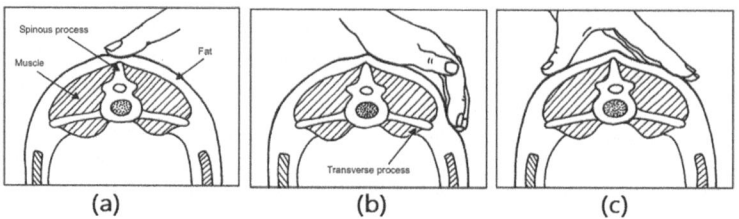

Fig. 1. Procedure for assessing muscle and fat deposition over the vertebrae in the loin region of a sheep: (a) palpation of the spinous process along the midline, located between the last rib and the hip bone; (b) palpation of the tips of the transverse processes; (c) evaluation of muscle fullness and fat coverage. [25]

Despite the recognized importance of BCS as a key indicator of sheep health and productivity, its assessment still relies on manual palpation, which is subjective and labor-intensive. Although computer vision techniques have shown promise in automating BCS evaluation for dairy cattle and goats using deep learning and depth imaging technologies [2,19,26], no studies have specifically addressed their application to sheep. This raises the critical question: can sheep BCS be accurately predicted without human touch? Wool coverage in sheep obscures key anatomical landmarks, making visual assessment particularly difficult. Combined with data scarcity and body morphology variability, these factors complicate automation. This gap highlights the need to develop computer vision-based tailored BCS systems for sheep.

Understanding how an AI model makes its predictions is also crucial to building trust between experts and nonexperts. Indeed, as models become more complex, their decision-making often becomes opaque, raising concerns about reliability. This issue is particularly important in agriculture, where decisions based on AI outputs can directly impact animal welfare, productivity, and farm operations. Explainable Artificial Intelligence (XAI) addresses this by making AI outputs more transparent through techniques such as SHAP (SHapley Additive exPlanations) [16], LIME (Local Interpretable Model-agnostic Explanations) [18], and Grad-CAM (Gradient-weighted Class Activation Mapping) [20].

In this work, and given the absence of any publicly available dataset focused on BCS prediction in sheep, we propose as the first stage of our pipeline the construction of a custom, domain-specific dataset. The images were collected specifically to capture visual indicators of BCS and were carefully annotated by livestock experts to ensure scoring accuracy and consistency. This dataset was then expanded through augmentation techniques and prepared through standard preprocessing steps to support the subsequent stages of the pipeline. The second stage involves model development, in which sheep are detected using Florence 2 and classified into one of five BCS categories using a pre-trained classification model Vision Transformer (ViT). The final stage introduces post-hoc explainability, in which several XAI methods—Grad-CAM++, Score-CAM, SHAP, LIME, and hybrid variants—are applied to visualize and interpret the model's decisions.

Our paper is structured as follows: Sect. 2 presents related works, covering object detection techniques and explainable artificial intelligence (XAI) in livestock applications. Section 3 details our explainable AI approach for automated BCS. In Sect. 4, we discuss our findings, highlighting the performance of our approach and its implications. Finally, Sect. 5 concludes the paper, summarizing key contributions and potential future research directions.

2   Related Works

2.1   Animal Body Condition Scoring

Research focused on predicting BCS using AI is still relatively limited, and no data are accessible on the network, particularly for sheep, as shown in Table 1. In [21], applied machine learning algorithms, such as Random Forest and Gradient Boosting Decision Trees, to predict BCS in mature Romney ewes based on liveweight records. The authors used a dataset of 5761 ewes and achieved over 85% accuracy in predictions, but faced challenges with dataset imbalance and robust classification. Beyond sheep, other studies have focused on different species, leveraging AI for BCS assessment. [19] concentrated on dairy cows, using convolutional neural networks (CNNs) with transfer learning and ensembling methods to classify BCS from 1661 depth images captured with a Kinect V2 camera. Their model achieved 97% accuracy within a 0.5-unit BCS margin but required high-quality depth images, limiting practical deployment. Similarly, [4] used the YOLOv8 deep learning architecture to classify BCS in Holstein and Simmental cows, achieving an accuracy of 81% across five BCS classes using a dataset of 1270 images. The challenges included maintaining consistent accuracy across varying breeds and conditions. In [2], the authors developed a real-time classification system using a VGG19-based CNN for dairy cows, making the model mobile-compatible. The approach achieved 78% accuracy but was limited by the constraints of mobile device hardware. Furthermore, the work in [11] employed an improved YOLOX object detection network to classify pig BCS from 21,943 images. The approach reached an accuracy of 80.06% but faced deployment challenges in real-world environments with variable lighting and background farming conditions. In [23] developed a lightweight CNN for pose-independent classification of goat BCS from dorsal-view images, achieving 97.94% accuracy. However, the challenge was ensuring consistency across diverse Finally, [27] employed a Vision Transformer (ViT) for RGB+depth image-based BCS prediction in Raramuri Criollo cattle. Their approach achieved up to 92%

accuracy but highlighted the critical importance of dataset curation for robust model performance.

Table 1. Comparative Table of BCS Research Papers

Paper	Animal	Tools	Techniques	Dataset	Results
[19]	Dairy Cows	Kinect V2 depth camera	Transfer learning, Fourier transform preprocessing	1661 depth images (Kinect V2)	Improved accuracy within 0.25 BCS: 82%, within 0.50 BCS: 97%
[2]	Dairy cattle	Mobile-compatible CNN system	Pre-trained VGG19 for mobile devices	Dataset of back images collected on farms	Real-time classification with 78%
[21]	Sheep	Statistical and ML algorithms	Gradient Boosting, Random Forest	Ewe liveweight and BCS data from 5761 ewes	> 85% with ML
[11]	Pigs	YOLOX architecture	Dynamic loss weighting, attention mechanism	21,943 images (upper and rear views)	Improved AP of 80.06%, real-time at 29 FPS
[23]	Goats	Custom lightweight CNN	Pose-independent CNN training	5000 goat images (dorsal views)	97.94%
[4]	Dairy Cows	YOLOv8	Hyperparameter optimization in YOLOv8	1270 images from 20 farms (Holstein and Simmental cows)	81% across five classes
[27]	Cattle	RGB+ Depth camera	Masked image modeling for pretraining ViT	RGB+ Depth images of Raramuri Criollo cows	From 64 % to 92% with curated datasets

In summary, no prior research work has specifically addressed BCS in sheep despite the distinct morphological characteristics, mainly the variations in wool coverage. Therefore, no studies have addressed explainability, even though it plays a crucial role in building trust and understanding of AI models, especially for end users (i.e. farmers).

2.2   Explainable Artificial Intelligence in Livestock Applications

Explainable Artificial Intelligence (XAI) is a multidisciplinary field that aims to make AI systems transparent, interpretable, and trustworthy by generating understandable explanations tailored to diverse users [1,6]. It bridges the gap between complex black-box models and human understanding by clarifying the rationale behind AI decisions. XAI has been leveraged in various innovative ways across recent studies. For instance, the authors in [5] utilized LIME method to elucidate the factors influencing livestock price predictions, offering buyers and sellers increased confidence in an otherwise opaque pricing model. The work in [12] emphasized the role of explainability in video surveillance systems to detectving signs, integrating Grad-CAM, LIME, and SHAP to provide farmers with interpretable insights into cattle behavior improve decision-making. [15] highlighted explainability by using SHAP in conjunction with random forests and neural networks to analyse milk quality predictors in sheep and goat farms, enabling the identification of actionable farm management factors. [13] encountered challenges related to sensor data complexity when applying XAI techniques such as LIME and SHAP to interpret ML models that predict cattle behavior, particularly in balancing interpretability with prediction accuracy. Finally, [22] demonstrated the utility of XAI in correlating combined behavior of cattle with health conditions, such as mastitis, through the SHAP library, but noted difficulties in generalizing their health classification model across various datasets. However, none of these studies explicitly evaluated the quality or effectiveness of the XAI methods they employed.

3   Explainable AI Approach for Automated BCS

In collaboration with the Institut National de Recherche Agronomique de Tunis (INRAT), we propose a novel AI-based approach for predicting the BCS of sheep, specifically designed to address the needs of both domain experts and farmers. This solution aims to support researchers in advancing their studies by providing reliable, interpretable data and tools, while also empowering farmers with a better understanding of their herd to enable more informed and efficient livestock management.

As the first effort of its kind in the literature, this work is surrounded by several challenges, not least of which is the inherent difficulty in estimating BCS— even by eye observation or non-expert palpation. Indeed, our research team represents the first line of users, navigating these complexities firsthand. Therefore, our objective is to lay the groundwork for a reliable, interpretable, and inclusive BCS prediction pipeline. We begin by thoroughly exploring and evaluating the strengths and limitations of computer vision techniques and explainability methods in the face of these real-world challenges.

Our proposed approach is structured into three main stages, as illustrated in Fig. 2. The first stage focuses on preparing a high-quality dataset from scratch. The second stage involves object detection and classification, where several pretrained deep learning models are assessed for their effectiveness. The third stage integrates and compares several explainability techniques to interpret and justify the classification outcomes. The following subsections provide a detailed discussion of each stage.

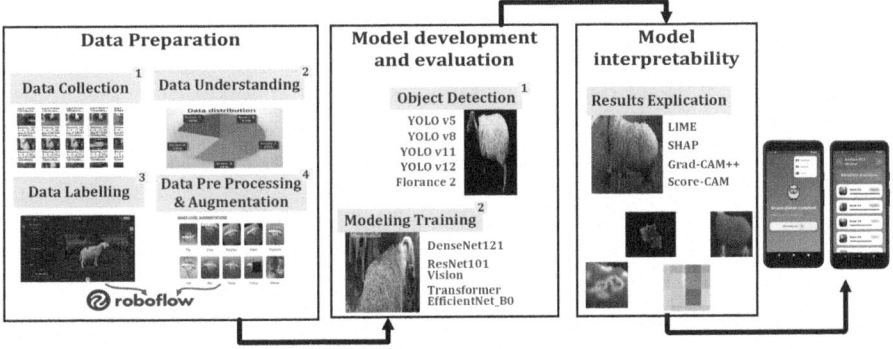

Fig. 2. Overview of the proposed pipeline for explainable BCS prediction in sheep.

3.1   Data Preparation

Data Collection. As there is no publicly available dataset specifically designed for Body Condition Score (BCS) prediction in sheep, we constructed our dataset from scratch. Image data were collected from the educational farm of the Higher School of Agriculture in Kef, Tunisia. A total of 204 short videos were recorded, capturing thin-tailed sheep with varying BCS levels under controlled conditions, with a focus on two key camera angles (side and rear) providing essential visual cues for assessing body shape and fat distribution. All videos were annotated on-site by an expert evaluator from the school, ensuring reliable and domain-informed labelling.

Data Understanding. The videos were sampled at 2 frames per second, and only images from side and rear views were retained. This resulted in a base dataset of 5,848 images, distributed across BCS scores as follows: 330 (Score_1), 1,044 (Score_2), 2,204 (Score_3), 1,510 (Score_4), and 760 (Score_5). During the image selection process, we aimed to ensure class balance while removing visually similar frames caused by the animals' slow movement.

Data Labelling. As previously mentioned, each image inherits its BCS label from expert annotations assigned at the video level, and this labeling was consistently preserved throughout frame extraction and augmentation to ensure annotation accuracy. We used the Smart Polygon (AI Labeling) tool provided by the Roboflow platform[1], which enabled precise polygon-based segmentation of body areas relevant to BCS prediction. After data cleaning and augmentation, the final annotated dataset comprised 5,848 images, evenly distributed across the five BCS classes. Our dataset was split into training (70%, 4,095 images), validation (20%, 1,169 images), and test (10%, 584 images) sets. To promote reproducibility and support future research, the dataset will be made publicly available on GitHub[2].

Image Processing and Data Augmentation. To ensure dataset consistency and improve class balance, we applied preprocessing and score-wise data augmentation using the Roboflow platform, which automatically propagates annotations to the newly generated images. Augmentation was selectively applied per BCS class to reduce imbalance: images from Score_1 were augmented 6× (330 × 6 = 1,980), Score_2 2× (1,044 × 2 = 2,088), while no augmentation was applied to Score_3 (2,204) or Score_4 (1,510). Score_5 was augmented 2× (760 × 2 = 1,520). Roboflow's augmentation tools included random horizontal flips and adjustments in hue (±23°), saturation (±33%), brightness (±24%), and exposure (±15%) to simulate natural variability. All images were auto-oriented and resized to 640 × 640 pixels using the stretch method.

3.2   Object Detection

To fully automate Body Condition Score (BCS) estimation, it is essential to first detect and localize the sheep in the input images. This step ensures that the clas-

---

[1] https://roboflow.com.
[2] https://github.com/nourelhoudahamoudaa/Sheep-Body-Condition-Score.

sification model receives clean, focused regions containing only the relevant animal, thereby reducing background noise and improving prediction accuracy. We therefore introduced a sheep detection model as the initial stage in our pipeline. To select the most effective detector, we compared Florence 2 with several YOLO variants (v5, v8, v11, v12). As shown in Table 2, Florence 2 achieved the highest accuracy (97.46%) and lowest error rate (1.81%), confirming its suitability for robust sheep detection.

Table 2. Comparison of Object Detection Models

Model	Florance_2	YOLOv5	YOLOv8	YOLOv11	YOLOv12
Accuracy (%)	**97.4638**	96.2924	88.6594	94.1495	89.2013
Misclassification Error Rate (%)	**1.8116**	4.7101	8.8768	2.8468	7.8157

3.3 Modeling Training

Four deep learning architectures: DenseNet121, EfficientNet_B0, ResNet101, and ViT were selected for training and evaluation on the BCS classification task. Each model was fine-tuned on the custom dataset and trained for 100 epochs using standard training procedures. Performance was assessed using accuracy, precision, recall, and F1-score, as summarized in Table 3.

Table 3. Performance comparison of the evaluated models.

Model/Metric	Accuracy	Precision	Recall	F1-Score
DenseNet121	0.6872	0.5804	0.6872	0.6190
EfficientNet_B0	0.7055	0.6082	0.7055	0.6384
ResNet101	0.7192	0.6121	0.7192	0.6495
ViT	**0.7237**	**0.6256**	**0.7237**	**0.6561**

3.4 Experimental Results

Figure 3 presents six representative examples from the test set, evaluated using the first two steps of our proposed approach: sheep detection followed by BCS classification. For each image, the detected sheep is enclosed in a bounding box, annotated with both the ground-truth label (GT) and the predicted label (Pred).

Most predictions closely align with the true scores. When misclassifications occur, the predicted class generally falls within a neighboring BCS category. This finding aligns with the earlier analysis of the confusion matrix, which indicated that most classification errors happened between adjacent scores. The overall classification accuracy reached 72%, considering exact class matches.

Fig. 3. Sample visual results of the proposed approach applied to six test images.

3.5   XAI Integration

To enhance the transparency of our BCS classification, we incorporated and compared multiple post-hoc explainability techniques applied to the ViT model, including Grad-CAM++, Score-CAM, SHAP, LIME, and two hybrid methods (Score-CAM + SPHAP and Score-CAM + LIME). Figure 4 illustrates visual explanations for three representative test samples from all applied methods. Each row shows the initial image, the sheep detection result (where the image background was removed beforehand to reduce noise and focus on the sheep's body), and the corresponding attention visualization from each technique.

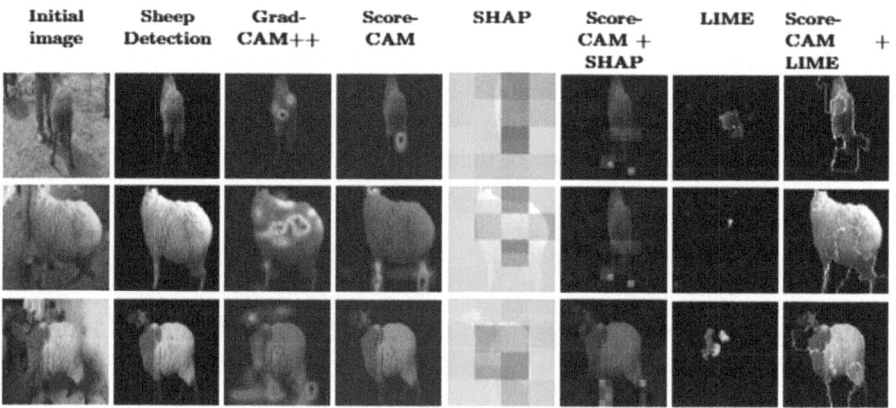

Fig. 4. Visualization of ViT-based explanations for sheep BCS classification using various XAI techniques.

Table 4 presents a quantitative comparison of various XAI methods using standard evaluation metrics that are widely used to assess the interpretability, fidelity, and localization ability of visual explanations.

- **Deletion AUC** [7]: measures how the model's confidence drops when the most important pixels are progressively removed.

$$\text{AUC}_{\text{deletion}} = \sum_{j=0}^{n} f(\tilde{x}(j)) - \frac{n+1}{2} \left[f(\tilde{x}(0)) + f(\tilde{x}(n)) \right]$$

where $\tilde{x}(j)$ is the input after inserting or deleting the j-th most important pixels, $f(\cdot)$ is the model prediction confidence, $\tilde{x}(0)$ is the original input, and $\tilde{x}(n)$ is the fully perturbed input.

- **Intersection over Union (IoU)** [17]: assesses the overlap between the explanation and ground truth region. Higher is better.

$$\text{IoU} = \frac{|\mathcal{E} \cap \mathcal{G}|}{|\mathcal{E} \cup \mathcal{G}|}$$

where $\mathcal{E}$ is the set of pixels in the explanation mask and $\mathcal{G}$ is the ground-truth region of interest.

- **Sparsity** [17]: indicates how focused or minimal the explanation is. Lower values mean fewer highlighted pixels.

$$\text{Sparsity} = \frac{\text{non-zero pixels in } \mathcal{E}}{\text{total pixels}}$$

Lower sparsity indicates a more focused and interpretable explanation.

Table 4. Quantitative comparison of XAI methods

Metric	Grad-CAM++	Score-CAM	SHAP	Score-CAM + SHAP	LIME	Score-CAM + LIME
AUC_D ↓	0.9631	0.9632	0.2679	0.2543	0.5147	**0.1768**
IoU ↑	0.0323	0.0085	0.1354	0.0195	0.0181	**0.1940**
Sparsity ↑	0.9677	**0.9915**	0.8646	0.9219	0.9819	0.8676

The results indicate that no method dominates across all metrics, reflecting trade-offs between faithfulness, localization, and compactness. Score-CAM + LIME offers the best localization (IoU), and also achieves the strongest deletion score. Grad-CAM++ performs moderately overall, while hybrid methods, especially Score-CAM + LIME, show a balanced profile suitable for interpretable BCS classification.

3.6   Discussion

Our work presents the first attempt to predict the Body Condition Score (BCS) of sheep directly from images. While expert annotations often use fine-grained decimal scores (e.g., 3.25 or 3.75) [24], classification models rely on discrete labels, which can lead to misclassification especially between adjacent classes. Using strict evaluation (exact match), our model achieved 72% accuracy; however, allowing for adjacent score tolerance (reflecting real-world inter-observer variability) boosts accuracy to 95%. Notably, our model remains aligned with practical needs, as BCS can be meaningfully grouped into three broad categories—low, moderate, and high—as supported by [3], who linked these ranges to reproductive performance outcomes. To support real-world deployment, we have developed a mobile application, as shown in Fig. 5, using Flutter

Fig. 5. Our mobile application interface for BCS prediction.

that can process images or extract frames from videos to predict the BCS across different camera angles (side and rear). Then we seek to provide the final decision accompanied by an explanation through it.

From the AI perspective, the application of various XAI methods—including Grad-CAM++, Score-CAM, SHAP, LIME, and hybrid versions—helped us identify the key body parts the model relies on to make predictions. This guided us to focus image collection on these areas (e.g., back, rump, tail base) to improve model performance. Since no single method satisfied all evaluation metrics, we shifted toward a semantic-based approach. Our recent work [10] proposes an ontology-based XAI framework grounded in our MoonCAB ontology [8], which has been enriched with over 200 expert-defined rules. This enrichment is designed to support the generation of explanations that are both scientifically rigorous for domain experts and intuitively understandable for non-experts. Following the enrichment process, MoonCAB was evaluated and validated using the MoOnEv tool [9] in collaboration with experts.

4   Conclusion

We addressed the challenge of AI-based Body Condition Score (BCS) prediction in sheep by proposing a structured, explainable pipeline. Given the lack of public datasets in this domain, we began by constructing a custom, expert-annotated dataset tailored specifically for this task. This served as the foundation for the second stage, where Florence-2 was used for sheep detection, followed by ViT model to classify BCS into five discrete classes. The proposed system achieved 72% accuracy under strict class-matching and up to 95% when accounting for adjacent class tolerance, reflecting alignment with expert scoring variability. To ensure transparency, the third stage incorporated multiple post-hoc explainability techniques—including Grad-CAM++, Score-CAM, SHAP, LIME, and hybrid methods—allowing us to analyze the model's focus across different visual regions. While each method provided valuable insights, our comparative evaluation showed that no single technique consistently satisfied all interpretability metrics, highlighting the need for more semantically grounded explanations.

Future work will focus on expanding the dataset, supporting finer-grained decimal scoring (e.g., 2.25, 3.75), and generating explanations that are both scientifically rigorous for livestock experts and easily interpretable for non-expert users, particularly farmers—based on the integration of our previously developed ontology into the XAI pipeline to produce semantically enriched justifications.

Acknowledgements. We would like to express our sincere gratitude to our collaborators, Dr Samir Smetti and Pr Moukhtar Mahouchi, from the Institut National de Recherche Agronomique de Tunis (INRAT), for their valuable support and for facilitating access to the experimental farm of the University of Jendouba, at the Higher School of Agriculture of Kef, which enabled the collection of videos used in our dataset.

References

1. Al-Ansari, N., Al-Thani, D., Al-Mansoori, R.S.: User-centered evaluation of explainable artificial intelligence (XAI): a systematic literature review. Hum. Behav. Emerg. Technol. **2024**(1), 4628855 (2024)
2. Çevik, K.K.: Deep learning based real-time body condition score classification system. IEEE Access **8**, 213950–213957 (2020)
3. Corner-Thomas, R., Ridler, A., Morris, S., Kenyon, P.: Ewe lamb live weight and body condition scores affect reproductive rates in commercial flocks. N. Z. J. Agric. Res. **58**(1), 26–34 (2015)
4. Dandıl, E., Çevik, K.K., Boğa, M.: Automated classification system based on yolo architecture for body condition score in dairy cows. Vet. Sci. **11**(9), 399 (2024)
5. Dave, D., Naik, H., Singhal, S., Dwivedi, R., Patel, P.: Towards designing computer vision-based explainable-ai solution: a use case of livestock mart industry. arXiv preprint arXiv:2103.03096 (2021)
6. Gunning, D., Stefik, M., Choi, J., Miller, T., Stumpf, S., Yang, G.Z.: XAI—explainable artificial intelligence. Sci. Robot. **4**(37), eaay7120 (2019)
7. Hama, N., Mase, M., Owen, A.B.: Deletion and insertion tests in regression models. J. Mach. Learn. Res. **24**(290), 1–38 (2023)
8. Hammouda, N., Mahfoudh, M., Boukadi, K.: Mooncab: a modular ontology for computational analysis of animal behavior. In: 2023 20th ACS/IEEE International Conference on Computer Systems and Applications, pp. 1–8. IEEE (2023)
9. Hammouda, N., Mahfoudh, M., Boukadi, K.: Moonev: modular ontology evaluation and validation tool. Procedia Comput. Sci. **246**, 3532–3541 (2024)
10. Hammouda, N., Mahfoudh, M., Boukadi, K.: Toward Semantic Explainable AI in Livestock: MoonCAB Enrichment for O-XAI to Sheep BCS Prediction. In: The 17th International Conference on Knowledge Engineering and Ontology Development (2025)
11. He, H., Chen, C., Zhang, W., Wang, Z., Zhang, X.: Body condition scoring network based on improved YOLOX. Pattern Anal. Appl. **26**(3), 1071–1087 (2023)
12. Hyodo, R., Saito, S., Nakano, T., Akabane, M., Kasuga, R., Ogawa, T.: Video surveillance system incorporating expert decision-making process: a case study on detecting calving signs in cattle. arXiv preprint arXiv:2301.03926 (2023)
13. Ibrahim, T., et al.: Interpretable machine learning techniques for predictive cattle behavior monitoring. In: 2024 2nd International Conference on Sustainable Computing and Smart Systems (ICSCSS), pp. 1219–1224. IEEE (2024)
14. Kenyon, P., Maloney, S., Blache, D.: Review of sheep body condition score in relation to production characteristics. N. Z. J. Agric. Res. **57**(1), 38–64 (2014)

15. Lianou, D.T., et al.: The use of explainable machine learning for the prediction of the quality of bulk-tank milk in sheep and goat farms. Foods **13**(24), 4015 (2024)
16. Lundberg, S.: A unified approach to interpreting model predictions. arXiv preprint arXiv:1705.07874 (2017)
17. Nauta, M., et al.: From anecdotal evidence to quantitative evaluation methods: a systematic review on evaluating explainable ai. ACM Comput. Surv. **55**(13s), 1–42 (2023)
18. Ribeiro, M.T., Singh, S., Guestrin, C.: Why should i trust you? Explaining the predictions of any classifier. In: Proceedings of the 22nd ACM SIGKDD International Conference on Knowledge Discovery and Data Mining, pp. 1135–1144 (2016)
19. Rodríguez Alvarez, J., et al.: Estimating body condition score in dairy cows from depth images using convolutional neural networks, transfer learning and model ensembling techniques. Agronomy **9**(2), 90 (2019)
20. Selvaraju, R.R., Cogswell, M., Das, A., Vedantam, R., Parikh, D., Batra, D.: Gradcam: visual explanations from deep networks via gradient-based localization. In: Proceedings of the IEEE International Conference on Computer Vision, pp. 618–626 (2017)
21. Semakula, J., Corner-Thomas, R.A., Morris, S.T., Blair, H.T., Kenyon, P.R.: Application of machine learning algorithms to predict body condition score from liveweight records of mature romney ewes. Agriculture **11**(2), 162 (2021)
22. Shi, Z., et al.: Classifying and understanding of dairy cattle health using wearable inertial sensors with random forest and explainable artificial intelligence. IEEE Sens. Lett. (2024)
23. Temenos, A., et al.: Goat-cnn: a lightweight convolutional neural network for pose-independent body condition score estimation in goats. J. Agric. Food Res. **16**, 101174 (2024)
24. Thompson, A.N., et al.: Additive impacts of liveweight and body condition score at breeding on the reproductive performance of merino and non-merino ewe lambs. Animals **14**(6), 867 (2024)
25. Thompson, J.M., Meyer, H.H., et al.: Body condition scoring of sheep (1994)
26. Vázquez-Martínez, I., et al.: Predicting body weight through biometric measurements in growing hair sheep using data mining and machine learning algorithms. Trop. Anim. Health Prod. **55**(5), 307 (2023)
27. Winkler, Z., Boucheron, L.E., Utsumi, S., Nyamuryekung'e, S., McIntosh, M., Estell, R.E.: Effects of dataset curation on body condition score (bcs) determination with a vision transformer (vit) applied to rgb+ depth images. Smart Agric. Technol. 100482 (2024)

What Does Gait Reveal About Health?
Investigating Human Motion
as an Indicator

Rafael Aguilar-Ortega[1]([⊠]) [iD], Shiqi Yu[2] [iD], Nuria Marin-Jimenez[3,4] [iD],
and Manuel J. Marin-Jimenez[1,5] [iD]

[1] Department of Computer Science and Artificial Intelligence, University of Córdoba,
Córdoba, Spain
{raortega,mjmarin}@uco.es

[2] Department of Computer Science and Engineering, Southern University of Science
and Technology, Shenzhen, China
yusq@sustech.edu.cn

[3] GALENO Research Group, Department of Physical Education,
Faculty of Education Sciences, University of Cadiz, Cádiz, Spain
nmjimenez@ual.es

[4] Instituto de Investigación e Innovación Biomédica de Cádiz (INiBICA),
Cádiz, Spain

[5] Instituto Maimónides de Investigación Biomédica de Córdoba (IMIBIC),
Córdoba 14004, Spain

Abstract. Gait analysis offers a non-invasive, scalable approach to infer individual health information using vision-based methods. In this work, we propose a multi-task learning framework that simultaneously estimates several health-related indicators grouped into four categories: biometric identification, body composition, body measures, and physical activity traits. Our model operates solely on silhouette gait sequences, avoiding the need for wearable sensors, depth cameras, or pose annotations. Unlike prior approaches that focus on isolated tasks or rely on multimodal inputs, our method leverages shared spatiotemporal representations to jointly predict diverse health factors from video alone. We conduct extensive experiments using the Health&Gait dataset, which includes 398 individuals walking naturally in indoor conditions with clinically relevant annotations. We show that grouping tasks by physiological correlation improves performance across model backbones, revealing structure in the health representation space. Results demonstrate that multi-task learning improves prediction accuracy for most tasks compared to single-task baselines, particularly benefiting from correlations between related attributes. These findings support the viability of gait-based health modeling as a contactless and privacy-conscious tool for comprehensive health profiling. Our work lays the groundwork for the development of generalist models for preventive health monitoring.

Keywords: Gait analysis · Health assessment · Multi-task learning · Biometrics

1   Introduction

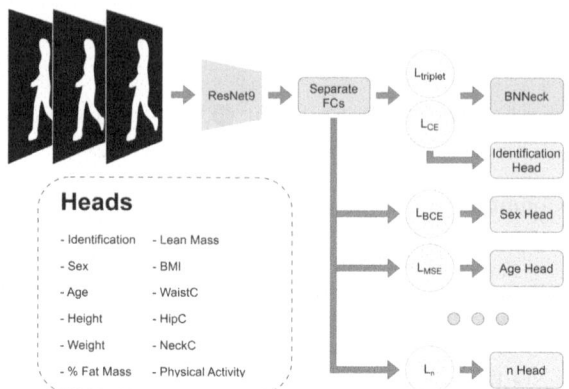

Fig. 1. Proposed framework for the estimation of multiple health-related measures. Input silhouettes are processed by a ResNet9 backbone, followed by separate fully connected layers (FCs) for each task. The model outputs predictions for multiple tasks, described in the "Heads" box. Task-specific losses such as Triplet loss ($L_{triplet}$), Cross-Entropy loss (L_{CE}), Binary Cross-Entropy loss (L_{BCE}), and Mean Squared Error loss (L_{MSE}) are used to optimize the respective task heads. The BNNeck layer is included for the identification head to improve performance.

Healthcare systems worldwide often face periods of saturation due to a high influx of patients with diverse and chronic conditions [2]. In this context, non-invasive, automated tools can assist medical professionals in preliminary health assessments and decision-making. Vision-based analysis offers a promising direction, enabling passive, scalable evaluation of individuals without physical contact or active participation. For instance, camera-based systems combined with predictive models can support early detection and reduce the burden on healthcare infrastructure [1].

Gait, as a behavioral biometric, encodes valuable information about an individual's physical characteristics and functional health. Gait analysis has traditionally been applied to identity recognition [10], but recent studies have demonstrated its potential for estimating demographic and medical traits such as age [14], sex [4], and even early-stage conditions such as scoliosis [23]. Importantly, silhouette-based gait models preserve privacy while still capturing discriminative spatiotemporal patterns. These advantages make gait a strong candidate for integration into contactless health screening pipelines in clinical and non-clinical environments.

Despite these benefits, most prior works address single-task settings and focus on a limited number of attributes, often requiring multi-view setups or wearable sensors. The estimation of a broader set of health-related indicators from vision alone remains an underexplored challenge. To our knowledge, this is the first work that extends gait-based analysis to jointly estimate a broad range of 12 health attributes using a single end-to-end framework trained exclusively on silhouette data. Prior studies focus on 2–4 attributes or require multimodal input. We show that appropriate task grouping enables accurate estimation of clinically relevant traits without additional sensors or pose estimation.

In this work, we propose a deep multi-task learning framework that predicts multiple demographic, anthropometric, and behavioral health indicators from silhouette gait sequences. Our method builds upon the OpenGait framework [6], integrating multi-task prediction heads into two backbone architectures: GaitBase and DeepGaitV2. We evaluate our approach on the Health&Gait dataset [20], which contains over 1,500 videos from 398 subjects annotated with up to twelve clinical and behavioral variables.

Our contributions are as follows: i) We design and implement a multi-task prediction framework over two gait analysis backbones to estimate a diverse set of health indicators from silhouette sequences. ii) We conduct comprehensive experiments using the Health&Gait dataset, demonstrating the feasibility of using gait to infer demographic, anthropometric, and behavioral traits. iii) We investigate feature combinations and task grouping strategies to identify effective learning synergies in multi-task settings. And, iv) we analyze the impact of backbone architecture on multi-task performance and generalization.

The remainder of this paper is organized as follows. Section 2 reviews relevant prior work on gait analysis and multi-task learning. Section 3 presents our proposed multi-task learning framework and training strategy. Section 4 describes the dataset and presents a comparative analysis of the tasks. Section 5 reports and discusses the results across multiple tasks and backbones. Finally, Sect. 6 concludes the paper and outlines directions for future work.

2 Related Works

Gait analysis has been widely explored as a means to assess physical and neurological health conditions. Tracking gait patterns over time enables the detection of subtle changes that may indicate underlying disorders, facilitating early intervention and improving long-term outcomes [3]. Among gait parameters, gait speed is one of the most frequently studied for the prevention and diagnosis of aging-related conditions such as mental health decline [8] and balance impairments [16]. These measurements are commonly obtained using environmental sensors, motion capture systems, wearable devices [13], or inertial measurement units (IMUs) [9]. In contrast, our work relies solely on monocular vision using silhouettes, offering a passive, privacy-conscious alternative that removes the need for specialized hardware.

Beyond general health assessment, gait information has also been leveraged in clinical contexts, supporting rehabilitation monitoring [11] and aiding in the treatment of various disorders [18]. In biomechanical studies, gait features are used to assess fall risk, especially in elderly populations [17], and to evaluate differences between post-surgery and healthy patients [7]. While these studies demonstrate the medical value of gait, they typically focus on predefined features or require clinical-grade setups. Our approach instead learns predictive representations directly from video, enabling broader and more scalable health profiling.

Several works have applied deep learning to estimate subject attributes using gait data. In [12], the authors proposed a multi-task model to predict identity, sex, and age simultaneously, demonstrating that joint training improves performance. Similarly, [21] estimated age and sex, while [22] extended the scope to include BMI alongside sex and age. However, these studies focus on a small number of output variables and do not explore the combined estimation of biometric, anthropometric, and behavioral traits. In contrast, our work targets more than ten health indicators across those categories, using a unified multi-task framework.

Other approaches rely on multi-stage pipelines or multimodal inputs. In [15], separate models are used in a sequential fashion to estimate sex and age. Meanwhile, [19] employed wearable sensors to predict sex, height, and weight, though this requires active user participation and is less suitable for passive, vision-based deployment. Our method avoids these limitations by using a single-stage model trained end-to-end from silhouette video, without the need for multiple models or sensor-based inputs.

In summary, gait analysis provides a non-invasive and versatile source of information for health monitoring. From rehabilitation assessment to fall risk prevention, the integration of objective gait features into clinical workflows offers valuable insights into an individual's condition. Continued progress in multi-task learning and standardized gait analysis protocols will further strengthen its role in early diagnosis and personalized healthcare. Our work contributes to this goal by demonstrating the feasibility of large-scale, privacy-aware estimation of diverse health indicators using silhouette-based gait analysis.

3 Methodology

We propose a generalizable multi-task learning framework that extends existing gait recognition models with task-specific heads for health-related prediction. The framework is designed to estimate a wide range of biometric, anthropometric, and behavioral indicators from gait silhouettes, using a shared gait encoder and multiple output branches.

Architecture Overview. As illustrated in Fig. 1, the model receives sequences of silhouette images as input and processes them through a backbone encoder, either GaitBase [6] or DeepGaitV2 [5]. These encoders extract high-level gait features, which are then fed into parallel task-specific heads. For each health-related target, we design a prediction head tailored to its output type.

The model supports 12 supervised tasks, as listed in Fig. 1. These include biometric attributes (e.g., sex, age), anthropometric measures (e.g., weight, BMI, waist circumference), and physical activity indicators (PA_level). Each head consists of fully connected (FC) layers whose depth and activation functions are adapted to the task type.

Specifically, classification tasks such as sex and physical activity level use a single FC layer with softmax activations. Regression tasks employ a deeper stack of three FC layers with linear activations. This design was selected empirically to stabilize training and reduce overfitting on continuous variables.

In the case of identification (not used during health prediction but shown in Fig. 1 for completeness), we include a BNNeck module and apply a Triplet loss alongside Cross-Entropy, following standard gait recognition practices.

Loss Formulation. Let T be the total number of supervised tasks. For each active task $t \in \{1, \ldots, T\}$, the model predicts either a continuous value or a categorical label. The total multi-task loss is computed as a weighted sum of task-specific losses: $\mathcal{L}_{\text{total}} = \sum_{t=1}^{T} \lambda_t \mathcal{L}_t$. Each $\mathcal{L}_t$ is selected based on the task type:

$$
\mathcal{L}_t =
\begin{cases}
-\frac{1}{N} \sum_{i=1}^{N} \left[y_i \log \sigma(\hat{y}_i) + (1 - y_i) \log(1 - \sigma(\hat{y}_i)) \right], & \text{Classification} \\
\frac{1}{N} \sum_{i=1}^{N} (y_i - \hat{y}_i)^2, & \text{Regression}
\end{cases}
\tag{1}
$$

where y_i is the ground truth, $\hat{y}_i$ the prediction, and $\sigma(\cdot)$ the sigmoid function for classification tasks. We use Binary Cross-Entropy for binary classification, and Mean Squared Error for regression targets.

The weight λ_t is set to 1 for each active task during training and 0 otherwise. Although this scheme treats all tasks equally, more advanced weighting strategies could be explored in future work.

A summary of the tasks and their mapping to loss types is shown in Fig. 1. Categorical targets include identification, sex and physical activity level, whereas all anthropometric variables are treated as continuous and predicted via regression.

Implementation Details. We implement our heads on top of two backbones from the OpenGait framework: (a) GaitBase [6]: a lightweight ResNet-like model optimized for robustness in constrained and unconstrained settings. (b) Deep-GaitV2 [5]: a deeper architecture that integrates 3D convolutions for temporal modeling of silhouette sequences.

Both models are initialized with random weights and extended with our modular head design. The input silhouettes are normalized and resized to a fixed shape of 64×64 pixels. Each input sequence contains 30 frames sampled at fixed intervals.

Training is conducted for 20,000 iterations using the AdamW optimizer with an initial learning rate of 0.001. The learning rate decays at 10k, 17k, and 19k iterations with a decay factor of 0.1. For comparison purposes, when multiple heads are active, all loss weights are set equally to 1 (on a scale from 0 to 1). Training and evaluation were conducted using the four data partitions provided by the authors of the Health&Gait dataset. Each partition defines a different train-test split to assess robustness across subject groupings. Final results are reported as the average performance across these four splits. All experiments are executed on a dual-GPU workstation with two NVIDIA Titan XP GPUs and 64 GB of RAM. Training takes approximately 8 h per fold.

4   Experimental Analysis

This section presents our experimental setup, including details of the dataset, a correlation analysis among prediction targets, and our task grouping strategy. The goal of these analyses is to explore whether related health indicators can be effectively predicted using a shared multi-task learning framework. We begin by describing the Health&Gait dataset, followed by a study of inter-task correlations, which informs our model design. Finally, we define task groups based on semantic and statistical relationships to support structured evaluation.

4.1   Dataset

We conduct all experiments on the Health&Gait dataset [20], which comprises 1,564 gait video sequences from 398 participants from 18 to 64 years old with a balanced ratio between females and males. Each participant performed walking trials along a 10-meter corridor at two gait speeds: usual gait speed (UGS) and fast gait speed (FGS), and under two clothing conditions: with jacket (WJ) and without jacket (WoJ). Videos were recorded in RGB format at a resolution of 1920×1080 pixels and 30 frames per second, using a lateral single-view setup.

Participants' demographic and physiological data were collected, including age, sex, height, weight, waist, hip, and neck circumferences, body fat percentage, and lean mass. Body mass index (BMI) was derived from height and weight. In addition, subjects self-reported their average physical activity level using a standardized questionnaire. Post-processing provided multiple data modalities per sequence, including silhouettes, pose estimates, optical flow, and body segmentation maps.

In this study, we use only the silhouette modality, which offers privacy-preserving input while retaining motion dynamics crucial for gait-based prediction. We aim to estimate the following 13 attributes: Identification (ID), Sex, Age, Height, Weight, Percentage of Fat Mass (%FatMass), Lean Mass (L_Mass), Body Mass Index (BMI), Waist Circumference (WaistC), Hip Circumference (HipC), Neck Circumference (NeckC), and Physical Activity Level (PA_level).

4.2   Task Correlation Analysis

To explore interdependencies between prediction targets, we compute pairwise Spearman correlation coefficients using the ground-truth labels in the Health&Gait dataset. As shown in Fig. 2, several strong correlations emerge across anthropometric variables. For instance, Sex is strongly correlated with Height ($\rho = 0.76$), Lean Mass (0.82), and Neck Circumference (0.79). Similarly, Weight correlates highly with Waist Circumference (0.87), Neck Circumference (0.85), and Lean Mass (0.87). BMI shows strong correlations with Waist (0.85) and Hip Circumference (0.77), reflecting its dependence on body fat distribution. Conversely, Fat Mass shows moderate to negative correlations with Lean Mass (-0.31) and Sex (-0.52).

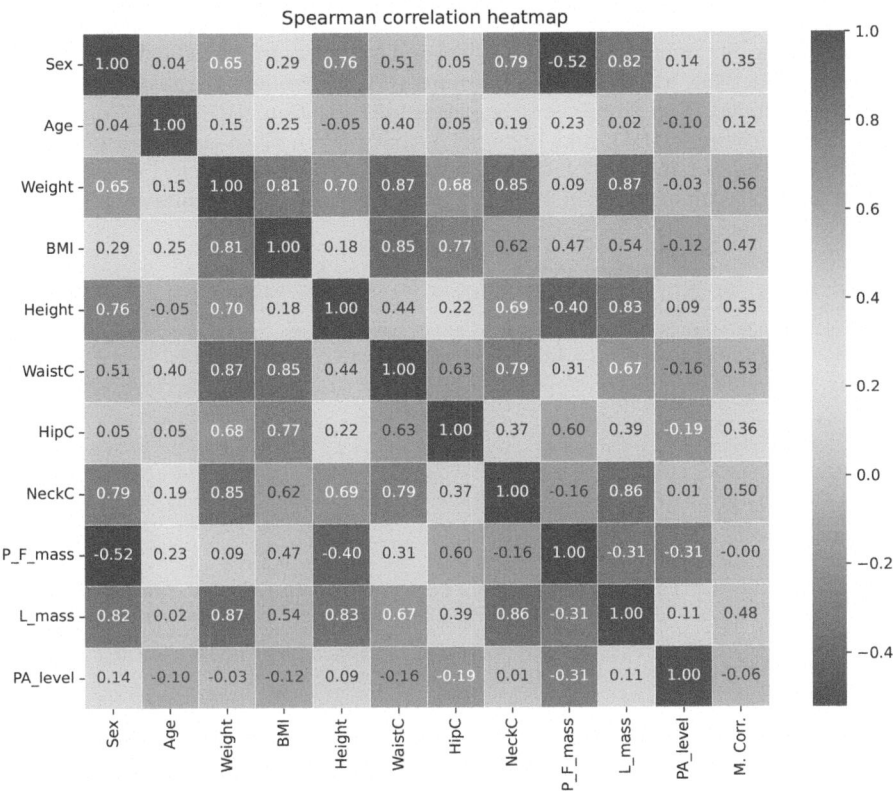

Fig. 2. Spearman correlation heatmap. Pairwise relationships between anthropometric and physical activity variables. Each cell represents the Spearman correlation coefficient, with positive correlations shown in red and negative in blue. (Best viewed in digital format)

Interestingly, Physical Activity Level exhibits low or negligible correlation with most anthropometric traits (e.g., BMI: -0.12; WaistC: -0.16), supporting its interpretation as a behaviorally independent variable. These insights guide the design of our multi-task prediction framework, where highly correlated tasks are grouped for shared representation learning, while weakly correlated ones like PA_level are modeled independently.

4.3 Task Grouping Strategy

Based on domain knowledge and correlation analysis, we organize the prediction targets into four semantically coherent categories. **Biometric Identification** includes traits tied to individual uniqueness, commonly used in person re-identification and demographic profiling. **Body Composition** focuses on physical attributes related to muscle-to-fat balance and metabolic health. **Body Measures** captures anthropometric dimensions that characterize body

Table 1. Comparison of the results of the single-task (ST) and multi-task (MT) models. M1 refers to GaitBase model and M2 refers to DeepGaitV2 model. The multi-task model was also divided into groups (G), where only the tasks of the same group were trained. The groups are Biometric Identification (BI), Body Composition (BC), Body Measures (BM), and Physical Activity (PA). Each row corresponds to a task. For each related metric, ↑ higher is better and ↓ lower is better. The magnitudes and loss functions can be found in Table 2. For each task, the best results are highlighted in the background.

G	Task	M1 ST	M1 MT (G)	M1 MT	M2 ST	M2 MT (G)	M2 MT
	ID ↑	99.41 ± 0.32	99.23 ± 0.73	98.64 ± 0.95	99.68 ± 0.39	99.74 ± 0.3	99.48 ± 0.56
BI	Sex ↑	94.04 ± 2.24	95.03 ± 1.69	94.78 ± 1.77	94.68 ± 2.18	94.71 ± 2.28	94.48 ± 2.57
	Age ↓	9.15 ± 1.47	8.49 ± 0.38	8.49 ± 0.6	8.48 ± 0.53	8.54 ± 0.63	8.47 ± 0.31
	He ↓	4.44 ± 1.54	3.78 ± 0.66	3.89 ± 0.50	4.78 ± 2.06	4.12 ± 0.93	4.65 ± 0.4
	We ↓	8.06 ± 1.72	7.16 ± 1.14	7.93 ± 1.74	10.18 ± 2.52	9.08 ± 1.83	8.52 ± 2
BC	FMa ↓	5.05 ± 1.03	5.16 ± 1.43	4.97 ± 1.44	6.02 ± 2.67	5.45 ± 1.56	4.89 ± 1.66
	LMa ↓	6.76 ± 2.15	5.59 ± 0.61	5.56 ± 0.97	6.58 ± 2.27	5.76 ± 1.89	6.19 ± 1.48
	BMI ↓	2.82 ± 0.75	2.44 ± 0.53	2.61 ± 0.53	3.63 ± 0.67	3.17 ± 0.99	2.61 ± 0.7
	WC ↓	6.4 ± 1.39	6.14 ± 1.37	6.22 ± 1.34	8.99 ± 2.08	9.43 ± 1.43	6.53 ± 1.48
BM	HC ↓	15.15 ± 11.52	12.84 ± 13.44	15.55 ± 12.02	15.48 ± 12.43	16.98 ± 14.18	16.63 ± 12.69
	NC ↓	1.75 ± 0.06	1.68 ± 0.09	1.64 ± 0.12	2.2 ± 0.27	2.36 ± 0.61	1.69 ± 0.17
P	PA ↑	56.8 ± 3.25	56.8 ± 3.25	58.03 ± 3.60	57.75 ± 5.1	57.75 ± 5.1	59.04 ± 1.17

morphology and structural traits. Finally, **Physical Activity** is treated as a standalone task due to its behavioral nature and weak correlation with other variables.

This grouping serves two goals: (i) it enables us to test the hypothesis that learning related tasks jointly yield performance gains, and (ii) it supports mod-

Table 2. Overview of the tasks considered for multi-task learning. Each task is associated with its evaluation magnitude and the specific loss function employed during training.

ID	Task Name	Metric
ID	Identification	Fast@R1 (%)
Sex	Sex	Accuracy (%)
Age	Age	MAE (years)
He	Height	MAE (cm)
We	Weight	MAE (Kg)
FMa	Fat Mass	MAE (%)
LMa	Lean Mass	MAE (Kg)
BMI	Body Mass Index	MAE (kg/m^2)
WC	Waist Circ.	MAE (cm)
HC	Hip Circ.	MAE (cm)
NC	Neck Circ.	MAE (cm)
PA	Physical Act. Lev.	Accuracy (%)

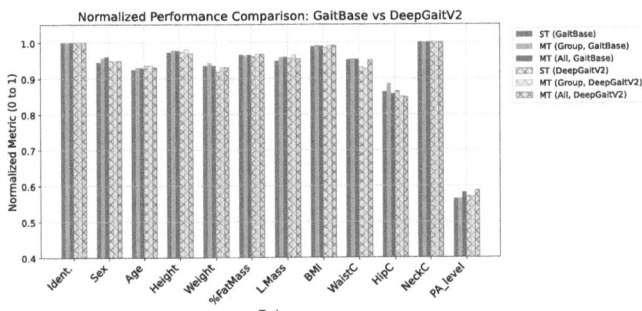

Fig. 3. Comparison of normalized metrics (scaled from 0 to 1) for single-task (ST) and multitask (MT) learning experiments across various tasks. The graph shows the performance of single-task mean (ST), multitask group mean (MT Group), and multitask overall mean (MT All) for both GaitBase (solid bars) and DeepGaitV2 (textured bars). (Best viewed in digital format)

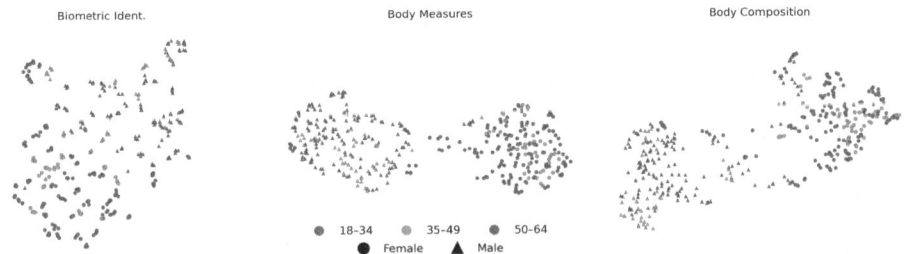

Fig. 4. UMAP projection of the feature embeddings obtained from models trained on three different task groups: *Biometric Identification* (ID, Sex, Age), *Body Measures* (Height, Weight, % Fat Mass, Lean Mass, BMI), and *Body Composition* (Waist, Hip, Neck Circumference). The plots are colored by age group (18–34, 35–49, 50–64) and shaped by sex (• Female, ▲ Male). The visualizations highlight how the feature spaces encode demographic attributes, revealing clustering patterns associated with age and sex. (Best viewed in digital format.)

ular training and ablation analysis. Section 5 presents comparative experiments on single-task, grouped-task, and full multi-task learning setups.

5 Results and Discussion

This section presents a comprehensive evaluation of our proposed multi-task learning (MTL) framework. We analyze prediction performance across 12 health-related tasks, compare two backbone architectures (GaitBase and DeepGaitV2), and explore the effects of task grouping on model accuracy. Our analysis includes both quantitative metrics and qualitative embedding visualizations.

5.1   Quantitative Evaluation

Table 1 compares the results for each task under three configurations: single-task (ST), grouped-task multitask (MT (G)), and full multitask (MT), using both GaitBase (M1) and DeepGaitV2 (M2) backbones. For each metric, higher or lower values indicate better performance depending on the task type. Best-performing values are highlighted.

Biometric Identification (BI). Identification (ID) achieves its highest accuracy in the single-task setting for both backbones (M2-ST: 99.68 ± 0.39), suggesting that isolated training is best suited for this classification task. However, both Age and Sex show clear gains under the grouped multitask setting. For instance, Age estimation improves from 9.15 ± 1.47 (M1-ST) to 8.49 ± 0.38 (M1-MT(G)). This supports the hypothesis that related demographic features benefit from joint representation learning.

Body Composition (BC). Performance varies across configurations. For Gait-Base, multi-task grouping consistently improves results for Weight, Lean Mass, and BMI. For example, Weight MAE drops from 8.06 ± 1.72 (ST) to 7.16 ± 1.14 (MT(G)). However, full multitask training (M1-MT) sometimes degrades performance slightly, likely due to task interference.

Body Measures (BM). Results show mixed trends. Some variables such as Neck Circumference are predicted equally well across all configurations (e.g., M1: $1.75 \rightarrow 1.68 \rightarrow 1.64$), whereas others such as Hip Circumference see improvements under M1-MT(G) (12.48 ± 13.44 vs. 15.55 ± 12.02 in M1-ST).

Physical Activity (PA). This task stands apart. While largely independent from other features (see correlation analysis in Sect. 4.2), PA_level estimation benefits slightly from full multitask learning, especially in M2-MT (59.04 ± 1.17 vs. 57.75 ± 5.1 in M2-ST).

5.2   Normalized Comparison

Figure 3 provides a normalized comparison across all tasks and models. Metrics were scaled to a 0–1 range to enable joint visualization. For metrics where lower values are better (e.g., MAE), scores were inverted.

This comparison highlights several trends. Most tasks benefit from multitask training, particularly under the grouped configuration (MT(G)). DeepGaitV2, despite its higher complexity, does not significantly outperform the simpler GaitBase model on average. Additionally, performance gains obtained through grouped multitask training tend to be more stable than those achieved with full multitask setups.

These findings confirm that shared learning among semantically and statistically related tasks enhances performance while reducing the risk of interference.

5.3   Embedding Visualization and Feature Space Analysis

To qualitatively assess the learned representations, we visualize the feature embeddings of GaitBase using UMAP projections in Fig. 4. Each subplot corresponds to a model trained on a specific task group: Biometric Identification,

Body Measures, and Body Composition. Embeddings are colored by age group (18–34, 35–49, 50–64) and shaped by sex (• Female, ▲ Male).

The feature space learned by the Body Measures model exhibits the clearest separation by sex. Distinct clusters of male and female subjects are visible, reflecting the sexual dimorphism inherent in traits such as waist, hip, and neck circumferences. Interestingly, this separation is more pronounced than in the Biometric Identification model, despite the latter being explicitly trained to classify sex. This suggests that predicting sex-correlated physical traits encourages the emergence of a feature space organized by sex, even without a dedicated classification objective.

The Biometric Identification embedding does reveal some structured patterns, particularly along the age axis, with age groups showing a degree of spatial grouping. However, sex-based separation is weaker, likely due to the model's focus on individual identity rather than demographic class boundaries.

The Body Composition embedding also shows meaningful structure, particularly along the sex axis. Male and female subjects tend to form distinct clusters, with a clear spatial separation between the two groups. This is likely due to the influence of sex on body composition variables such as fat mass, lean mass, and BMI, which exhibit well-documented physiological differences. While age-based clustering is less pronounced than in the Biometric Identification embedding, the model appears to learn sex-sensitive features, even without explicit supervision on demographic labels.

5.4 Error Analysis and Limitations

While multitask learning generally improves prediction, some tasks underperform in the full MT configuration compared to the grouped variant. For example, Weight and Hip Circumference slightly degrade under full multitask with GaitBase. This points to representational conflict between unrelated tasks.

Another observation is that more complex models such as DeepGaitV2 do not yield consistent performance gains. In some cases, they introduce higher variance (e.g., Lean Mass: 6.58 ± 2.27 in M2-ST vs. 6.19 ± 1.48 in M2-MT). This may reflect overfitting or inefficient feature use due to task interference. One plausible explanation is that DeepGaitV2, originally designed for person identification, is highly optimized for discriminative feature learning via mechanisms such as BNNeck and triplet loss. These components enforce tight intra-class clustering and inter-class separation, which benefit classification tasks like ID but may constrain the flexibility of the learned embeddings for regression-based health prediction. In contrast, GaitBase offers a simpler and more general architecture, which may allow it to adapt more easily to a broader set of prediction tasks under multi-task learning supervision.

Physical Activity remains the most challenging task, with relatively low performance across all configurations. This is likely due to its behavioral and self-reported nature, which makes it harder to estimate from visual cues alone.

6   Conclusions

This study presents a comprehensive multi-task learning framework for estimating a wide range of health-related measures from gait silhouette videos. Unlike prior work that focuses primarily on age or sex estimation, our approach predicts up to 12 variables spanning biometric traits, body composition, and physical activity levels. We demonstrate that task grouping improves learning stability and accuracy, and that lightweight models such as GaitBase are well-suited for these multi-output estimations, sometimes outperforming deeper alternatives like DeepGaitV2.

Our adaptation of the Health&Gait dataset to the OpenGait framework provides a reproducible experimental benchmark, enabling future research in vision-based health monitoring. The experimental findings suggest that gait-based prediction can offer privacy-preserving, non-invasive insights into a patient's physical condition, supporting early detection and preventative healthcare in clinical and ambient settings.

For future work, we will investigate dynamic loss weighting strategies, more flexible task grouping schemes, and architecture modifications tailored to regression-heavy multitask settings. We also plan to evaluate generalization under real-world variability, including different camera views, walking surfaces, and clothing occlusions by testing in different datasets. Additionally, we aim to revisit our frame sampling strategy to explore the benefits of using longer input sequences.

Acknowledgments. This work was supported by the Spanish Ministry of Economy, Industry and Competitiveness (Grant No. PID2023-147296NB-I00), and the National Natural Science Foundation of China (Grant No. 62476120).

Disclosure of Interests. The authors have no competing interests to declare that are relevant to the content of this article.

References

1. Albahri, A.S., et al.: Iot-based telemedicine for disease prevention and health promotion: state-of-the-art. J. Netw. Comput. Appl. **173** (2021)
2. Badr, S., Nyce, A., Awan, T., Cortes, D., Mowdawalla, C., Rachoin, J.: Measures of emergency department crowding, a systematic review. How to make sense of a long list. Open Access Emergency Med. **14**, 5–14 (2022)
3. Cai, Y., Qian, X., Cao, H., Zheng, J., Xu, W., Huang, M.C.: mHealth technologies toward active health information collection and tracking in daily life: a dynamic gait monitoring example. IEEE Internet Things J. **9**(16), 15077–15088 (2022)
4. Do, T.D., Nguyen, V.H., Kim, H.: Real-time and robust multiple-view gender classification using gait features in video surveillance. Pattern Anal. App. **23**(1), 399–413 (2020)
5. Fan, C., Hou, S., Huang, Y., Yu, S.: Exploring deep models for practical gait recognition. CoRR **abs/2303.03301** (2023)

6. Fan, C., Liang, J., Shen, C., Hou, S., Huang, Y., Yu, S.: Opengait: revisiting gait recognition towards better practicality. In: Proceedings of the IEEE/CVF Conference on CVPR, pp. 9707–9716 (2023)
7. Fuchs, S., Flören, M., Skwara, A., Tibesku, C.: Quantitative gait analysis in unconstrained total knee arthroplasty patients. Int. J. Rehabil. Res. **25**(1), 65–70 (2002)
8. Hahm, K.S., Chase, A.S., Dwyer, B., Anthony, B.W.: Indoor human localization and gait analysis using machine learning for in-home health monitoring. In: Proceedings of the IEEE EMBC, pp. 6859–6862 (2021)
9. Kim, M.J., Han, J.H., Shin, W.C., Hong, Y.S.: Gait pattern identification using gait features. Electronics **13**(10) (2024)
10. Lee, L., Grimson, W.: Gait analysis for recognition and classification. In: Proceedings of Fifth IEEE International Conference on Automatic Face Gesture Recognition, pp. 155–162 (2002)
11. Lianzhen, C., Hua, Z.: Athlete rehabilitation evaluation system based on internet of health things and human gait analysis algorithm. Complexity **2021** (2021)
12. Marín-Jiménez, M., Castro, F., Guil, N., de la Torre, F., Medina-Carnicer, R.: Deep multi-task learning for gait-based biometrics. In: IEEE ICIP, pp. 106–110 (2017)
13. Saboor, A., et al.: Latest research trends in gait analysis using wearable sensors and machine learning: a systematic review. IEEE Access **8**, 167830–167864 (2020)
14. Sakata, A., Makihara, Y., Takemura, N., Muramatsu, D., Yagi, Y.: Gait-based age estimation using a DenseNet. In: Carneiro, G., You, S. (eds.) ACCV 2018. LNCS, vol. 11367, pp. 55–63. Springer, Cham (2019). https://doi.org/10.1007/978-3-030-21074-8_5
15. Sakata, A., Takemura, N., Yagi, Y.: Gait-based age estimation using multi-stage convolutional neural network. IPSJ Trans. Comput. Vis. Appl. **11**(1), 1–10 (2019). https://doi.org/10.1186/s41074-019-0054-2
16. Schniepp, R., Möhwald, K., Wuehr, M.: Key gait findings for diagnosing three syndromic categories of dynamic instability in patients with balance disorders. J. Neurol. **267**(1), 301–308 (2020). https://doi.org/10.1007/s00415-020-09901-5
17. Silva, J., Atalaia, T., Abrantes, J., Aleixo, P.: Gait biomechanical parameters related to falls in the elderly: a systematic review. Biomechanics **4**(1), 165–218 (2024)
18. Werner, C., Awai Easthope, C., Curt, A., Demko, L.: Towards a mobile gait analysis for patients with a spinal cord injury: a robust algorithm validated for slow walking speeds. Sensors **21**(21) (2021)
19. Xing, H., Ren, L., Zhao, Y.: The gender, height and weight prediction based on gait using mems sensor. In: IEEE 11th Annual International Conference on CYBER Technology in Automation, Control, and Intelligent Systems, pp. 830–834 (2021)
20. Zafra-Palma, J., Marín-Jiménez, N., Castro-Piñero, J., Cuenca-García, M., Muñoz-Salinas, R., Marín-Jiménez, M.J.: Health & gait: a dataset for gait-based analysis. Sci. Data **12**(1) (2025)
21. Zhang, S., Wang, Y., Li, A.: Gait-based age estimation with deep convolutional neural network. In: 2019 International Conference on Biometrics (ICB), pp. 1–8 (2019)
22. Zhang, Y., Huang, Y., Wang, L., Yu, S.: A comprehensive study on gait biometrics using a joint CNN-based method. Pattern Recogn. **93**, 228–236 (2019)
23. Zhou, Z., Liang, J., Peng, Z., Fan, C., An, F., Yu, S.: Gait patterns as biomarkers: a video-based approach for classifying scoliosis. In: MICCAI, pp. 284–294 (2024)

Enhancing Synthetic CT from CBCT via Multimodal Fusion and End-To-End Registration

Maximilian Tschuchnig[1,3]([🖂]) [iD], Lukas Lamminger[2], Philipp Steininger[2], and Michael Gadermayr[1] [iD]

[1] Salzburg University of Applied Sciences, Puch bei Hallein, Austria
[2] MedPhoton GmbH, Salzburg, Austria
[3] University of Salzburg, Salzburg, Austria
`maximilian.tschuchnig@fh-salzburg.ac.at`

Abstract. Cone-Beam Computed Tomography (CBCT) is widely used for intraoperative imaging due to its rapid acquisition and low radiation dose. However, CBCT images typically suffer from artifacts and lower visual quality compared to conventional Computed Tomography (CT). A promising solution is synthetic CT (sCT) generation, where CBCT volumes are translated into the CT domain. In this work, we enhance sCT generation through multimodal learning by jointly leveraging intraoperative CBCT and preoperative CT data. To overcome the inherent misalignment between modalities, we introduce an end-to-end learnable registration module within the sCT pipeline. This model is evaluated on a controlled synthetic dataset, allowing precise manipulation of data quality and alignment parameters. Further, we validate its robustness and generalizability on two real-world clinical datasets. Experimental results demonstrate that integrating registration in multimodal sCT generation improves sCT quality, outperforming baseline multimodal methods in 79 out of 90 evaluation settings. Notably, the improvement is most significant in cases where CBCT quality is low and the preoperative CT is moderately misaligned.

Keywords: Synthetic CT · Registration · Multimodal Learning · Deep Learning

1 Introduction

Mobile robotic imaging systems, such as cone-beam computed tomography (CBCT) [11] provide real-time, intraoperative imaging, facilitating guidance during medical procedures. This is especially useful, for example, in radiation therapy [13]. CBCT uses a cone-shaped X-ray beam to acquire 3D images in a single rotation, reducing acquisition time and radiation exposure. However, CBCT images typically suffer from lower contrast-to-noise ratios and increased artifacts compared to CT scans [17].

M. Castrillón-Santana et al. (Eds.): CAIP 2025, LNCS 15622, pp. 62–73, 2026.
https://doi.org/10.1007/978-3-032-05060-1_6

One approach to mitigating CBCT artifacts, as highlighted by Altalib et al. [1], is the conversion of high-artifact CBCT volumes into a low-artifact CT domain using image-to-image translation. sCTs [1,2] aim to leverage domain advantages, here the acquisition speed and mobility of CBCT while improving image quality by reducing artifacts. The importance of robust sCT was underscored by the SynthRad Grand Challenge 2023 [13], which exposed key limitations of existing methods, especially their sensitivity to CBCT artifacts and anatomical misalignment, and motivated the need for techniques that are more robust to artifacts and anatomical misalignment.

However, sCT generation faces two key limitations. The methods cannot create new anatomical details beyond those present in the input data, and the effectiveness of training models generating sCT depends on the availability of high-quality training data. While this second limitation can be partially addressed by data augmentation, addressing the first limitation requires integrating additional information such as high-quality preoperative CT scans into the sCT generation process [1–3,14]. Tschuchnig et al. [14], for example, combine high quality, preopera-tivce CT with intraoperative CBCT for multimodal sCT generation. These high-quality, preoperative CT scans are typically acquired before a procedure to aid in treatment planning. For instance, in radiotherapy, treatment plans are initially based on high-resolution preoperative CT scans [6]. However, since intraoperative CBCT and preoperative CT are typically acquired at different times and under varying anatomical conditions, accurately aligning them remains an essential component of any multimodal sCT pipeline.

As shown by Tschuchnig et al., the fusion of CBCT-CT for multimodal sCT improves on unimodal sCT approaches in most cases [14,18]. This fusion allows the model to learn from the complementary strengths of both modalities: CBCT provides real-time anatomical information during the procedure, while CT offers high-quality structural detail acquired during planning. Multimodal learning, effectively leverages this complementary information. However, a key challenge in multimodal sCT generation is the inherent misalignment between intraoperative CBCT and preoperative CT, caused by anatomical changes and differences in patient positioning. While traditional multimodal fusion approaches often rely on external pre-registration, Tschuchnig et al. [14] show that U-Net based multimodal approaches improve on unimodal, even in misaligned scenarios. However, they still reported a strong gap between their observed implicit registration and potential additional registration. We focus on integrating a learnable registration component directly into the multimodal sCT generation pipeline. Specifically, we employ Spatial Transformer Networks (STN) [7] to enable end-to-end optimization of both the registration and synthesis tasks, allowing the network to jointly optimize registration and synthesis, and adaptively align preoperative CT to intraoperative CBCT during training.

The main contributions of this work are manifold. First, we propose an end-to-end multimodal sCT generation framework that integrates a STN to explicitly address anatomical misalignment between intraoperative CBCT and preoperative CT. Second, we conduct an extensive evaluation on the effect of the STN component, comparing performance against conventional unimodal and multimodal baselines without registration. Third, using a synthetic dataset with controlable quality parameters, we investigate the effects of CBCT quality and CBCT-CT alignment on the effect of STN. Fourth, we demonstrate the robustness and reproducibility of our approach across two real-world clinical datasets.

2 Methodology and Materials

The proposed 3D multimodal sCT generation method combines deep learning with multimodal learning by adapting a 3D image reconstruction model, specifically 3D U-Net [4], with early fusion multimodal learning [10]. To facilitate CBCT-CT alignment, a STN module is added before the first layer of the 3D U-Net, aligning the preoperative CT with the intraoperative CBCT. The full architecture is illustrated in Fig. 1. The model F takes as input a combination of preoperative, unaligned CT volumes U_{CT} and intraoperative CBCT volumes V_{CBCT}. U_{CT} is aligned to V_{CBCT} using the STN. To accomplish this the STN takes the concatenation of $[U_{CT}, V_{CBCT}]$ to estimate affine registration parameters, aimed at registering U_{CT} to V_{CBCT}, resulting in V_{CT}. Then, the volumes V_{CBCT} and V_{CT} are concatenated along a fourth dimension, analogous to how RGB channels are treated in 2D images, to form the combined volume $V = [V_{CT}, V_{CBCT}]$. This volume is then fed into a reconstruction U-Net to generate the corresponding sCT $\hat{Y}$. Using $\hat{Y}$ and the preoperative, aligned CT, Y, the loss is calculated with $L(\hat{Y}, Y)$.

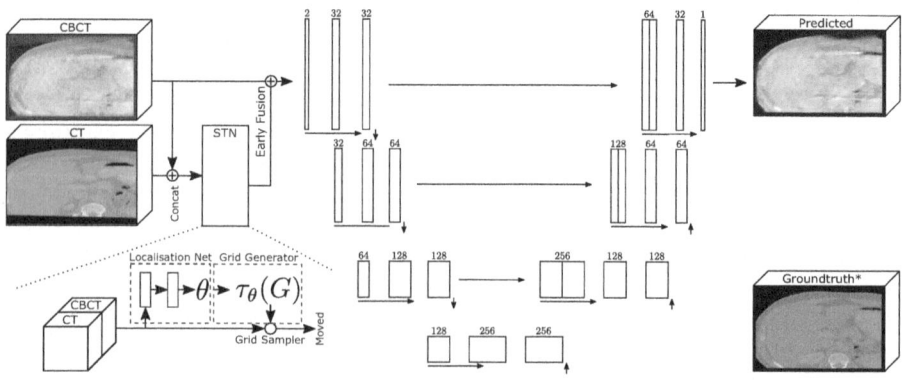

Fig. 1. Proposed 3D, (early fusion) multimodal sCT model, based on U-Net.

Specifically, the model begins with an STN component, followed by an encoder consisting of three double convolutional blocks, each with $3 \times 3 \times 3$ kernels, connected by 3D max pooling layers. The latent space contains one additional double convolutional block, followed by a decoder that mirrors the encoder structure. Each encoder block is connected to its corresponding decoder block via skip connections. To facilitate final image reconstruction, a $1 \times 1 \times 1$ 3D convolutional layer with a single output channel is added at the end of the decoder. Since no transfer function is applied to this layer, it outputs logits directly. The number of feature maps at each level is $\{32, 64, 128, 256\}$, and batch normalization is applied after each convolutional layer.

The STN module is implemented as a CNN with three 3D convolutional layers containing $\{16, 32, 64\}$ filters and kernel sizes of $\{7, 5, 3\}$, respectively. Max pooling is applied after the first convolutional layer. Following the convolutional stack, adaptive average pooling is used to compress the spatial dimensions before passing the features to the parameter estimation head. This head consists of two fully connected layers with output sizes $\{32, 12\}$. The 12 predicted parameters represent an affine transformation, which is then converted into a sampling grid and applied to the moving (preoperative) CT using a grid sampler. Both the U-Net and STN use ReLU activation functions throughout.

To train the model we use a weighted sum of losses. Specifically, voxel based, patch based and perceptual losses, as typically used in image reconstruction and image-to-image based models [1,2,8]. As a voxel based loss we apply mean absolute error (MAE) to focus on preserving image structure and robustness to noise. As a patch based loss, the structured similarity index measure (SSIM) is applied to also focus on the patch information luminance, contrast and structure. To convert the similarity into a loss function, we utilize $1 - SSIM(\hat{Y}, Y)$. Additionally, a perceptual loss [8] based on vgg16 [12], pretrained on the imagenet dataset [5] is added to preserve perceptual information. In detail, we use the pretrained vgg16 to extract mid-level feature maps (vgg16 up to and including layer 23) from the sCT and compare these feature maps to the original CTs vgg16 feature maps using MAE. Therefore, our perceptual loss is defined as $MAE(vgg_{:23}(\hat{Y}), vgg_{:23}(Y))$. The final U-Net loss is given by $\alpha_1 \cdot MAE(\hat{Y}, Y) + \alpha_2 \cdot 1 - SSIM(\hat{Y}, Y) + \alpha_3 \cdot MAE(vgg_{:23}(\hat{Y}), vgg_{:23}(Y))$ with $\alpha_1 = 0.2$, $\alpha_2 = 0.1$ and $\alpha_3 = 0.7$. To tune the STN, an additional registration loss is added, defined as $10^{-3} \cdot MSE(V_{CT}, Y)$.

Loss weights were empirically chosen based on validation metrics and qualitative assessment. The perceptual loss was prioritized to improve sharpness and anatomical detail. In qualitative observations, relying solely on perceptual loss led to hallucinated structures, which were mitigated by incorporating the voxel-wise MAE term. SSIM further complemented both by encouraging local structural coherence and contrast preservation. To facilitate reproducibility, all source code used for these experiments is publicly available on GitHub at: https://github.com/MaxTschuchnig/Enhan cingSyntheticCTfromCBCTviaMultimodalFusionandEnd-To-EndRegistration.

As baselines, unimodal sCT models are applied using only CBCT as input and multimodal models without STN [14]. To assess whether the multimodal model primarily relies on the unaligned CT while disregarding the intraoperative CBCT, we introduce an additional baseline (CT-only), which uses only the unaligned CT as input and compares the output to the perfectly aligned CT.

However, due to the inherent unalignment of CBCT and CT, this baseline is imperfect in real-world datasets.

2.1   Dataset

To investigate the effect of the STN, three datasets, one synthetic and two real-world datasets are used:

1. CBCT Liver Tumor Segmentation Benchmark (CBCTLiTS) dataset [16]. CBCTLiTS consists of 131 synthetic CBCT and corresponding real CT images of the abdominal region. The paired volumes are perfectly aligned and available in five different quality levels, enabling a controlled study of intraoperative image quality (α_{np}).
2. Pancreatic-CT-CBCT-SEG dataset [6] (pancreas dataset) The pancreas dataset provides 40 CBCT-CT pairs of the abdominal region, serving as a challenging benchmark to validate our findings in a clinical setting. Unlike CBCTLiTS, the pancreas dataset includes naturally occurring misalignment and relatively uniform CBCT quality.
3. SynthRAD2023 [13] dataset (synthrad). Synthrad provides 360 paired CT-CBCT and CT-MR volumes across brain and pelvis anatomies, acquired from three clinical centers. The dataset includes both rigidly registered CBCT and MRI images to planning CTs, enabling sCT generation tasks across varying acquisition protocols and image qualities. For our experiments, we chose the target of pelvis sCT generation from CBCT.

All datasets used are publicly available: CBCTLiTS under CC BY-NC-ND 4.0, Pancreatic-CT-CBCT-SEG under CC BY 4.0, and SynthRAD under CC BY-NC 4.0.

In the real-world datasets (pancreas and synthrad), CBCT-CT alignment is not perfect (even if performed manually or semi-automatically) due to patient movement and respiratory variations. Even with controlled breathing techniques designed to minimize motion-induced misalignment, perfect CBCT-CT alignment remains difficult to achieve [6]. This is an important detail which potentially affects both, training (training with imperfect pairs) and evaluation (applying measures on imperfectly aligned pairs).

To assess the impact of alignment on multimodal sCT reconstruction, the parameter α_a is introduced. This factor, defined as α_a, describes how strongly the preoperative CT is artificially unaligned. α_a reduces the number of parameters controlling unalignment by combining multiple affine unalignments (rotation, scaling and translation) into a single parameter. Artificial unalignment was performed using TorchIO RandomAffine. In detail, affine misaligned was performed using random (non-isotropic) scaling, with the scaling parameter sampled from $\mathcal{U}(1 - 0.5 \cdot \alpha_a, 1 + 0.5 \cdot \alpha_a)$, rotation, parameters sampled from $\mathcal{U}(-22.5 \cdot \alpha_a, 22.5 \cdot \alpha_a)$, and translation, with the parameter sampled from $\mathcal{U}(0, 0.05 \cdot \alpha_a)$ with tri-linear interpolation. Sample results of this unalignment on the CBCTLiTS dataset are shown in Fig. 2. All datasets were unaligned in the same way.

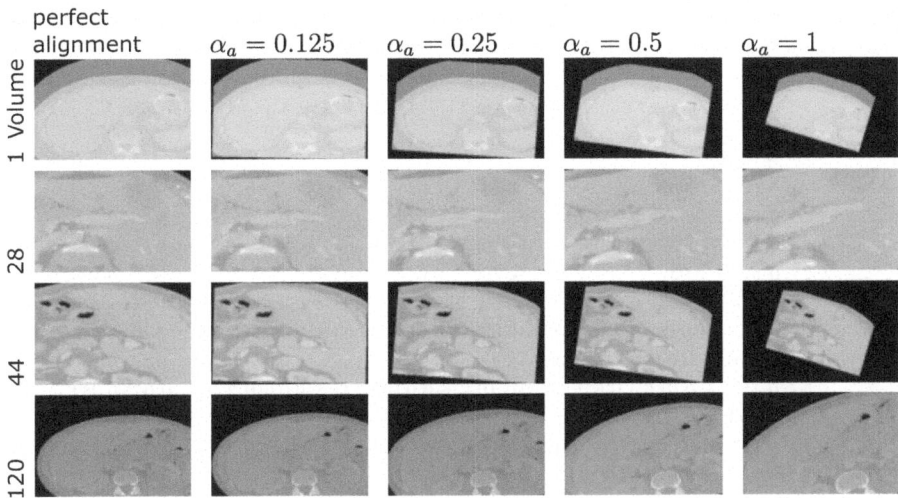

Fig. 2. Sample results, showing the original CT as well as synthetically unaligned version of the same CT (CBCTLiTS dataset). Four different volumes are shown with increasing α_a (unalignment), displaying rotation, scaling and minor translation.

To quantify the unalignment effect of α_a, the average voxel distance between the original and unaligned volumes was calculated using the perfectly paired CBCTLiTS dataset. The resulting average voxel-wise displacement (in voxels) for different α_a values were $\alpha_a = 0 \rightarrow 0, \alpha_a = 0.125 \rightarrow 2.6, \alpha_a = 0.25 \rightarrow 3.9, \alpha_a = 0.5 \rightarrow 6, \alpha_a = 1 \rightarrow 7.8$, corresponding to physical distances of approximately $0\,\mathrm{mm}, 2.1\,\mathrm{mm}, 3.1\,\mathrm{mm}, 4.8\,\mathrm{mm}, 6.3\,\mathrm{mm}$, respectively. These values cover a realistic range from mild misalignment (typical of breathing-induced motion) to more substantial, clinically relevant misalignments, in line with previously reported registration errors of up to $10\,\mathrm{mm}$ for lung SBRT patients during CBCT-CT alignment [9].

2.2 Experimental Details

All models were trained on a Ubuntu server using NVIDIA RTX A6000 graphics cards, using Pytorch. Due to the large data size and 48 GB VRAM memory limit, CBCTLiTS, pancreas data and synthrad volumes were isotropically downscaled by a factor of two [15]. The dataset was split into training, validation, and testing subsets using a ratio of 0.7 (training), 0.2 (validation), and 0.1 (testing). All experiments were conducted and evaluated using four different random splits to ensure stable results and to report averaged metrics and their standard deviation. To enable comparability, the same data splits and random CT misalignments were applied across all model configurations. All model variants were trained for 100 epochs using the Adam optimizer, with a learning rate of $1 \cdot 10^{-4}$, a weight decay of $1 \cdot 10^{-5}$ for regularization, and gradient accumulation over 8 steps (effective batch size 8) with a per-step batch size of 1.

Table 1. Evaluation results (mean ± std) on the Pancreas Dataset. The results shown include unimodal baselines (Base), multimodal model results [14] (MM) and the proposed MM+STN model results. Values in bold show improvements towards all baselines.

	α_a	MAE	CT-only	1-SSIM	CT-only	Perceptual	CT-only
Base		0.111 ± 0.037		0.094 ± 0.026		0.225 ± 0.025	
MM	1	0.114 ± 0.038	0.243 ± 0.070	0.108 ± 0.020	0.169 ± 0.049	0.227 ± 0.028	0.350 ± 0.065
	0.5	0.114 ± 0.039	0.150 ± 0.044	0.105 ± 0.018	0.135 ± 0.038	0.232 ± 0.024	0.272 ± 0.054
	0.25	0.105 ± 0.038	0.088 ± 0.027	0.100 ± 0.027	0.101 ± 0.028	0.213 ± 0.022	0.195 ± 0.038
	0.125	0.087 ± 0.033	0.051 ± 0.016	0.091 ± 0.020	0.068 ± 0.022	0.172 ± 0.020	0.136 ± 0.027
	0	0.045 ± 0.027	0.000 ± 0.000	0.031 ± 0.013	0.000 ± 0.000	0.096 ± 0.012	0.000 ± 0.000
MM+STN	1	**0.110 ± 0.044**	0.246 ± 0.060	0.124 ± 0.058	0.164 ± 0.048	0.228 ± 0.023	0.359 ± 0.042
	0.5	**0.107 ± 0.036**	0.155 ± 0.029	**0.104 ± 0.032**	0.134 ± 0.030	**0.221 ± 0.025**	0.289 ± 0.029
	0.25	**0.094 ± 0.028**	0.093 ± 0.015	**0.090 ± 0.018**	0.104 ± 0.020	**0.205 ± 0.023**	0.206 ± 0.023
	0.125	**0.080 ± 0.027**	0.053 ± 0.009	**0.071 ± 0.019**	0.071 ± 0.015	**0.155 ± 0.015**	0.142 ± 0.016
	0	**0.041 ± 0.027**	0.000 ± 0.000	**0.018 ± 0.010**	0.000 ± 0.000	**0.082 ± 0.005**	0.000 ± 0.000

3   Results

Tables 3, 1, and 2 summarize the quantitative results of sCT generation across three datasets: synthetic CBCTLiTS, pancreas, and synthrad. The data reported are mean values over four different train-val-test splits, with added standard deviations. Both unimodal and CT-only baselines are given. Overall, the proposed method improves on the baseline multimodal method in 79 out of 90 experimental setups. On the unimodal baseline, the STM+MM method achieves improved results in 83 out of 90 cases and against the CT-only baseline in 6 out of 72 cases. Overall, CT-only baselines show little difference between MM+STN and MM.

On the CBCTLiTS dataset, our proposed MM+STN model consistently outperforms both the multimodal (MM) setup without STN as well as the unimodal and CT-only baselines. The addition, MM+STN has the biggest positive effects on low to moderately aligned setups with low intraoperative CBCT visual quality. In the pancreas dataset, MM+STN improves over MM in all but two cases. Compared to the unimodal and CT-only baselines applied to the pancreas dataset, STM+MM improve results in 12 out of 15 (unimodal) and 8 out of 12 (CT-only) setups respectively. The synthrad dataset shows inconclusive results. While MAE and SSIM metrics worsen in most (70%) cases, Perceptual metrics improve in four out of five cases. In detail, compared to MM, MM+STN improves results in seven out of 15 cases, 11 out of 15 cases compared to the unimodal baseline and 10 out of 12 cases compared to CT-only.

Figure 3 qualitatively evaluates a randomly sampled sCT, generated from different source quality CBCT, CBCT-CT alignment and sCT generation method (multimodal as MM and multimodal with registration as MM+STN).

Table 2. Evaluation results (mean ± std) on the SynthRad Dataset. The results shown include unimodal baselines (Base), multimodal model results [14] (MM) and the proposed MM+STN model results. Values in bold show improvements towards all baselines.

	α_a	MAE	CT-only	1-SSIM	CT-only	Perceptual	CT-only
Base		0.127 ± 0.059		0.149 ± 0.138		0.188 ± 0.039	
MM	1	0.120 ± 0.067	0.304 ± 0.091	0.136 ± 0.107	0.285 ± 0.110	0.198 ± 0.038	0.420 ± 0.068
	0.5	0.121 ± 0.061	0.187 ± 0.053	0.119 ± 0.102	0.194 ± 0.071	0.196 ± 0.038	0.350 ± 0.056
	0.25	0.111 ± 0.050	0.114 ± 0.032	0.131 ± 0.116	0.131 ± 0.046	0.196 ± 0.035	0.257 ± 0.044
	0.125	0.101 ± 0.047	0.069 ± 0.020	0.128 ± 0.109	0.086 ± 0.032	0.184 ± 0.028	0.181 ± 0.034
	0	0.055 ± 0.028	0.000 ± 0.000	0.077 ± 0.111	0.000 ± 0.000	0.101 ± 0.021	0.000 ± 0.000
MM+STN	1	0.127 ± 0.063	0.312 ± 0.083	0.142 ± 0.122	0.289 ± 0.109	0.198 ± 0.037	0.429 ± 0.058
	0.5	0.121 ± 0.055	0.194 ± 0.047	0.146 ± 0.122	0.198 ± 0.071	**0.191 ± 0.037**	0.358 ± 0.049
	0.25	0.114 ± 0.054	0.117 ± 0.029	**0.123 ± 0.103**	0.135 ± 0.048	**0.192 ± 0.037**	0.264 ± 0.042
	0.125	**0.097 ± 0.043**	0.071 ± 0.018	0.130 ± 0.114	0.089 ± 0.034	**0.163 ± 0.024**	0.186 ± 0.032
	0	**0.053 ± 0.025**	0.000 ± 0.000	0.077 ± 0.109	0.000 ± 0.000	**0.096 ± 0.018**	0.000 ± 0.000

4 Discussion

Our results on synthetic data demonstrate that the addition of STN into the MM model systematically improves alignment between preoperative CT and intraoperative CBCT, generally enhancing sCT generation (79 out of 90 cases). In nearly all setups with moderate to low misalignment, the inclusion of STN led to consistent improvements over the multimodal model without STN, the unimodal CBCT baseline, and the CT-only baseline. This confirms that end-to-end learning with STN is effective for compensating moderate anatomical discrepancies, particularly when training data closely resemble the synthetic setup where near-perfect CBCT-CT alignment exists.

However, performance declines at high misalignment levels ($\alpha_a = 1$), suggesting that STN struggles with extreme spatial transformations. This is especially noticable in real-world data scenarios with fuzzy ground truth registration targets. In such cases, external pre-registration, possibly informed by clinical expertise, may be necessary to bring inputs into a roughly aligned space before fine-tuning with STN. Also, for extreme misalignment, reverting to unimodal sCT models may be more effective due to their simplicity and consistency.

Results on real-world clinical datasets largely mirror trends from synthetic data, indicating strong generalizability. While the SynthRad dataset yielded mixed results (improvements in 47% of cases), the pancreas dataset showed notable improvements (improvements in 87% of cases) over the multimodal method without STN. Interestingly, STN also provided benefits in well-aligned settings, potentially due to its ability to capture subtle anatomical variations not addressed by conventional registration or static model assumptions present in the used real-world datasets.

Despite its benefits, STN occasionally led to decreases in MAE and 1-SSIM on the SynthRad dataset. However, perceptual metrics generally improved across

Table 3. Evaluation results (mean ± std) on the CBCTLiTS Dataset. The results shown include unimodal baselines (Base), multimodal model results [14] (MM) and the proposed MM+STN model results. Values in bold show improvements towards all baselines.

	α_{np}	α_a	MAE	CT-only	1-SSIM	CT-only	Perceptual	CT-only
Base	32		0.306 ± 0.128		0.399 ± 0.064		0.391 ± 0.056	
MM	32	1	0.302 ± 0.116	0.764 ± 0.216	0.396 ± 0.070	0.668 ± 0.097	0.392 ± 0.054	0.666 ± 0.056
	32	0.5	0.280 ± 0.101	0.525 ± 0.124	0.361 ± 0.066	0.561 ± 0.079	0.388 ± 0.055	0.594 ± 0.061
	32	0.25	0.241 ± 0.075	0.348 ± 0.076	0.354 ± 0.075	0.458 ± 0.074	0.380 ± 0.054	0.475 ± 0.063
	32	0.125	0.206 ± 0.057	0.219 ± 0.050	0.320 ± 0.066	0.341 ± 0.066	0.353 ± 0.055	0.343 ± 0.057
	32	0	0.060 ± 0.020	0.000 ± 0.000	0.086 ± 0.044	0.000 ± 0.000	0.187 ± 0.019	0.000 ± 0.000
MM+STN	32	1	**0.281 ± 0.114**	0.847 ± 0.267	**0.357 ± 0.070**	0.696 ± 0.104	**0.380 ± 0.055**	0.670 ± 0.081
	32	0.5	**0.253 ± 0.095**	0.575 ± 0.158	**0.323 ± 0.062**	0.582 ± 0.086	**0.378 ± 0.055**	0.617 ± 0.082
	32	0.25	**0.207 ± 0.068**	0.380 ± 0.092	**0.303 ± 0.088**	0.477 ± 0.077	**0.348 ± 0.059**	0.504 ± 0.079
	32	0.125	**0.164 ± 0.049**	0.241 ± 0.055	**0.234 ± 0.086**	0.363 ± 0.069	**0.316 ± 0.050**	0.372 ± 0.071
	32	0	0.062 ± 0.019	0.000 ± 0.000	**0.081 ± 0.042**	0.000 ± 0.000	**0.185 ± 0.018**	0.000 ± 0.000
Base	64		0.289 ± 0.122		0.358 ± 0.061		0.380 ± 0.057	
MM	64	1	0.282 ± 0.112	0.764 ± 0.216	0.357 ± 0.060	0.668 ± 0.097	0.380 ± 0.055	0.666 ± 0.056
	64	0.5	0.262 ± 0.093	0.525 ± 0.124	0.326 ± 0.057	0.561 ± 0.079	0.377 ± 0.056	0.594 ± 0.061
	64	0.25	0.233 ± 0.082	0.348 ± 0.076	0.316 ± 0.074	0.458 ± 0.074	0.371 ± 0.056	0.475 ± 0.063
	64	0.125	0.200 ± 0.062	0.219 ± 0.050	0.316 ± 0.073	0.341 ± 0.066	0.343 ± 0.054	0.343 ± 0.057
	64	0	0.069 ± 0.016	0.000 ± 0.000	0.096 ± 0.049	0.000 ± 0.000	0.194 ± 0.017	0.000 ± 0.000
MM+STN	64	1	**0.268 ± 0.118**	0.847 ± 0.267	**0.322 ± 0.064**	0.696 ± 0.104	**0.370 ± 0.057**	0.670 ± 0.081
	64	0.5	**0.243 ± 0.093**	0.575 ± 0.158	**0.300 ± 0.067**	0.582 ± 0.086	**0.365 ± 0.055**	0.617 ± 0.082
	64	0.25	**0.200 ± 0.069**	0.380 ± 0.092	**0.276 ± 0.074**	0.477 ± 0.077	**0.346 ± 0.054**	0.504 ± 0.079
	64	0.125	**0.161 ± 0.049**	0.241 ± 0.055	**0.227 ± 0.082**	0.363 ± 0.069	**0.313 ± 0.052**	0.372 ± 0.071
	64	0	**0.062 ± 0.015**	0.000 ± 0.000	**0.087 ± 0.048**	0.000 ± 0.000	**0.185 ± 0.016**	0.000 ± 0.000
Base	128		0.281 ± 0.122		0.328 ± 0.068		0.372 ± 0.059	
MM	128	1	0.275 ± 0.113	0.764 ± 0.216	0.331 ± 0.072	0.668 ± 0.097	0.369 ± 0.055	0.666 ± 0.056
	128	0.5	0.250 ± 0.102	0.525 ± 0.124	0.296 ± 0.070	0.561 ± 0.079	0.368 ± 0.057	0.594 ± 0.061
	128	0.25	0.219 ± 0.087	0.348 ± 0.076	0.293 ± 0.069	0.458 ± 0.074	0.362 ± 0.055	0.475 ± 0.063
	128	0.125	0.191 ± 0.063	0.219 ± 0.050	0.280 ± 0.076	0.341 ± 0.066	0.343 ± 0.051	0.343 ± 0.057
	128	0	0.074 ± 0.021	0.000 ± 0.000	0.097 ± 0.048	0.000 ± 0.000	0.192 ± 0.016	0.000 ± 0.000
MM+STN	128	1	**0.263 ± 0.104**	0.847 ± 0.267	**0.301 ± 0.073**	0.696 ± 0.104	**0.361 ± 0.053**	0.670 ± 0.081
	128	0.5	**0.242 ± 0.103**	0.575 ± 0.158	**0.267 ± 0.079**	0.582 ± 0.086	**0.361 ± 0.058**	0.617 ± 0.082
	128	0.25	**0.203 ± 0.079**	0.380 ± 0.092	**0.249 ± 0.077**	0.477 ± 0.077	**0.349 ± 0.056**	0.504 ± 0.079
	128	0.125	**0.160 ± 0.051**	0.241 ± 0.055	**0.211 ± 0.081**	0.363 ± 0.069	**0.309 ± 0.054**	0.372 ± 0.071
	128	0	**0.065 ± 0.021**	0.000 ± 0.000	**0.094 ± 0.054**	0.000 ± 0.000	**0.188 ± 0.014**	0.000 ± 0.000
Base	256		0.275 ± 0.126		0.313 ± 0.072		0.360 ± 0.062	
MM	256	1	0.273 ± 0.118	0.764 ± 0.216	0.312 ± 0.068	0.668 ± 0.097	0.363 ± 0.060	0.666 ± 0.056
	256	0.5	0.250 ± 0.097	0.525 ± 0.124	0.268 ± 0.069	0.561 ± 0.079	0.357 ± 0.059	0.594 ± 0.061
	256	0.25	0.222 ± 0.087	0.348 ± 0.076	0.264 ± 0.068	0.458 ± 0.074	0.353 ± 0.057	0.475 ± 0.063
	256	0.125	0.184 ± 0.063	0.219 ± 0.050	0.260 ± 0.070	0.341 ± 0.066	0.339 ± 0.050	0.343 ± 0.057
	256	0	0.070 ± 0.023	0.000 ± 0.000	0.095 ± 0.049	0.000 ± 0.000	0.197 ± 0.014	0.000 ± 0.000
MM+STN	256	1	**0.267 ± 0.116**	0.847 ± 0.267	**0.295 ± 0.074**	0.696 ± 0.104	**0.354 ± 0.058**	0.670 ± 0.081
	256	0.5	**0.239 ± 0.100**	0.575 ± 0.158	**0.259 ± 0.076**	0.582 ± 0.086	**0.350 ± 0.058**	0.617 ± 0.082
	256	0.25	**0.203 ± 0.073**	0.380 ± 0.092	**0.231 ± 0.070**	0.477 ± 0.077	**0.338 ± 0.058**	0.504 ± 0.079
	256	0.125	**0.151 ± 0.046**	0.241 ± 0.055	**0.198 ± 0.072**	0.363 ± 0.069	**0.303 ± 0.049**	0.372 ± 0.071
	256	0	**0.065 ± 0.016**	0.000 ± 0.000	**0.091 ± 0.051**	0.000 ± 0.000	**0.188 ± 0.013**	0.000 ± 0.000

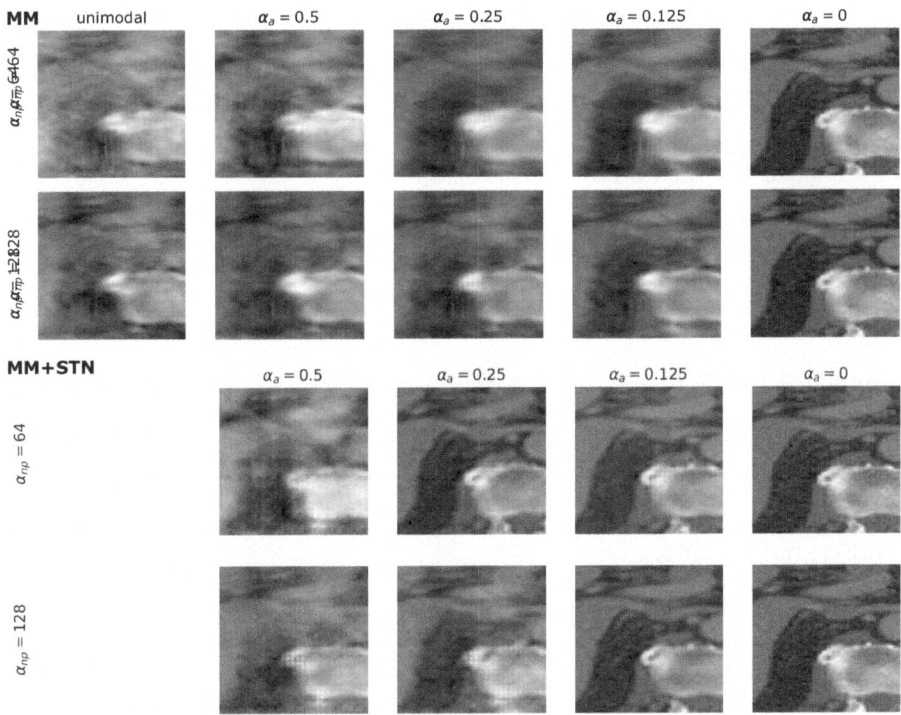

Fig. 3. Qualitative comparison of sCT reconstructions on the CBCTLiTS dataset. Rows correspond to different CBCT quality levels (top: low, bottom: high). Columns compare methods: a unimodal CBCT baseline (left) and multimodal reconstructions conditioned on preoperative CTs with increasing alignment (right). Images are normalized for visualization and shown at increased zoom to better highlight anatomical structures. Results of both multimodal (MM) and multimodal with STN registration (MM+STN) are shown.

all datasets and setups. This suggests that STN may enhance spatial coherence and anatomical realism even if pixelwise similarity does not improve uniformly. This qualitative improvement is especially visible in the presented CBCTLiTS qualitative results figure, where STN clearly enhances fine structural detail.

To conclude, we demonstrated how the STN can be seamlessly integrated into a U-Net based sCT generation framework to address anatomical misalignment in multimodal imaging and how this additional STN reacts to the parameters of intraoperative CBCT quality and CBCT-CT alignment. Our findings suggest that STN-based alignment consistently enhances sCT quality across both synthetic and clinical datasets, particularly under moderate misalignment. We find STN to be especially useful if the intraoperative volume is of low visual quality. These insights highlight the potential of learnable registration to improve intraoperative imaging workflows and motivate further research into robust, adaptive multimodal learning strategies for clinical deployment.

Acknowledgments. This project was partly funded by the Austrian Research Promotion Agency (FFG) under the bridge project "CIRCUIT: Towards Comprehensive CBCT Imaging Pipelines for Real-time Acquisition, Analysis, Interaction and Visualization" (CIRCUIT), no. 41545455 and by the county of Salzburg under the project AIBIA and the Salzburg University of Applied Sciences under the project FHS-trampoline-8 (Applied Data Science Lab).

Disclosure of Interests. Not applicable.

References

1. Altalib, A., McGregor, S., Li, C., Perelli, A.: Synthetic CT image generation from CBCT: a systematic review. IEEE Trans. Radiat. Plasma Med. Sci. (2025)
2. Chen, L., Liang, X., Shen, C., Jiang, S., Wang, J.: Synthetic CT generation from CBCT images via deep learning. Med. Phys. **47**(3), 1115–1125 (2020)
3. Chen, L., Liang, X., Shen, C., Nguyen, D., Jiang, S., Wang, J.: Synthetic CT generation from CBCT images via unsupervised deep learning. Phys. Med. Biol. **66**(11), 115019 (2021)
4. Çiçek, Ö., Abdulkadir, A., Lienkamp, S.S., Brox, T., Ronneberger, O.: 3D U-net: learning dense volumetric segmentation from sparse annotation. In: Medical Image Computing and Computer-Assisted Intervention, pp. 424–432 (2016)
5. Deng, J., Dong, W., Socher, R., Li, L.J., Li, K., Fei-Fei, L.: ImageNet: a large-scale hierarchical image database. In: IEEE Conference on Computer Vision and Pattern Recognition, pp. 248–255 (2009)
6. Hong, J., et al.: CT and cone-beam CT of ablative radiation therapy for pancreatic cancer with expert organ-at-risk contours. Sci. Data **9**(1), 637 (2022)
7. Jaderberg, M., Simonyan, K., Zisserman, A., et al.: Spatial transformer networks. In: Advances in Neural Information Processing Systems, vol. 28 (2015)
8. Johnson, J., Alahi, A., Fei-Fei, L.: Perceptual losses for real-time style transfer and super-resolution. In: Leibe, B., Matas, J., Sebe, N., Welling, M. (eds.) ECCV 2016. LNCS, vol. 9906, pp. 694–711. Springer, Cham (2016). https://doi.org/10.1007/978-3-319-46475-6_43
9. Oechsner, M., Chizzali, B., Devecka, M., Combs, S.E., Wilkens, J.J., Duma, M.N.: Registration uncertainties between 3D cone beam computed tomography and different reference CT datasets in lung stereotactic body radiation therapy. Radiat. Oncol. **11**, 1–10 (2016)
10. Podobnik, G., Strojan, P., Peterlin, P., Ibragimov, B., Vrtovec, T.: Multimodal CT and MR segmentation of head and neck organs-at-risk. In: Medical Image Computing and Computer-Assisted Intervention, pp. 745–755 (2023)
11. Rafferty, M.A., et al.: Intraoperative cone-beam CT for guidance of temporal bone surgery. Otolaryngol.-Head Neck Surg. **134**(5), 801–808 (2006)
12. Simonyan, K., Zisserman, A.: Very deep convolutional networks for large-scale image recognition. In: International Conference on Learning Representations (2015)
13. Thummerer, A., et al.: SynthRAD 2023 grand challenge dataset: generating synthetic CT for radiotherapy. Med. Phys. **50**(7), 4664–4674 (2023)
14. Tschuchnig, M., Lamminger, L., Steininger, P., Gadermayr, M.: Enhancing synthetic CT from CBCT via multimodal fusion: a study on the impact of cbct quality and alignment. arXiv preprint arXiv:2506.08716 (2025)

15. Tschuchnig, M.E., Coste-Marin, J., Steininger, P., Gadermayr, M.: Multi-task learning to improve semantic segmentation of CBCT scans using image reconstruction. In: BVM Workshop, pp. 243–248 (2024)
16. Tschuchnig, M.E., Steininger, P., Gadermayr, M.: CBCTLiTS: a synthetic, paired CBCT/CT dataset for segmentation and style transfer. In: Eurographics Workshop on Visual Computing for Biology and Medicine (2024)
17. Wei, C., et al.: Reduction of cone-beam CT artifacts in a robotic CBCT device using saddle trajectories with integrated infrared tracking. Med. Phys. (2024)
18. Zhao, Z., et al.: Equivariant multi-modality image fusion. In: IEEE Conference on Computer Vision and Pattern Recognition, pp. 25912–25921 (2024)

Increasing Resolution of MRI Volumes Through Interleaved Slice Synthesis Based on Generative Adversarial Networks

Giuseppe Rauso[1], Mara Sangiovanni[1], Silvio Barra[1], Daniel Riccio[1], Mariacarla Staffa[2], Lorenzo D'Errico[1], and Francesco Longobardi[1]([⊠])

[1] DIETI - Department of Electrical Engineering and Information Technology, University of Naples Federico II, Via Claudio 21, 80125 Naples, Italy
{giuseppe.rauso,mara.sangiovanni,silvio.barra,daniel.riccio,lorenzo.derrico, francesco.longobardi3}@unina.it
[2] University of Naples Parthenope, Centro Direzionale di Napoli, Isola C4, 80143 Naples, Italy
mariacarla.staffa@uniparthenope.it

Abstract. In modern medical diagnostics, Magnetic Resonance Imaging (MRI) has emerged as an indispensable tool, playing a fundamental role in diagnosing several diseases. However, the widespread use of dated diagnostic machinery often results in MRI scans of limited slice count and suboptimal quality. Such limitations can considerably impede diagnostic accuracy, particularly in complex cases like small lesion detection, early-stage tumor identification, and fine structural anomalies. Enhancing the resolution and overall image quality of these MRI scans holds the potential to boost diagnostic precision for an array of conditions significantly. Furthermore, the rise of deep learning AI techniques for segmentation and classification has amplified the need for large datasets, ensuring that models can effectively generalize across diverse imaging scenarios. In recent times, Generative Adversarial Networks (GANs) have garnered attention for their ability to generate high-fidelity synthetic images, which could potentially bridge the gap between existing diagnostic needs and the current limitations of MRI scans. This paper introduces a new GAN architecture designed to enhance the resolution of MRI volumes for detailed structure analysis, synthesizing new slices from existing adjacent ones in the MRI volume. Such a system increases data quantity and broadens the model's adaptability, enhancing modeling accuracy. The experiments demonstrate superior accuracy, robustness, and computational efficiency compared to traditional methods.

Keywords: GANs · MRI · Super-resolution · Data augmentation

1 Introduction

In modern medical diagnostics, Magnetic Resonance Imaging (MRI) has emerged as an indispensable tool for diagnosing several diseases. However, the widespread

M. Castrillón-Santana et al. (Eds.): CAIP 2025, LNCS 15622, pp. 74–85, 2026.
https://doi.org/10.1007/978-3-032-05060-1_7

reliance on outdated diagnostic equipment frequently leads to MRI scans with fewer slices and reduced image quality. Such limitations can considerably impede diagnostic accuracy, particularly in complex cases like small lesion detection, early-stage tumor identification, and fine structural anomalies. Moreover, the paucity of training data often hampers the application of deep learning techniques in the medical image domain. Data augmentation is a standard approach to increase the number of images used for the training phase. In many domains, standard techniques include cropping, padding, Gaussian noise addition, flipping, and deformation of the original images. However, these approaches are unsuitable for medical images: the basic operations may produce the disruption of the logical image structure, unreal image deformations, and the generation of aberrant data. These limitations, in turn, adversely affect the performance of image processing models. To address the issue, several data augmentation techniques have been devised to increase diversity, reduce imbalance, and improve learning performances while preserving medical image inner coherence [6]. Among them, deep learning based approaches are emerging for several classification tasks and different image modalities [8]. Generative approaches, such as Generative Adversarial Networks (GAN), Variational AutoEncoders (VAE), and diffusion models, are gaining momentum due to their ability to produce realistic images while being more adherent to the true distribution of the data [13]. Among them, GANs have proved to be an effective approach to synthetic image generation in the medical domain [24].

Here, we present a new GAN architecture to enhance the resolution of MRI volumes, synthesizing new slices from existing adjacent ones. The contribution of the proposed methodology is twofold: i) super-resolution: the method produces a new image constrained between two consecutive slices. When done in a volume for all the slices along a dimension, the newly generated images could be interleaved with the existing ones, thus generating a new 3D volume of higher resolution; ii) augmentation: generating new 2D images might be useful to widen existing datasets.

1.1 Related Works

Post-processing of MR images has become an essential step, with a particular emphasis on image reconstruction (IR) and super-resolution (SR). SR addresses the problem of building high-resolution (HR) images starting from low-resolution (LR) ones. However, the higher the scaling factor, the more complex and error-prone the restoration process. The problem has been addressed using a wide range of methodologies [20] that, in more recent years, include deep learning [15,19]. SR strategies can be divided into two broad categories: 2D, which aims to increase the resolution of single images, whereas in 3D, the information recovered is volumetric and distributed across several slices along different dimensions [23]. Among the various approaches to SR, including Convolutional Neural Networks (CNNs) [28], Diffusion models [29], and Transformers [10], Generative Adversarial Networks (GANs) [27] are showing great promise. This is mainly due to their adversarial learning process, which guides GANs in producing high-fidelity synthetic images without the need for as much data as the Transformers, without the computational demands of Diffusion models [14], and without suffering the depth/gradient vanishing problems that affect CNNs [26]. GANs have been used in 2D SR tasks to obtain HR images from LR images, such as in [12]. However, all those approaches use only a single slice and do not exploit the information coming from adjacent slices. Few works are going in this direction: in [22] an unsupervised optical flow approach was used to build 3D isotropic cardiac MR reconstructions. The generation of HR images is guided by newly generated slices obtained by training a GAN to generate intermediate slices from adjacent ones. In [21] the authors use a transformer-based architecture to learn intermediate slices from the upper and lower adjacent HR slices. In [25], the problem of generating complex 3D volumes is decomposed into successive cross-sectional 2D mappings across single rectilinear dimensions. The authors of [18] train an autoencoder to compress and reconstruct high-resolution slices taken from highly anisotropic volumes. The authors of [16] propose a hybrid model trained on consecutive LR axial slices. In [17], the authors propose a multi-slice 2D network that combines a GAN with feature fusion and a pre-trained slice interpolation network to obtain 3D super-resolution. These works differ from the proposed approach, which relies only on two consecutive HR images to generate a new HR slice. Moreover, it does not need a vast amount of training data, as is the case with transformer architectures, which are notoriously data-greedy. Concerning 3D volume augmentation, some works have been proposed, which often suffer from the scarcity of volumetric training data.

The synthetic slice generation approach proposed here considerably differs from state-of-the-art approaches under three different aspects: i) generation approach. The method focuses on enhancing resolution by synthesizing new MRI slices from adjacent ones, aiming for comprehensive tissue characterization [5]; ii) purpose and application. Aimed at improving resolution and details in MRI volumes, these synthetic MRIs support quantitative analysis and standardized imaging, enhancing diagnostic precision; iii) data quantity and adaptability. By increasing resolution and data quantity, this GAN-based approach offers greater model adaptability and accuracy, supporting tailored research needs.

2   The Proposed Architecture

The underlying idea of this work is to devise a generative adversarial network to generate an intermediate MRI slice from two consecutive ones. This way, the dataset's volume increases with greater accuracy as the network produces non-existent frames that could be intermediate scans. The basic GAN framework consists of two neural networks: a Generator G and a Discriminator D, trained together in a minimax game. The proposed architecture takes in input 256×256 pixel images, dividing each into overlapping 128×128 patches used for GAN-based synthesis. These patches, treated as independent training units, are then generated and reassembled into the full image. This reassembly corrects overlaps by averaging pixel values from intersecting patches, ensuring consistent and accurate image reconstruction. The generator adopts a U-Net as the backbone network. It has been selected due to its capability to transfer a vast amount of low-level information directly through the network from input to output.

In our setting, the generator receives two consecutive images, applies compression to both, concatenates the images post-transformation, and proceeds along the expansion path. In the latter, the outputs from deconvolutions are concatenated with corresponding encoding outputs (skip-connections). Each encoding and decoding block includes batch normalization levels after the convolution levels. The LeakyReLU activation function is utilized, preventing saturation—a situation where the network can no longer learn due to 'gradient vanishing' or excessive noise from approximations. The compression path consists of three encoding blocks with a filter number of 32-64-128. Each block applies a sequence of Conv2D-BatchNorm-LeakyReLU twice, followed by a 2×2 max pooling operation. The network's central part concatenates the two images post-compression and applies two sequences of Conv2D-BatchNorm-LeakyReLU, using eight kernels for the convolutions. The result is then passed to the decoder, which consists of 3 upsampling blocks that sequentially apply UpSampling2D-Conv2D-Concatenate (skip connections) and then Conv2D-BatchNorm-LeakyReLU twice. All kernels used for convolutions have a size of 3×3. The encoder outputs used as skip-connections are the concatenation of the intermediate outputs of the encoder, with the first image passed as input, and the same intermediate outputs passed with the second image as input.

The discriminator, which is a critic in this case, takes in three images and assesses whether the middle one could be a plausible 'intermediate' image between the first and third. Based on the original publication about the Wasserstein GANs (WGANs) [2], the last layer's activation function is linear rather than sigmoidal. The three input images are overlaid in a Concatenate layer. The result enters the downsampling path. The first downsampling block uses 32 filters for convolutions, the second and third use 64, and the fourth uses 128. Each block applies a sequence Conv2D(stride $= 2 \times 2$)-LeakyReLU-Conv2D-LeakyReLU. In this case, downsampling is performed using a 2×2 stride in the first convolutional layer of each block, while the filters have a size of 5×5. After the last downsampling block, the result is flattened and fed into a fully connected layer with a single neuron and the identity function as activation. Since we use gradient penalty [9] in the objective function, batch normalization is not helpful because it alters the discriminator problem's shape by mapping a batch of inputs to a batch of outputs, but using this penalty, the norm of the critic's gradient is penalized with respect to each input independently, rather than the entire batch, as highlighted in the paper.

The Generator creates data mimicking real data; the Discriminator differentiates real from synthetic data. Our architecture addresses GAN issues like mode collapse and training instability by using Wasserstein distance (from WGANs), improving stability, diversity, and reliability in medical imaging, compared to standard GANs.

The training objective is then defined as

$$L_D = L + P \qquad (1)$$

where

$$L = \mathbb{E}_{x_1,x_3}[D(x_1, G(x_1, x_3), x_3)] - \mathbb{E}_{x_1,x_2,x_3}[D(x_1, x_2, x_3)]$$
$$P = \alpha \mathbb{E}_{\hat{x}}[(\|\nabla_{\hat{x}} D(\hat{x})\|_2 - 1)^2]$$

with

$$\hat{x} = \epsilon x_2 + (1 - \epsilon)G(x_1, x_3), \quad 0 \le \epsilon \le 1$$

D and G are the critic and the generator, respectively, x_1, x_2, and x_3 are three consecutive slices of a volume with respect to a fixed timestep and α is the weight of the gradient penalty. The generator is updated using the following loss:

$$L_G = -\mathbb{E}_{x_1,x_3}[D(x_1, G(x_1, x_3), x_3)] + \lambda_p L_{L1}$$

where $L_{L1} = \mathbb{E}_{x_1,x_2,x_3}[\|x_2 - G(x_1, x_3)\|_1]$ and λ_p is its weight. The decision to incorporate the L1 loss was inspired by the Pix2Pix architecture [11]. The value of λ_p is the same as in the referenced work.

2.1 Recomposition Protocol for GAN-Enhanced Volumes

In the proposed method, high-resolution reconstruction of medical volumes is achieved by applying a GAN, trained to synthesize realistic intermediate slices along a single anatomical direction (axial, coronal, or sagittal). For each direction, an enhanced volume is produced in which the resolution is doubled along the target axis, producing three complementary versions of the same anatomical volume, each with greater structural detail along one direction. While effective at producing locally plausible structures, radiometric and spatial inconsistencies are introduced.

Intra-volume discrepancies in grayscale tone arise due to the difference between real slices, acquired via medical imaging and synthetic slices, which are generated by learned interpolation. GANs typically optimize local visual fidelity but do not enforce global intensity consistency, resulting in histogram shifts, banding, and tonal discontinuities. Inter-volume discrepancies emerge because each volume is independently enhanced along a single axis. The lack of cross-directional consistency constraints implies that the same spatial location (i, j, k) may receive inconsistent voxel values across the three enhanced volumes, causing speckle-like artifacts and a structural incoherence in the reconstructed 3D models, especially near anatomical boundaries or intersections.

To address these issues, the proposed method applies a volumetric interpolation followed by voxel-wise fusion. Each volume $V^{(d)}$ is first interpolated onto a common $512 \times 512 \times 512$ grid using a finite-support Lanczos-3 filter [7]. This convolutional kernel is defined as $L(x) = \mathrm{sinc}(x) \cdot \mathrm{sinc}(x/a)$ for $|x| < a$

with $a = 3$, enabling band-limited reconstruction that minimizes aliasing while preserving local structures. The interpolated representation $\tilde{V}^{(d)}$ is given by:

$$\tilde{V}^{(d)}(i, j, k) = \sum_{m,n,p} V^{(d)}(m, n, p) \cdot L(i - m) \cdot L(j - n) \cdot L(k - p), \qquad (2)$$

where the index d denotes the direction of super-resolution. The fusion of the three interpolated volumes is then performed via voxel-wise summation:

$$V_{\text{init}}(i, j, k) = \tilde{V}^{(\text{Axial})}(i, j, k) + \tilde{V}^{(\text{Coronal})}(i, j, k) + \tilde{V}^{(\text{Sagittal})}(i, j, k). \qquad (3)$$

This choice is justified by the complementary nature of the input data: each GAN-generated volume is expected to offer reliable structure along a specific axis, while being less informative along the others. The voxel-wise fusion thus integrates the best local information from each view, reinforcing anatomical content and mitigating directional uncertainty. To finalize the fusion process and enhance radiometric consistency, an edge-aware filtering step is applied independently along each orthogonal directions. This operation aims to suppress inconsistencies. Filtering is performed slice-by-slice, treating each two-dimensional section S independently to improve inter-slice continuity and reduce noise caused by learned interpolation. A smoothed version S_{smooth} is computed using a non-centered mean filter over the 8 neighboring pixels (a 3×3 kernel with zero weight at the center). This filter preserves the central voxel's original value while providing a robust local estimate to suppress high-frequency noise without degrading anatomical details. An edge-aware weight map W is then computed to guide an adaptive combination of the original and smoothed images. This map is derived from the gradient magnitude $\|\nabla S\|$, estimated using Sobel horizontal G_x and vertical G_y operators:

$$\|\nabla S\| = \sqrt{(G_x * S)^2 + (G_y * S)^2}, \qquad W = \frac{\log(1 + \|\nabla S\|)}{\max \log(1 + \|\nabla S\|)}. \qquad (4)$$

The logarithmic transformation emphasizes boundaries while maintaining a smooth response even in regions of low gradient. The normalization ensures that $W \in [0, 1]$, where high values indicate strong structural confidence. The final filtered slice is then computed as a weighted interpolation:

$$S_{\text{out}} = W \cdot S + (1 - W) \cdot S_{\text{smooth}}. \qquad (5)$$

In this formulation, high-gradient regions ($W \to 1$), typically corresponding to anatomical boundaries or salient structures, retain the original content S, thus preserving structural fidelity. Conversely, in flat or noisy regions ($W \to 0$), where GANs are prone to introducing artifacts, the smoothed content S_{smooth} dominates, providing regularization. The combination of volumetric Lanczos-3 interpolation, multi-directional fusion, and edge-aware slice-wise filtering results in a high-resolution volume that is radiometrically uniform, structurally coherent, and free from speckling artifacts—well-suited for downstream medical analysis or accurate 3D visualization (see Fig. 1).

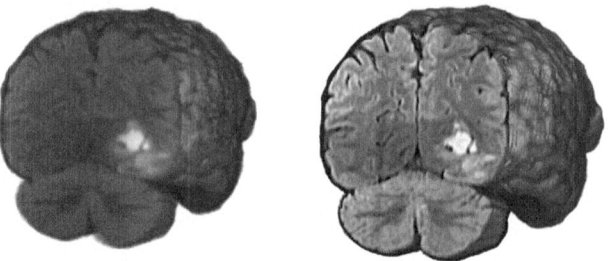

Fig. 1. Original volume (left) vs reconstructed (right).

2.2 Dataset and Experiments

We used T1 brain images from the ADNI dataset's normal control group (CN) [1]. Specifically, the preprocessed version has been considered in which non-brain areas have been removed using the FreeSurfer software. The dataset is composed of 997 scans across 231 different subjects. Each scan is a $256 \times 256 \times 256$ volume. For the training, we used five volumes from 5 different subjects, and we reduced them by removing images with more than 93% of black pixels. Moreover, each image is divided into 25 overlapping patches of size 128×128 with a step of 32, and the values are scaled to $[-1, 1]$. Hence, the training data consists of triplets (x_i, x_{i+t}, x_{i+2t}) where x_i is a patch of the i-th image within a volume, and t (timestep) is the distance between slices within a volume. To train the WGAN, triplets were selected for values of $t \in \{1, 2, 3\}$. Different distances between slices were used to make the network more robust and less sensitive to differences in slice density in MRI scans. To the best of our knowledge, no prior work deals with generating MRI intermediate sequences starting from two slices of a scan. Thus, we compare our method with a standard-trained UNet, retaining the same architecture as the generator. Specifically, we trained two UNets, namely UNet$_1$ and UNet$_2$. The former is able to generate x_{i+t} given as input x_i, while the latter outputs x_i from x_{i+t}. Given a triplet (x_i, x_{i+t}, x_{i+2t}), we calculate a metric f as follows: $f(G(x_i, x_{i+2t}), x_{i+t})$, $f(\text{UNet}_1(x_i), x_{i+t})$, $f(\text{UNet}_2(x_{i+2t}), x_{i+t})$, $f(Mean(\text{UNet}_1(x_i), \text{UNet}_2(x_{i+2t})), x_{i+t})$.   To assess the quality and accuracy of MRI images enhanced by the proposed GAN architecture, various metrics are employed [3]. These include HVS L1 and L2 for pixel differences, PSNR and SSIM for image fidelity, IoU for segmentation precision, and FSIM and VIF for perceptual accuracy. Additionally, GMSD and MSE evaluate edge sharpness and detail, while CPBD assesses overall image sharpness. This comprehensive evaluation covers various aspects of image quality, from basic pixel differences to complex perceptual features. Metrics have been computed by randomly sampling 50 subjects, generating 2 samples from each patient, obtaining 100 random triplets (x_i, x_{i+t}, x_{i+2t}) with $t = 1, 3$ across two experiments. Results are averaged over 10 runs with fixed seed for reproducibility.

Results in Table 1 highlight that while some metrics exhibit higher values for UNet generated images, they appear visually more blurred than those pro-

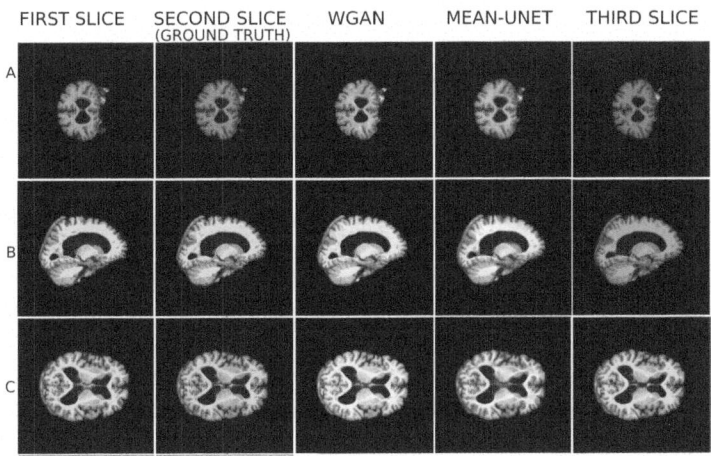

Fig. 2. Reconstructed slices from the T1 modality. A: axial view, B: coronal view, C: sagittal view.

Table 1. Comparative evaluation on timesteps $t = 1$ and 3. $CPBD$ was calculated only on the predicted image, while the other metrics were calculated using the predicted image and ground truth (middle image in the triplet). Here, $\uparrow$ means higher is better, $\downarrow$ means lower is better.

(a) Comparative evaluation on timesteps $t=1$.

	G (GAN Generator)	$UNet_1$	$UNet_2$	Mean($UNet_1$, $UNet_2$)
HVS L1 $\downarrow$	0.371 ± 0.004	0.486 ± 0.006	0.543 ± 0.003	0.425 ± 0.003
HVS L2 $\downarrow$	0.401 ± 0.004	0.525 ± 0.005	0.587 ± 0.004	0.457 ± 0.002
PSNR $\uparrow$	32.738 ± 0.194	29.402 ± 0.259	25.989 ± 0.441	30.088 ± 0.366
SSIM $\uparrow$	0.982 ± 0.001	0.961 ± 0.001	0.945 ± 0.002	0.969 ± 0.001
IoU $\uparrow$	0.997 ± 0	0.954 ± 0	0.997 ± 0	0.988 ± 0
FSIM $\uparrow$	0.834 ± 0.001	0.738 ± 0.003	0.752 ± 0.004	0.798 ± 0.002
VIF $\uparrow$	0.536 ± 0.004	0.395 ± 0.006	0.333 ± 0.006	0.456 ± 0.007
GMSD $\downarrow$	0.032 ± 0.001	0.059 ± 0.001	0.087 ± 0.003	0.055 ± 0.003
CPBD$^{(*)}$ $\uparrow$	0.332 ± 0.006	0.267 ± 0.005	0.290 ± 0.005	0.236 ± 0.002
ESMSE $\downarrow$	2.433 ± 0.043	4.418 ± 0.118	5.101 ± 0.115	3.353 ± 0.079

(b) Comparative evaluation on timesteps $t = 3$

	G (GAN Generator)	$UNet_1$	$UNet_2$	Mean($UNet_1$, $UNet_2$)
HVS L1 $\downarrow$	0.648 ± 0.009	0.652 ± 0.006	0.651 ± 0.006	$\mathbf{0.621 \pm 0.005}$
HVS L2 $\downarrow$	0.699 ± 0.008	0.701 ± 0.005	0.692 ± 0.006	$\mathbf{0.661 \pm 0.005}$
PSNR $\uparrow$	25.649 ± 0.216	23.957 ± 0.139	24.032 ± 0.171	25.381 ± 0.173
SSIM $\uparrow$	0.934 ± 0.001	0.905 ± 0.002	0.903 ± 0.002	0.918 ± 0.001
IoU $\uparrow$	0.997 ± 0	0.994 ± 0	0.992 ± 0	0.992 ± 0
FSIM $\uparrow$	0.719 ± 0.002	0.664 ± 0.003	0.659 ± 0.002	0.687 ± 0.002
VIF $\uparrow$	0.287 ± 0.004	0.197 ± 0.003	0.203 ± 0.003	0.246 ± 0.003
GMSD $\downarrow$	0.085 ± 0.001	0.116 ± 0.001	0.118 ± 0.001	0.100 ± 0.001
CPBD$^{(*)}$ $\uparrow$	0.303 ± 0.004	0.097 ± 0.004	0.109 ± 0.004	0.11 ± 0.004
ESMSE $\downarrow$	5.701 ± 0.078	8.077 ± 0.127	8.254 ± 0.139	7.691 ± 0.122

duced by the Generative Adversarial Network. This discrepancy highlights a divergence between quantitative evaluations and qualitative image perception. Images obtained with the WGAN and the mean UNET approach are shown in Fig. 2. A significant advantage of the GAN is its ability to be trained once, using triplets at variable timesteps, unlike UNet, which requires retraining for each specific timestep. This difference is also reflected in the stability of the Cumulative Probability of Blur Detection (CPBD) measurements across various timesteps; for GAN, CPBD values remain nearly constant, indicating consistency in the sharpness of the generated images, whereas UNet shows a considerable degradation in quality even when retrained for different timesteps. Although the PSNR measurements are similar between the two techniques, the analysis of CPBD reveals a marked advantage of the GAN in terms of clarity and image definition. Table 2 reports the mean and standard deviation of the segmentation metrics (Dice, Precision, Recall, Accuracy) computed over 16 test subjects, comparing the results obtained using the original volumes with those obtained after the proposed GAN-based volumetric augmentation. The BraTS2021 dataset [4] has been used in this comparative evaluation as it provides 3D brain tumor segmentation masks.

Table 2. Comparison of segmentation metrics (mean ± standard deviation) on 16 test subjects, for original and GAN-augmented volumes.

Class	Dice	Precision	Recall	Accuracy
Original volumes				
1	0.545 ± 0.248	0.576 ± 0.294	0.739 ± 0.263	0.998 ± 0.002
2	0.533 ± 0.250	0.439 ± 0.249	0.809 ± 0.128	0.988 ± 0.014
4	0.737 ± 0.162	0.724 ± 0.219	0.812 ± 0.132	0.998 ± 0.001
Augmented volumes				
1	0.602 ± 0.249	0.735 ± 0.265	0.651 ± 0.267	0.999 ± 0.001
2	0.717 ± 0.164	0.625 ± 0.182	0.883 ± 0.093	0.996 ± 0.001
4	0.825 ± 0.111	0.871 ± 0.117	0.802 ± 0.146	0.999 ± 0.001

The results reported in Table 2 highlight the impact of GAN-based volumetric augmentation on the performance of tumor segmentation across the three BraTS classes. In particular, the Dice coefficient increases for all classes. These improvements reflect a better overall overlap between the predicted and ground truth segmentations, especially for smaller and more challenging tumor subregions such as the tumor core (Class 2). Precision also benefits significantly from augmentation, implying a strong reduction in false positives, particularly important in medical settings where over-segmentation can lead to misdiagnosis or unnecessary interventions. Conversely, recall shows a mixed trend. While it improves substantially for Class 2 and Class 4 is marginally reduced, Class 1 experiences a slight decrease. This can be attributed to the augmented data being sharper but possibly with more selective boundaries, occasionally leading to under-segmentation in elongated or infiltrative regions. Nevertheless, the overall accuracy remains high across all classes ($\geq$0.996), confirming that the proposed approach maintains a reliable voxel-wise classification performance. From a clinical perspective, improved Dice and precision enhance tumor delineation accuracy, reducing overestimation, while recall gains—especially for the tumor core—boost sensitivity to critical areas. Overall, despite minor recall limitations, the method improves volumetric representation, aligning better with radiological expectations and aiding diagnostic and therapeutic tasks.

3   Conclusions

In this study, we proposed a novel Generative Adversarial Network architecture specifically designed to enhance the resolution of MRI volumes. This approach diverges significantly from traditional methods by synthesizing new slices from adjacent ones to improve overall image detail and resolution, showcasing the novelty of using GANs to improve MRI data quality directly. Despite similar PSNR values, the superior performance of the GAN is evident in terms of image sharpness and definition, as reflected by the Cumulative Probability of Blur Detection (CPBD) metric. This underscores the GAN's advantage in producing less blurred, more clinically useful MRI images than UNet. Furthermore, a critical advantage of the GAN architecture is its training efficiency—it requires only a single training session with triplets at variable timesteps, as opposed to UNet's need for retraining at each timestep. This not only simplifies the workflow but also ensures consistency in the quality of the generated images across different timesteps, as evidenced by the stability of the GAN's CPBD values. Looking ahead, this model's potential applications extend beyond traditional MRI to include other diagnostic methods, such as functional MRI (fMRI) and Computed Tomography (CT), where volume resolution is even more critical.

Acknowledgements. We acknowledge financial support from the project PNRR MUR project PE0000013-FAIR.

References

1. Alzheimer's disease neuroimaging initiative. https://adni.loni.usc.edu/
2. Arjovsky, M., Chintala, S., Bottou, L.: Wasserstein GAN (2017)
3. Avcıbaş, I.S., Sankur, B.L., Sayood, K.: Statistical evaluation of image quality measures. J. Electron. Imaging **11**(2), 206–223 (2002)
4. Baid, U., Ghodasara, S., et al.: The RSNA-ASNR-MICCAI BraTS 2021 benchmark on brain tumor segmentation and radiogenomic classification. arXiv preprint arXiv:2107.02314 (2021)
5. Blystad, I., Warntjes, J., Smedby, O., Landtblom, A.M., Lundberg, P., Larsson, E.M.: Synthetic MRI of the brain in a clinical setting. Acta Radiol. **53**(10), 1158–1163 (2012). pMID: 23024181

6. Chlap, P., Min, H., Vandenberg, N., Dowling, J., et al.: A review of medical image data augmentation techniques for deep learning applications. J. Med. Imaging Radiat. Oncol. **65**(5), 545–563 (2021)

7. Duchon, C.E.: Lanczos filtering in one and two dimensions. J. Appl. Meteorol. **1962–1982**, 1016–1022 (1979)

8. Goceri, E.: Medical image data augmentation: techniques, comparisons and interpretations. Artif. Intell. Rev. 1–45 (2023)

9. Gulrajani, I., Ahmed, F., Arjovsky, M., Dumoulin, V., Courville, A.C.: Improved training of wasserstein GANs (2017)

10. Huang, S., et al.: TransMRSR: transformer-based self-distilled generative prior for brain MRI super-resolution. Vis. Comput. **39**(8), 3647–3659 (2023)

11. Isola, P., Zhu, J., Zhou, T., Efros, A.A.: Image-to-image translation with conditional adversarial networks (2016)

12. Jiang, M., et al.: FA-GAN: fused attentive generative adversarial networks for MRI image super-resolution. Comput. Med. Imaging Graph. **92**, 101969 (2021)

13. Kebaili, A., Lapuyade-Lahorgue, J., Ruan, S.: Deep learning approaches for data augmentation in medical imaging: a review. J. Imaging **9**(4), 81 (2023)

14. Li, G., Rao, C., Mo, J., Zhang, Z., Xing, W., Zhao, L.: Rethinking diffusion model for multi-contrast MRI super-resolution. In: Proceedings of the IEEE/CVF Conference on Computer Vision and Pattern Recognition, pp. 11365–11374 (2024)

15. Li, Y., Sixou, B., Peyrin, F.: A review of the deep learning methods for medical images super resolution problems. Irbm **42**(2), 120–133 (2021)

16. Lu, Z., et al.: Two-stage self-supervised cycle-consistency transformer network for reducing slice gap in MR images. IEEE J. Biomed. Health Inform. **27**(7), 3337–3348 (2023)

17. Nimitha, U., Ameer, P.: MRI super-resolution using similarity distance and multi-scale receptive field based feature fusion GAN and pre-trained slice interpolation network. Magn. Reson. Imaging **110**, 195–209 (2024)

18. Sander, J., de Vos, B.D., Išgum, I.: Autoencoding low-resolution MRI for semantically smooth interpolation of anisotropic MRI. Med. Image Anal. **78**, 102393 (2022)

19. Umirzakova, S., Ahmad, S., Khan, L.U., Whangbo, T.: Medical image super-resolution for smart healthcare applications: a comprehensive survey. Inf. Fusion **103**, 102075 (2024)

20. Van Reeth, E., Tham, I.W., Tan, C.H., Poh, C.L.: Super-resolution in magnetic resonance imaging: a review. Concepts Magn. Reson. Part A **40**(6), 306–325 (2012)

21. Wang, L., Zhu, H., He, Z., Jia, Y., Du, J.: Adjacent slices feature transformer network for single anisotropic 3D brain MRI image super-resolution. Biomed. Signal Process. Control **72**, 103339 (2022)

22. Xia, Y., Ravikumar, N., Greenwood, J.P., Neubauer, S., Petersen, S.E., Frangi, A.F.: Super-resolution of cardiac MR cine imaging using conditional GANs and unsupervised transfer learning. Med. Image Anal. **71**, 102037 (2021)

23. Yang, H., Wang, Z., Liu, X., Li, C., Xin, J., Wang, Z.: Deep learning in medical image super resolution: a review. Appl. Intell. **53**(18) (2023)

24. Yi, X., Walia, E., Babyn, P.: Generative adversarial network in medical imaging: a review. Med. Image Anal. **58**, 101552 (2019)

25. Yurt, M., Özbey, M., Dar, S.U., Tinaz, B., Oguz, K.K., Çukur, T.: Progressively volumetrized deep generative models for data-efficient contextual learning of MR image recovery. Med. Image Anal. **78**, 102429 (2022)

26. Zeng, W., Peng, J., Wang, S., Liu, Q.: A comparative study of CNN-based super-resolution methods in MRI reconstruction and its beyond. Signal Process.: Image Commun. **81**, 115701 (2020)
27. Zhang, K., et al.: SOUP-GAN: super-resolution MRI using generative adversarial networks. Tomography **8**(2), 905–919 (2022)
28. Zhao, C., Dewey, B.E., Pham, D.L., Calabresi, P.A., Reich, D.S., Prince, J.L.: SMORE: a self-supervised anti-aliasing and super-resolution algorithm for MRI using deep learning. IEEE Trans. Med. Imaging **40**(3), 805–817 (2020)
29. Zhao, K., Pang, K., Hung, A.L.Y., Zheng, H., Yan, R., Sung, K.: MRI super-resolution with partial diffusion models. IEEE Trans. Med. Imaging (2024)

Model Robustness and Generalization

Simultaneous Robustness and Generalization Using Nearest Neighbor Classifiers

O. Deniz$^{(\boxtimes)}$ ⓘ, G. Bueno ⓘ, A. Pedraza ⓘ, and H. Singh ⓘ

VISILAB, University of Castilla-La Mancha, ETSI Industrial,
Avda Camilo Jose Cela 3, 13071 Ciudad Real, Spain
Oscar.Deniz@uclm.es

Abstract. While most attention in modern machine learning is devoted to accuracy gains, there is ample scope for research into fail cases. The phenomenon of adversarial examples shows in the most striking fashion how brittle the behavior of our models is. Adversarial examples are not exclusive of deep learning, they can equally appear with other machine learning methods too. Since they became popular, a multitude of attack and defense methods have been proposed in the literature to both generate adversarials and protect models from them. In this context, it has been shown that there is a general trade-off between robustness to adversarial examples and generalization: making a model more robust to adversarials causes it to lose generalization in the standard test set. This has been observed in so many ways (both theoretical and empirical) that several authors have argued that the trade-off is inescapable. In this work we use nearest neighbors to show that an algorithm can have optimal generalization and robustness, suggesting that the trade-off is not inescapable in that case.

Keywords: Adversarial examples · Nearest Neighbors

1 Introduction

Despite the power of modern machine learning methods based on deep learning, it is unclear how such methodologies can mimic human learning and behavior without making mistakes. Humans also make mistakes, and while most of the general attention in modern machine learning revolves around performance in terms of accuracy (as measured on average in certain datasets), comparatively less is known about fail cases. In this respect, the phenomenon of adversarial examples (henceforth AEs) is arguably the most striking negative feature of such methodologies [6]. AEs refer to specially crafted inputs that are intentionally designed to deceive machine learning models, causing them to make incorrect predictions or decisions. Figure 1 shows the phenomenon. It is important to note that adversarial examples are not exclusive to deep learning models, they can

ⓒ The Author(s), under exclusive license to Springer Nature Switzerland AG 2026
M. Castrillón-Santana et al. (Eds.): CAIP 2025, LNCS 15622, pp. 89–100, 2026.
https://doi.org/10.1007/978-3-032-05060-1_8

Fig. 1. Adversarial Example. From left to right: original image (classified as a "chair"), adversarial noise (amplified for visualization purposes), adversarial image obtained after adding the noise (classified as a "panda").

affect other types of machine learning models as well (SVMs, random forests, nearest neighbors, etc.).

More specifically, an adversarial example X' can be obtained from a legitimate unperturbed image X by adding a calculated noise δX, which should be the minimum that alters the class $f(X)$ predicted for the original image:

$$X' = X + \arg\min_{\delta X}\{||\delta X||\ \ s.t.\ \ f(X + \delta X) \neq f(X)\} \qquad (1)$$

Several of the so-called *attack* methods have been proposed to find the particular noise of Eq. 1. In turn, *defense* methods aim at detecting AEs or else they directly change the classifier decision into the correct one. For a recent survey see [7].

Adversarial examples represent a captivating area of study that may shed light on the inner workings of machine learning models and the challenges involved in building Artificial Intelligence systems. Apart from the arm's race between defense and attack methods, several works have aimed at a more exploratory research in order to gain more knowledge about the phenomenon.

In the AE literature it has been generally established that there is a trade-off between robustness (to AEs) and generalization: making a classifier robust against AEs implies accuracy drops in the legitimate test set. One of the first works to show this was [16], where the authors showed that a trade-off between the standard accuracy of a model and its robustness to adversarial perturbations provably exists in a simplified setting of binary classification of normal distributions. They concluded that this phenomenon is a consequence of robust classifiers learning fundamentally different feature representations than standard classifiers. In [8] the authors considered generalization to random spatial transformations (e.g., translations, rotations). They showed that: (a) as the spatial robustness of models improves by training augmentation with progressively larger transformations, their adversarial robustness worsens progressively, and (b) as state-of-the-art robust models are adversarially trained with progressively larger pixel-wise perturbations, their spatial robustness drops progressively.

A similar result is described in [9], where the authors showed, in a restricted scenario of binary classification with ReLU networks and well-separated data, that there is an exponential separation between the network size for achieving robustness and generalization. In [15] the experiments showed a trade-off between robustness accuracy, with robustness scaling linearly in the logarithm of the classification error. In [12] and [3] classification was performed in two regimes: normal training (which optimizes generalization in a test set) and overfitting (which optimizes error in the training set). The results consistently showed greater robustness to adversarial examples in the overfitting regime.

While there is larger evidence to the contrary, it should be noted that a few works have suggested that there is no clear robustness-generalization trade-off. In this respect, [5] described how better generalization helps robustness on both toy and large-scale datasets. Similarly, [20] argued that adversarial examples can be used to improve image recognition if harnessed in the right manner.

Given this panorama of an unavoidable robustness-generalization trade-off the question arises: are there any cases in which such trade-off does not actually exist? In this paper we use a canonical non-parametric classifier, K-Nearest Neighbors, to demonstrate that. In particular, we introduce a nearest-neighbor classifier that has the generalization of a K-NN classifier (with K > 1) and the robustness of a 1-NN classifier. As far as the authors know, this is the first time that optimal robustness and optimal generalization within a single family of classifiers is achieved simultaneously, showing that there is no trade-off in that case.

This paper is organized as follows. In Sect. 2 we describe the proposed methods. Experiments are described and discussed in Sect. 3. Finally, the main conclusions are outlined in Sect. 4.

2   Adversarial Examples in Nearest Neighbor Classifiers

While most of the work around adversarial examples has considered the prevailing methodologies of deep networks, some authors have studied them with simpler classifiers such as nearest neighbors (NNs). Authors have proposed several NN attacks, see for example [10] and [14].

Using nearest neighbors, in [18] the authors showed that the K-NN robustness properties depend critically on the value of K, with the classifier being inherently non-robust for small K, and robustness increasing for growing K. We note that this is in apparent contradiction with previously published work (in particular [3]), in which regularized methods (a K-NN with large K) help generalization but (due to the aforementioned trade-off) hurt robustness. Below we will test this in the experiments.

Apart from representing a simpler experimental setting, nearest neighbors present other interesting particularities that can be useful in the study of AEs. An example is: for a test sample that is correctly classified by 1-NN, we can expect that the space around it will belong to the same class (unless the test sample happens to fall exactly at a class boundary). This means that in principle, given a test sample we can calculate the minimum distance to its AE, i.e.

the minimal perturbation that is necessary to get an AE from that sample. That is precisely what was mathematically calculated by the authors of [17]. In that work, the authors formulate the task as a list of convex quadratic programming (QP) problems that can be efficiently solved for 1-NN models. For K-NN models with larger K, they showed that the same formulation can help efficiently compute upper and lower bounds of that minimum adversarial perturbation. Despite the efficient solvers proposed, the computation is still quite compute intensive, taking over 1 s per 28×28 image of the MNIST dataset.

While the work [17] showed the mathematical expression for the minimal AEs and attacks using those AEs, it fell short of proposing a defense mechanism. Based on that work, we first introduce a Monte-Carlo algorithm that can act as a "perfect" defense. To illustrate the basic principle, consider a two-class example in Fig. 2. The basic idea is to obtain the nearest neighbor of test sample x and attack it to obtain the minimal AE, which gives the minimal distance e. Then we apply a Monte Carlo sampling inside a ball of radius e around x. If most of the samples inside the ball are classified to the same class as x then x can be considered a legit image. Otherwise it is considered an adversarial example.

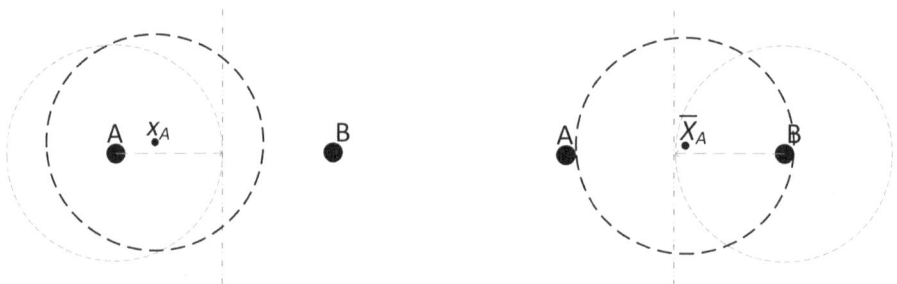

Fig. 2. Example for two classes A and B, the vertical line is the class boundary. Left: X_A is a legit test sample of class A. Right: $\overline{X}_A$ is an AE of class A. Points A and B are their nearest neighbors (which are samples of the training set). Since X_A is close to A most of samples inside the thick black circle (which is centered on X_A) will be of that same class. For $\overline{X}_A$ that ratio will be lower.

The detailed procedure is given in Algorithm 1.

Constant F controls the ratio of detection versus false positives. For values of F close to 1 the algorithm has a higher AE detection rate but also more false positives (legit images considered AEs) and vice versa.

Now that we have a "perfect" defense, we turn our attention to the following question: is it possible to have a classifier with good generalization and robustness *simultaneously*? In our case, is it possible to have a classifier with the robustness of K1-NN and the generalization of K2-NN (where K2 > K1)? The answer is yes, and we can apply Algorithm 2, a straightforward variant of Algorithm 1, to determine this.

Algorithm 1. Monte Carlo algorithm for a "Perfect" Defense

---

1. Let x = input test image
2. Let e = norm of the minimum AE, which is obtained in the following way:
   Classify x with 1-NN to get its nearest neighbor x_N
   Attack x_N with method of [17] to obtain value of e
3. Randomly generate P samples $x + d$, where d is the bounded perturbation, such that $|d| \leq e$
4. If at least some fraction $F(0.0 \leq F \leq 1.0)$ of those P samples give the same class as x
   Then x is a legit image
   Else x is an AE

Algorithm 2. Classifier with robustness of K1-NN and generalization of K2-NN

---

1. Let x = input test image
2. Let e = norm of the minimum AE, which is obtained in the following way:
   Classify x with K1-NN to get its nearest neighbor x_N
   Attack x_N with method of [17] to obtain value of e
3. Randomly generate P samples $x + d$, where d is the bounded perturbation, such that $|d| \leq e$
4. If at least some fraction $F(0.0 \leq F \leq 1.0)$ of those P samples give the same class as x
   Classify it with K2-NN
   Else Classify it with K1-NN

---

3   Experiments

Experiments were conducted on four different public datasets:

- MNIST [2]: a widely used benchmark for machine learning, consisting of 28×28 grayscale images of handwritten digits (0 to 9) with corresponding labels. The dataset contains 60,000 images for training and 10,000 for test.
- FMNIST [19]: a popular alternative to MNIST, containing 28×28 grayscale images of fashion items (e.g., clothes) with corresponding labels. The dataset contains 60,000 images for training and 10,000 for test.
- EMNIST [1]: an extension of MNIST, comprising 28×28 grayscale images of handwritten characters (both uppercase and lowercase letters, as well as digits) with corresponding labels. We used the EMNIST Balanced configuration: 131,600 images and 47 balanced classes.
- SVHN [11]: The SVHN (Street View House Numbers) dataset is a large-scale dataset containing color images of house numbers captured from Google Street View, used for digit recognition tasks. Images are 32×32, there are 73,257 images for training and 26,032 for test.

Figure 3 shows examples of the images in the four datasets considered. For the SVHN dataset we first converted all the images to grayscale.

In the first experiment we analyzed how robustness depends on K. As mentioned above, [18] argued that the classifier is less robust for small K, and more

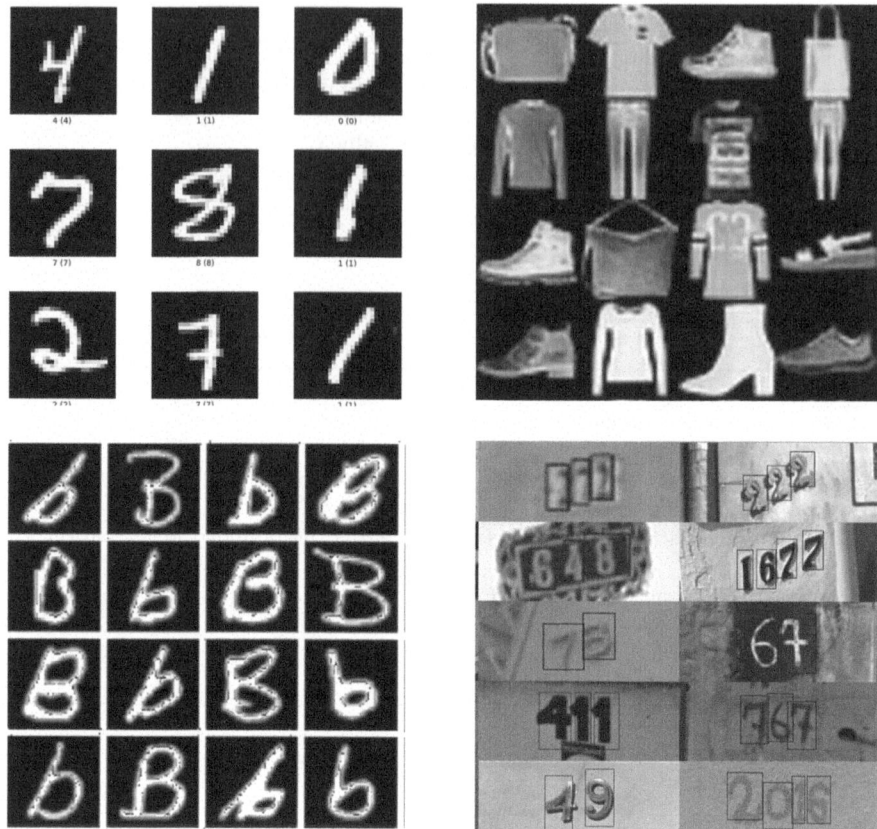

Fig. 3. From top to bottom, left to right: MNIST, FMNIST, EMNIST and SVHN datasets.

robust for large K. We noted that this is in apparent contradiction with previously published work (in particular [3]), in which less complex methods (a K-NN with large K) help generalization but (due to the aforementioned trade-off) hurt robustness. Figure 4 shows the result of this first experiment, where we used [17] to obtain the minimal distance to AEs. The figure shows minimum, maximum and average distance for a set of 40 adversarial examples obtained from the test set (accuracy was measured in the whole test set), with all values being the average of 8 runs. As can be observed, the distances tend to decrease with K. This means that the closest adversarials that can be obtained from legit images are increasingly closer to them, which represents less and less robustness (see Fig. 5). This first experiment shows that, as opposed to what is discussed in [18], but consistent with [3] and in general with the existence of a robustness-generalization trade-off, robustness decreases with K.

In a second experiment we show results for Algorithm 1 for AE detection. We used $P = 16$ and $F = 0.77$. The box plots of Fig. 6 show the mean accuracies for

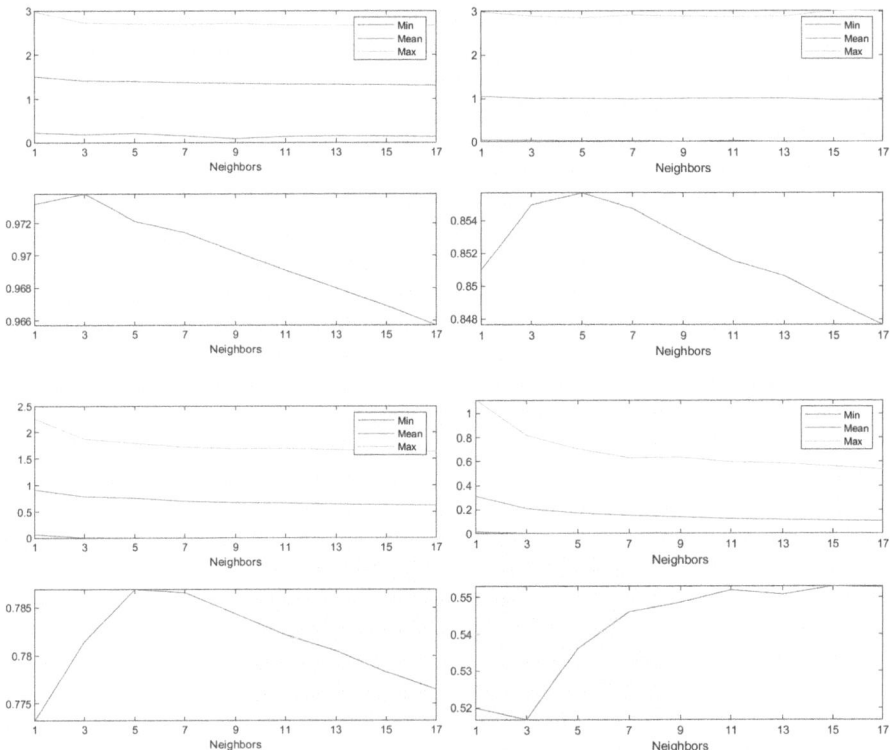

Fig. 4. From top to bottom, left to right: MNIST, FMNIST, EMNIST and SVHN datasets. For each figure, top: minimum, average and maximum distance to the AEs (shown in logarithmic scale), bottom: test accuracy.

both AE detection and legit image detection, that are practically 100% in both cases. We used the following $K2$ values: MNIST: 3, FMNIST: 3, EMNIST: 7 and SVHN: 15 (these values were chosen because test accuracy was the highest). 30 legit samples and 30 AEs were used in each run.

Despite the high computational cost (2.2 min per MNIST image), this result shows that Algorithm 1 provides a "perfect" detection mechanism, whereby all adversarial examples can be detected as such (and no legit images are considered adversarial examples).

In a third experiment we show results for Algorithm 2. Our objective in this case is to check if that classification algorithm has the same robustness of K1-NN (we used K1=1 in all the experiments) and the same generalization of K2-NN. Since we wanted to demonstrate this with statistically significant results, hypothesis testing was used. In our particular case, however, that test is not straightforward. Note that the typical case in research is when we want to check for statistically different means of two variables (one for the baseline algorithm and the other for a proposed method). The two-sample t-test is typically used

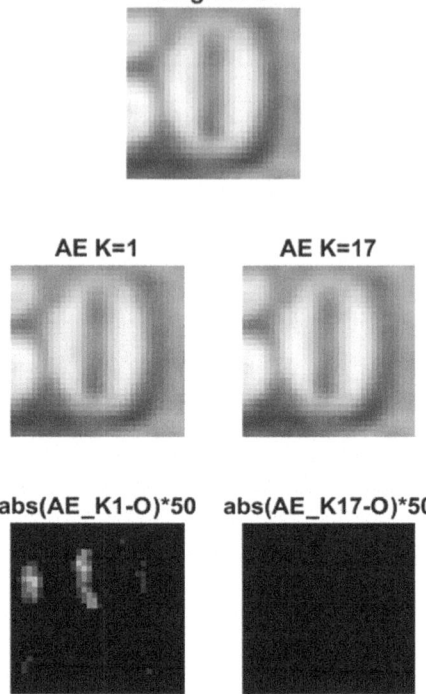

Fig. 5. Minimal adversarial examples generated for an image in the SVHN dataset, for K = 1 and K = 17.

for this. The null hypothesis is 'the means of the two variables are equal' and we can reject it if the p-value is below the alpha (where typically $alpha = 0.05$)). However, in our case we do not want to check for statistically different values but for statistically equivalent values. This cannot be achieved with a standard t-test, and we need to use a 'statistical equivalence test', where the null hypothesis is 'the means of the two variables are different'. One of the most used statistical equivalence tests is the 'two one-sided t-tests' (TOST) procedure [13].

In TOST, upper (U) and lower (L) equivalence thresholds are specified based on the smallest effect size of interest. Then two composite null hypotheses are tested: $HO1 : (\mu_1 - \mu_2) \leq L$ and $H02 : (\mu_1 - \mu_2) \geq U$. These tests produce two p-values, $p1$ and $p2$. When both tests can be statistically rejected (i.e. when $max(p1, p2) \leq \alpha$), we can conclude that the two variables are not different and therefore they are statistically equivalent.

Unfortunately, statistical equivalence tests like TOST require a user-defined equivalence threshold which is a problem-dependent value. To circumvent this we used the so-called 'minimal detectable effect' (MDE) [4]. This represents the effect size which, if it truly exists, can be detected with a given probability with a statistical test of a certain significance level. It is a characteristic of the

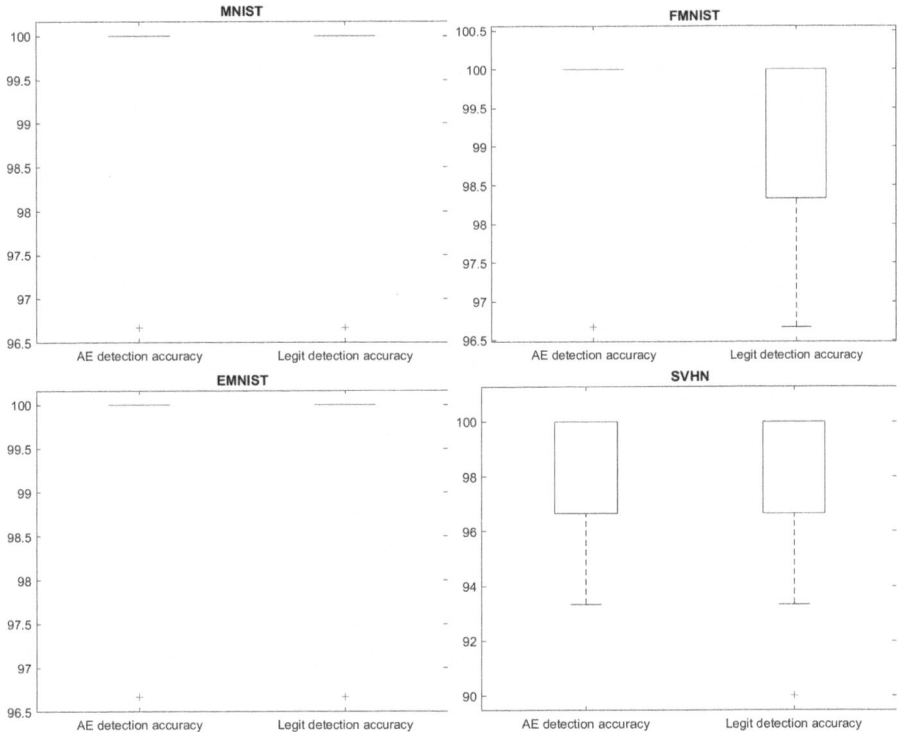

Fig. 6. From top to bottom, left to right: MNIST, FMNIST, EMNIST and SVHN datasets.

statistical hypothesis test used and can be fully determined by the choice of the test's sample size and significance threshold.

In the experiments we compared test accuracies (as a measure of generalization) and distance to the minimal AE (as a measure of robustness). We executed 10 runs, with 40 legit test samples and 40 AEs used in each run. Table 1 shows the results. In the hypothesis tests we compared our method (i.e. algorithm 2) with the K1-NN and K2-NN classifiers. It can be observed that generalization is statistically equivalent to that of K2-NN while the minimal AE distance is statistically equivalent to that of K1-NN. Thus, for nearest neighbors we 'break' the robustness-generalization trade-off. This is an encouraging result, since it shows that the trade-off is not necessarily unavoidable and can, under certain circumstances, be altogether removed.

Table 1. Results of Algorithm 2 on the four datasets. Statistically significant p-values (for $\alpha = 0.05$) are shown in bold. The test accuracy shown here for K1-NN was measured in a set of 40 test samples.

	MNIST	FMNIST	EMNIST	SVHN
Test Accuracy of K1-NN	96.50	86.75	75.75	47.25
Test Accuracy of proposed algorithm	95.75	87.00	79.00	55.25
Test Accuracy of K2-NN	96.00	87.25	79.75	55.25
$max(p1, p2)$	**0.0332**	**0.0234**	**0.0411**	**0.0158**
Minimal AE distance of proposed algorithm	1.1635	0.7126	0.7104	0.2668
Minimal AE distance of K1-NN	1.1653	0.7208	0.7175	0.2687
$max(p1, p2)$	**0.0211**	**0.0291**	**0.0341**	**0.0225**

4   Conclusions

In the context of research in adversarial examples, this paper makes two contributions. First, we show that, contrary to a previously published work, for K-nearest neighbors robustness tends to decrease with K. This behavior is in turn consistent with the prevailing notion of a general trade-off between robustness and generalization. Second, based on the possibility of obtaining minimal-distance adversarial examples for nearest neighbors classifiers, we show a "perfect" defense algorithm that has near 100% AE detection rate (and near 100% legit detection rate). A variation of this defense allows us to propose another algorithm that has the robustness of 1-NN and the generalization of K-NN (for K > 1). The latter result shows that, for nearest neighbors at least, there is no inescapable robustness-generalization trade-off.

The results are encouraging in the sense that they show how a family of classifiers can be made both robust and accurate, with no trade-off involved. Adapting the principles used here to other learning methodologies and, in particular, forms of deep learning, is a logical step for future work. In this respect, note that the algorithm introduced here is essentially relying on two steps, first the search for a nearest neighbor x_N in the training set and then a Monte Carlo sampling with a radius given by the minimal AE of x_N. Both steps can be conceivably implemented for a deep network. The nearest neighbor would be the training sample that provides an output closest to that of the test image (or alternatively the closest from an embedding -latent representation- layer). In turn, the minimal AE can be estimated by using known attack methods on the nearest neighbor.

Acknowledgements. This work was partially funded by project SBPLY/21/180501/000025 by the Autonomous Government of Castilla-La Mancha.

References

1. Cohen, G., Afshar, S., Tapson, J., van Schaik, A.: EMNIST: an extension of MNIST to handwritten letters (2017). http://arxiv.org/abs/1702.05373
2. Deng, L.: The MNIST database of handwritten digit images for machine learning research. IEEE Signal Process. Mag. **29**(6), 141–142 (2012)
3. Deniz, O., Pedraza, A., Vallez, N., Salido, J., Bueno, G.: Robustness to adversarial examples can be improved with overfitting. Int. J. Mach. Learn. Cybern. **11**(4), 935–944 (2020). https://doi.org/10.1007/s13042-020-01097-4
4. Georgiev, G.: Statistical methods in online A/B testing (2019)
5. Gilmer, J., et al.: Adversarial spheres (2018). https://openreview.net/forum?id=SyUkxxZ0b
6. Goodfellow, I.J., Shlens, J., Szegedy, C.: Explaining and harnessing adversarial examples. In: ICLR (2015)
7. Han, S., Lin, C., Shen, C., Wang, Q., Guan, X.: Interpreting adversarial examples in deep learning: a review. ACM Comput. Surv. **55** (2023). https://doi.org/10.1145/3594869
8. Kamath, S., Deshpande, A., Kambhampati Venkata, S., N Balasubramanian, V.: Can we have it all? On the trade-off between spatial and adversarial robustness of neural networks. In: Advances in Neural Information Processing Systems, vol. 34, pp. 27462–27474. Curran Associates, Inc. (2021). https://proceedings.neurips.cc/paper_files/paper/2021/file/e6ff107459d435e38b54ad4c06202c33-Paper.pdf
9. Li, B., Jin, J., Zhong, H., Hopcroft, J.E., Wang, L.: Why robust generalization in deep learning is difficult: perspective of expressive power (2022)
10. Li, X., Chen, Y., He, Y., Xue, H.: AdvKnn: adversarial attacks on k-nearest neighbor classifiers with approximate gradients (2019). https://doi.org/10.48550/arXiv.1911.06591
11. Netzer, Y., Wang, T., Coates, A., Bissacco, A., Wu, B., Ng, A.: Reading digits in natural images with unsupervised feature learning (2011)
12. Pedraza, A., Deniz, O., Bueno, G.: On the relationship between generalization and robustness to adversarial examples. Symmetry **13**(5) (2021). https://doi.org/10.3390/sym13050817, https://www.mdpi.com/2073-8994/13/5/817
13. Schuirmann, D.J.: A comparison of the two one-sided tests procedure and the power approach for assessing the equivalence of average bioavailability (1987). https://doi.org/10.1007/bf01068419
14. Sitawarin, C., Wagner, D.A.: Minimum-norm adversarial examples on KNN and KNN-based models. In: 2020 IEEE Security and Privacy Workshops, pp. 34–40 (2020)
15. Su, D., Zhang, H., Chen, H., Yi, J., Chen, P.Y., Gao, Y.: Is robustness the cost of accuracy? – a comprehensive study on the robustness of 18 deep image classification models. In: Proceedings of the ECCV (2018)
16. Tsipras, D., Santurkar, S., Engstrom, L., Turner, A., Madry, A.: Robustness may be at odds with accuracy. In: International Conference on Learning Representations (2019). https://openreview.net/forum?id=SyxAb30cY7
17. Wang, L., Liu, X., Yi, J., Zhou, Z.H., Hsieh, C.J.: Evaluating the robustness of nearest neighbor classifiers: a primal-dual perspective. ArXiv abs/1906.03972 (2019)

18. Wang, Y., Jha, S., Chaudhuri, K.: Analyzing the robustness of nearest neighbors to adversarial examples. In: Dy, J., Krause, A. (eds.) Proceedings of the of the 35th International Conference on Machine Learning, vol. 80, pp. 5133–5142 (2018)
19. Xiao, H., Rasul, K., Vollgraf, R.: Fashion-MNIST: a novel image dataset for benchmarking machine learning algorithms (2017). https://github.com/zalandoresearch/fashion-mnist
20. Xie, C., Tan, M., Gong, B., Wang, J., Yuille, A., Le, Q.V.: Adversarial examples improve image recognition (2020). https://doi.org/10.48550/arXiv.1911.09665

Adaptive Model Selection for Expanded Post Hoc Debiasing and Mitigating Varying Degrees of Spurious Correlations

Jan Blunk$^{(\boxtimes)}$ (ID), Paul Bodesheim (ID), and Joachim Denzler (ID)

Computer Vision Group, Friedrich Schiller University Jena, Jena, Germany
{jan.blunk,paul.bodesheim,joachim.denzler}@uni-jena.de
https://inf-cv.uni-jena.de

Abstract. Deep neural networks are prone to shortcut bias, where models rely on features that are statistically associated with the target label but lack causal relevance, leading to poor generalization under distribution shifts. To address this, debiasing methods aim to improve robustness by reducing reliance on these spurious features. Unfortunately, existing approaches typically assume unbiased test distributions, an idealized scenario that rarely holds in practice. As a result, they often underperform on the original biased distribution when compared with standard empirical risk minimization (ERM) models. We propose a novel *Adaptive Model SELection* approach for expanding post hoc debiasing called *AMSEL*, which maintains strong performance across test distributions with varying strength of spurious correlation. Using the fixed feature extractor of the biased model, AMSEL trains a family of lightweight classifier heads on simulated distributions ranging from the original biased data to a fully balanced version. At test time, it estimates the degree of spurious correlation in the test data and selects the most suitable classifier. We validate AMSEL on CelebA and ChestX-ray14, demonstrating that it matches the performance of debiased models under unbiased conditions while preserving the accuracy of the original biased model when spurious correlations are prevalent. AMSEL thus offers an adaptive solution to mitigate the impact of spurious correlations when their strength is either unknown or varies across application environments. Code and models are publicly available at https://github.com/debiasing/AMSEL.

Keywords: Post Hoc Debiasing · Shortcut Bias · Spurious Correlations · Dynamic Classifier Selection

1 Introduction

Deep neural networks often struggle to generalize when facing shifts in test distributions [10,36]. One reason for this is *shortcut bias*, where the model relies

Supplementary Information The online version contains supplementary material available at https://doi.org/10.1007/978-3-032-05060-1_9.

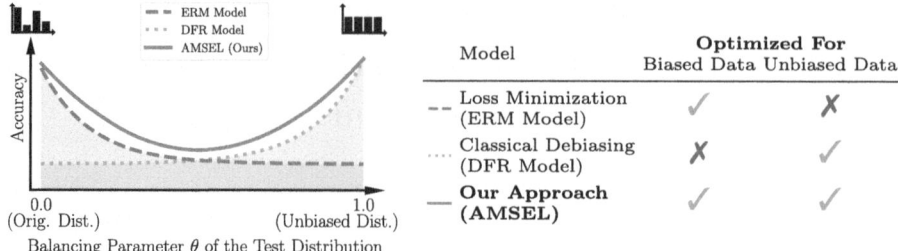

Fig. 1. Expanding Post Hoc Debiasing by Adaptive Model SELection (AMSEL): The ERM model, trained on biased data, performs well in biased settings, whereas classical debiasing models are optimized for unbiased data and thus perform best in unbiased settings. Our method **AMSEL** bridges this gap by (i) estimating test distribution characteristics and (ii) adaptively selecting the best classifier head, achieving strong performance across the entire spectrum.

on statistical associations present in the training data that are not causally warranted [10]. For instance, in natural image classification, models may erroneously associate background textures with certain classes rather than focusing on intrinsic object properties [3,28,34] while in medical imaging, they can rely on hospital-specific artifacts rather than clinically relevant features [7,14,37]. Consequently, when these *spurious correlations* no longer hold, the shortcut bias leads to significant performance drops.

To address this, a variety of *debiasing* methods have been proposed [2,12, 16,18,21,25,28,36], with some even avoiding a complete retraining. One such post hoc approach is *Deep Feature Reweighting (DFR)* [18], which observes that models extract both meaningful and spurious features when trained via standard loss minimization for empirical risk minimization (ERM) [31]. The authors of DFR thus argue that the classifier disproportionately relies on the spurious features and propose freezing the feature extractor while retraining the classifier head on unbiased data. Despite its simplicity and low computational overhead, DFR substantially improves performance on unbiased data [17,18].

However, classical debiasing methods like DFR tend to underperform on the original biased distribution compared to the ERM model, visualized in Fig. 1. This drawback stems from optimizing for an *unbiased* test distribution. This assumption is particularly strong given biased training data, as it requires test data to be free of the spurious correlations present in the training set. In real-world applications, assuming a *fixed* level of spurious correlation is further problematic, as test data often exhibits varying levels of spurious correlation due to changes in acquisition conditions or site-specific characteristics [3,7,14].

We reject the assumption of a fixed test distribution and instead frame the problem of post hoc debiasing as dynamic classifier selection [1,5,6]. Specifically, we assume a continuum of candidate test distributions, each characterized by a different strength of spurious correlation, ranging from the original biased setting to an ideal, unbiased scenario. Building on the observations of DFR, our method trains a family of classifier heads, each optimized for a different candidate test distribution while keeping the feature extractor fixed. At test time, we select the classifier best suited to the actual conditions by comparing the predictions of the classifiers trained on the original biased data and unbiased data. Crucially, we assume that they primarily agree on simple, bias-aligned samples, allowing us to select the head optimized for the corresponding level of spurious correlation.

We validate AMSEL on CelebA [20] and ChestX-ray14 [33], a real-world medical imaging dataset, and find that it matches DFR's benefits on unbiased data while avoiding the performance degradation observed with conventional debiasing methods on biased data.

Our contributions are twofold: (i) We provide a framework for modeling varying test environments by training a family of classifier heads optimized for different levels of spurious correlation. (ii) We propose a test-time mechanism to select the estimated best classifier based on the degree of bias in the input data. Finally, we provide empirical evidence that our approach robustly adapts to varying test conditions, bridging the gap between the original level of spurious correlation and scenarios with minimal spurious correlation.

2   Related Work

Real-world datasets often exhibit *spurious correlations*, which are features that are statistically predictive in the training data but lack a causal link to the task [22,36]. Models trained via empirical risk minimization (ERM) [31] tend to exploit these correlations, in part due to the inherent *simplicity bias* of deep networks [29,30]. When simplicity bias aligns with spurious features, models develop a *shortcut bias* [10], also sometimes referred to as *Clever Hans phenomenon* [19]. Instead of relying on task-relevant features, they focus on these simple yet irrelevant attributes, leading to poor generalization when the spurious correlations do not persist at test time.

To mitigate shortcut bias, various debiasing strategies have been proposed, often requiring retraining of the entire model. These approaches are categorized by the availability of *subgroup labels* that provide information about spurious attributes [35]. Approaches leveraging subgroup labels involving full-model retraining include reweighting and resampling techniques [2,16] as well as worst-case loss minimization frameworks such as GroupDRO [28]. For a comprehensive comparison of debiasing techniques, we refer the reader to [35,36].

In contrast, *post hoc* debiasing techniques modify a pre-trained ERM model with minimal adjustments [12,18,25,26]. For instance, Deep Feature Reweighting (DFR) [17,18] retrains only the classifier on a group-balanced validation set to recover robust performance on unbiased test distributions. This approach is motivated by the observation that while ERM-trained feature extractors encode both biased and unbiased information, the classifier tends to over-rely on spurious features [18,26]. Similarly, He et al. [12] directly adjust classifier weights to suppress the influence of spurious correlations.

Unlike these methods, which optimize for a fixed, unbiased test setting, our approach considers a spectrum of test distributions characterized by varying strengths of spurious correlations. This leads us to draw connections to *dynamic classifier selection (DCS)* [1,5,6], where the classifier is chosen based on the input data. However, unlike previous work in DCS focusing on feature space competence estimation and ensemble generation, our approach is tailored for debiasing. We constrain test distributions to a spectrum between the original biased and ideal unbiased settings, enabling a competence measure based on the estimated ratio of bias-conflicting samples.

3   Method

Preliminaries and Notation. We consider a standard supervised learning setting with input space $\mathcal{X}$ and class labels $\mathcal{Y}$. The corresponding ERM model [31] $m : \mathcal{X} \rightarrow \mathcal{Y}$ is decomposed as $m = h \circ e$, where $e : \mathcal{X} \rightarrow \mathcal{F}$ is the feature extractor and $h : \mathcal{F} \rightarrow \mathcal{Y}$ is the classifier. The training dataset $\mathcal{D}$ of size $n \in \mathbb{N}_{\geq 0}$ is given by $\mathcal{D} = \{(x_i, y_i)\}_{i=1}^{n} \subseteq \mathcal{X} \times \mathcal{Y}$, and is *biased* due to the spurious correlations.

Following standard practice in debiasing [2,17,18,28], we formalize this bias as an imbalance between subgroups, which the model may exploit to make predictions based on the subgroup rather than the label. Formally, we define a surjective subgroup labeling function $g : \mathcal{X} \times \mathcal{Y} \rightarrow \mathcal{G}$, where $\mathcal{G} = \{1, 2, \ldots, l\}$ indexes the subgroups defined by combinations of (class label, bias attribute). Probabilistically, the subgroup imbalance is equivalent to an imbalance in the joint distribution $P(Y, B)$ of class labels Y and bias attributes B. In line with other post hoc debiasing approaches [12,18], we assume that subgroup labels are available only on the validation dataset.

Method Overview. AMSEL addresses robust test-time generalization by constructing a family of candidate models, each optimized for a different test distribution, and by adaptively selecting the appropriate model based on the input data. Our method consists of three main steps:

(i) *Candidate Model Construction:* With a fixed feature extractor e, construct a family of classifier heads $\{h_\theta\}_{\theta \in \Theta}$ by training on subsets $\mathcal{D}_\theta$, where θ controls the degree of spurious correlation between class label Y and bias attribute B. Thus, we also refer to θ as the *balancing parameter*.

(ii) *Estimation of Test Distribution Characteristics:* Given a score function $s(\cdot)$ that measures aspects of spurious correlation from an n-tuple of inputs, learn a mapping $M : s(\mathcal{X}^n) \rightarrow \Theta$ from these scores to the corresponding balancing parameter $\theta \in \Theta$.

(iii) *Adaptive Inference:* At test time, estimate the balancing parameter $\theta_{\text{test}} = M(s(\mathcal{D}_{\text{test}}))$ from the test data and select the corresponding model $m_{\theta_{\text{test}}} = h_{\theta_{\text{test}}} \circ e$ for making predictions.

Following standard practice, we train the feature extractor e on $\mathcal{D}$ via ERM [31], and then learn the candidate classifier heads $\{h_\theta\}$ and the mapping M on disjoint validation set halves, denoted by $\mathcal{D}_{\text{candidate}}$ and $\mathcal{D}_{\text{mapping}}$, respectively.

Candidate Model Construction. Standard DFR [18] retrains the classifier on a balanced validation subset, implicitly assuming an unbiased test distribution. We instead assume a parameterized family of candidate joint distributions $\{P_\theta\}_{\theta \in \Theta}$, where θ controls the degree of spurious correlation between class label Y and bias attribute B. In our experiments, we use the following instantiation:

$$P_\theta(Y, B) = \theta\, P_{\text{bal}}(Y, B) + (1 - \theta)\, P_{\text{orig}}(Y, B), \quad \theta \in [0, 1], \tag{1}$$

where P_{orig} is the original (biased) joint distribution and P_{bal} is the uniform (unbiased) joint distribution. Note that due to our definition of subgroups, Equation (1) is equivalent to interpolating between subgroup distributions, thereby allowing θ to control the degree of subgroup balance.

For each θ, we sample a subset $\mathcal{D}_\theta \subseteq \mathcal{D}_{\text{candidate}}$ without replacement to approximate the distribution P_θ. For binary labels and bias attributes, assuming class balance, the strength of spurious correlation in $\mathcal{D}_\theta$ decreases monotonically with increasing θ, vanishing at $\theta = 1$ (see Appendix A2 for details).

With the fixed feature extractor e, AMSEL then trains the corresponding classifier heads h_θ via logistic regression [4] on features extracted from $\mathcal{D}_\theta$:

$$h_\theta = \arg\min_{h \in \mathcal{H}} \frac{1}{|\mathcal{D}_\theta|} \sum_{(x,y) \in \mathcal{D}_\theta} \ell\big(h(e(x)), y\big) + R(h), \tag{2}$$

where $\ell(\cdot, \cdot)$ denotes the cross-entropy loss, $\mathcal{H}$ is the set of classification functions obtainable via logistic regression, and $R(h)$ is a regularizer. Following DFR [18], we average the coefficients over multiple runs to reduce sensitivity to initialization. The final candidate model for parameter θ is given by $m_\theta = h_\theta \circ e$. This decoupled strategy enables efficient training of multiple candidate models using a single, fixed feature extractor.

Adaptive Model Selection. At test time, the test data may exhibit varying levels of spurious correlation, and the true joint distribution of class labels and bias attributes, $P_{\text{test}}(Y, B)$, is unknown. Rather than estimating the bias level directly, AMSEL identifies which candidate test distribution from the predefined set best approximates P_{test}. To this end, we define a score function s, which captures statistical properties indicative of the number of bias-conflicting samples from a given dataset of variable size, i.e., an ordered sequence of n input samples:

$$s : \bigcup_{n \in \mathbb{N}} \mathcal{X}^n \to \mathbb{R}. \tag{3}$$

In our implementation, we compare the predictions of the two most extreme candidate models: m_0 (trained on the original distribution) and m_1 (trained on an unbiased subset). Inspired by Ho et al. [13], we assume that these models agree on simple, bias-aligned samples while diverging on more challenging, bias-conflicting ones. For an input $x \in \mathcal{X}$, we define the binary consensus function:

$$\text{consensus}(x) = \begin{cases} 1, & \text{if } \arg\max(m_0(x)) \\ & \quad = \arg\max(m_1(x)) \\ 0, & \text{otherwise.} \end{cases} \tag{4}$$

For a given n-tuple of inputs $X = (x_1, x_2, \ldots, x_n) \in \mathcal{X}^n$, the consensus score is defined as the percentage of top-1 prediction agreement:

$$s_{\text{consensus}}(X) = \frac{1}{n} \sum_{i=1}^{n} \text{consensus}(x_i). \tag{5}$$

AMSEL learns a mapping $M : s(\mathcal{X}^n) \rightarrow \Theta$ from this score to the balancing parameter θ on $\mathcal{D}_{\text{mapping}}$. Thus, we estimate the balancing parameter θ_{test} of the test set as $\theta_{\text{test}} = M\big(s(\mathcal{D}_{\text{test}})\big)$. Finally, for any test input $x \in \mathcal{X}$, the prediction is generated using the corresponding candidate model: $\hat{y} = m_{\theta_{\text{test}}}(x) = h_{\theta_{\text{test}}}(e(x))$. This direct model selection based on the estimated θ_{test} avoids the need to evaluate all candidate models at test time, enabling efficient adaptation to varying levels of spurious correlation.

4   Experiments

We evaluate AMSEL on both a widely used debiasing benchmark (CelebA [20]) and a real-world medical imaging dataset (ChestX-ray14 [33]) and compare it against standard ERM [31] as well as existing post hoc debiasing methods [12, 18].

4.1   Implementation Details

Datasets and ERM Models. We evaluate our approach on the following two datasets. The *CelebA* dataset [20] is a popular debiasing benchmark, which comprises celebrity face images. Following [28], we focus on the binary target attribute *blond hair*, which is spuriously correlated with gender. In contrast, *ChestX-ray14* [33] is a real-world dataset containing chest X-rays of patients with various diseases. We consider the binary classification task of detecting *pneumothorax* (diagnosed vs. not diagnosed), where the presence of chest drains (a treatment artifact) acts as a spurious feature [23,27]. We use automatically generated chest drain labels from [22]. For both datasets, we first train an ERM model [31], which serves as the baseline and provides the fixed feature extractor e for subsequent debiasing. For CelebA, we follow the training procedure of [17] using a ResNet50 [11] pre-trained on ImageNet [8]. For ChestX-ray14, we adopt the approach described in [22] with a DenseNet121 [15] architecture.

Conventional Debiasing and AMSEL. In practice, we implement the mapping $M : s(\mathcal{X}^n) \rightarrow \Theta$ via linear regression and clip its predictions to $[0, 1]$. The parameter space is discretized as $\Theta = \{0.0, 0.05, 0.1, \ldots, 1.0\}$ to reduce the number of classifier heads. Following Kirichenko et al. [18], both the standard DFR models and our candidate classifier heads are trained via logistic regression using the `liblinear` solver [9] with an ℓ_1 penalty (see Appendix A1). To reduce sensitivity to initialization, each classifier head is obtained by averaging coefficients over 20 independent runs. The inverse regularization strength c is selected

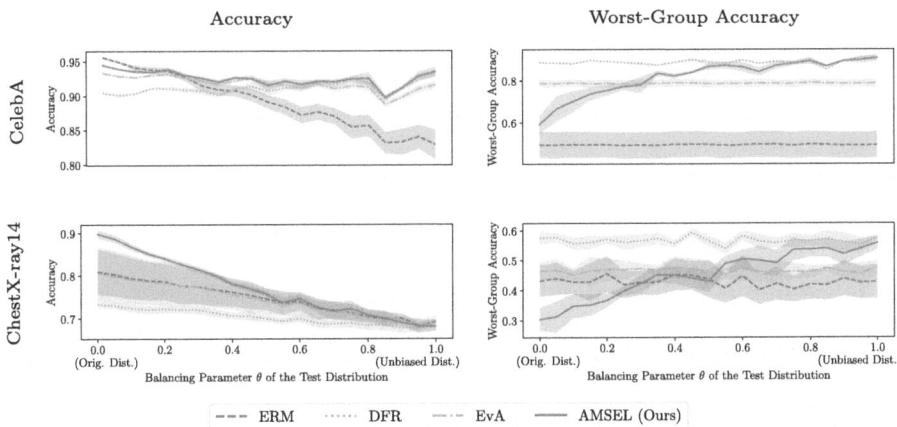

Fig. 2. Performance Trade-Off Across Test Distributions: We evaluate accuracy ($\uparrow$, *left*) across test sets with varying spurious correlation (cf. Equation (1)), with ERM [31] performing best on biased data, while conventional post hoc debiasing methods (DFR [18] and EvA [12]) outperform the ERM model on balanced data. In contrast, AMSEL maintains competitive accuracy across all distributions by adaptively selecting the estimated best classifier. The adaptive nature of AMSEL is further evident in its worst-group accuracy [28] ($\uparrow$, *right*), a proxy for model bias, demonstrating an increased model bias for biased distributions. Shaded regions around each line represent standard deviations.

by training on $\mathcal{D}_{\text{candidate}}$ and tuning based on the worst-group accuracy [28] on $\mathcal{D}_{\text{mapping}}$, considering $c \in \{a \cdot 10^b \mid a \in \{1, 3, 7\}, b \in \{-3, -2, -1, 0, 1, 2, 3, 4\}\}$. For EvA [12], we follow the original procedure and retrain the classifier head using SGD (see Appendix A3 for additional details). Sample-wise reweighting is applied to all classifiers to address class imbalance. All reported results include standard deviations computed over five independent runs.

4.2   Results

Trade-Offs Between ERM and Debiasing Methods. To motivate the need for adaptive selection, Fig. 2 compares the ERM model [31] with the debiased models obtained via DFR [18] and EvA [12], reporting both overall accuracy and worst-group accuracy [28] across test sets with varying spurious correlation strength (see Equation (1)). A clear trade-off emerges: the ERM model excels on biased data by exploiting shortcuts, whereas debiased models, by ignoring these shortcuts, outperform it on balanced sets where such correlations no longer hold. The debiased models' insensitivity to spurious correlations is reflected in their higher worst-group accuracy, indicating more uniform performance across subgroups. Since the test set's level of spurious correlation is typically unknown, the optimal model choice is ambiguous. AMSEL addresses this by adaptively selecting the classifier head based on the estimated test distribution.

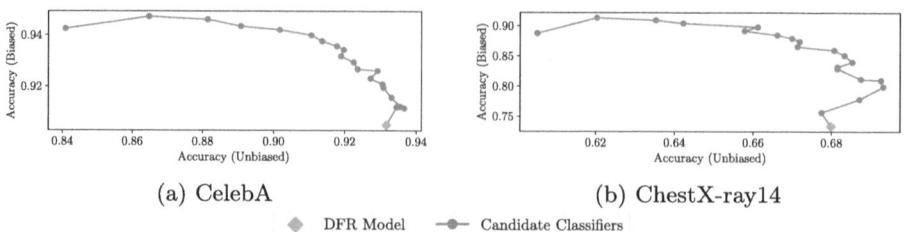

(a) CelebA   (b) ChestX-ray14

◆ DFR Model   —●— Candidate Classifiers

Fig. 3. Comparison of Candidate Classifiers: AMSEL trains classifier heads on a range of candidate test distributions ranging from the original, biased distribution to the unbiased distribution, which corresponds to standard DFR. When comparing their accuracy (↑) on the original (biased) and the unbiased distribution, we observe that there is no single uniformly dominant candidate classifier.

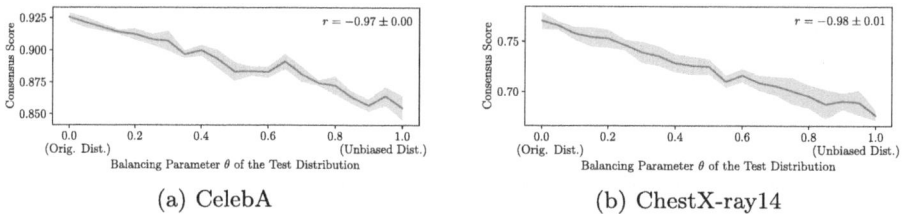

(a) CelebA   (b) ChestX-ray14

Fig. 4. Informative Value of the Score Function: Our consensus-based score function $s_{consensus}$ exhibits a near-linear relationship with the balancing parameter θ (Pearson correlation coefficients denoted as r), which controls the strength of spurious correlation, making it suitable for test distribution identification. Shaded regions around each line represent standard deviations.

Candidate Classifier Heads. Fig. 3 depicts the performance trade-offs among the candidate classifier heads $\{h_\theta\}_{\theta \in [0,1]}$ (see Equation (2)), where $\theta = 1.0$ corresponds to the standard DFR setting [18]. The candidate models exhibit an almost smooth transition in performance, shifting their focus from biased to unbiased data. As no single candidate model is uniformly optimal, adaptive model selection is essential to navigate this performance trade-off.

Estimation of Test Distribution Characteristics. The effectiveness of AMSEL depends on accurately estimating the balancing parameter θ to select the estimated best classifier for a given test distribution. Figure 4 shows that our consensus-based score function $s_{consensus}$ is almost linearly related to the balancing parameter θ, as confirmed by the Pearson correlation coefficients of -0.97 ± 0.00 for CelebA [20] and -0.98 ± 0.01 for ChestX-ray14 [33]. This strong linear dependency confirms $s_{consensus}$ as a suitable score function and justifies our design choice of implementing the mapping M for estimating the test distribution's bias characteristics via linear regression.

Adaptive Model Selection (AMSEL). Finally, we combine the balancing parameter estimation with the candidate classifier heads to evaluate the full AMSEL approach. For a test set $\mathcal{D}_{\text{test}}$, we estimate $\theta_{\text{test}} = M(s(\mathcal{D}_{\text{test}}))$ and select the corresponding model $m_{\theta_{\text{test}}} = h_{\theta_{\text{test}}} \circ e$ for prediction. As shown in Fig. 2, for CelebA [20], AMSEL nearly matches the performance of the ERM model [31] on test distributions with strong spurious correlation (benefiting from shortcut bias exploitation) while also capturing the improvements of debiasing methods on balanced test distributions where spurious correlations no longer apply. For ChestX-ray14 [33], the adaptive shift from a biased to an unbiased model is most apparent when considering worst-group accuracy [28], indicated by an improved worst-group performance for higher balancing parameters. To capture overall performance across the entire spectrum of considered joint distributions, we compute the area under the accuracy curve (AuC; see Appendix A4). On CelebA [20] AMSEL achieves an AuC of 0.925 ± 0.002, outperforming DFR [18] (0.911 ± 0.003), EvA [12] (0.917 ± 0.003), and ERM [31] (0.891 ± 0.010). On ChestX-ray14 [33], similar trends are observed with AMSEL (0.771 ± 0.007) outperforming the other approaches by a margin of 0.025. These results confirm that AMSEL is highly beneficial when the strength of the spurious correlation of the test set is unknown. When it is known, AMSEL also retains high performance.

5   Conclusion

We propose a lightweight yet effective approach to spurious feature mitigation that expands debiasing with *adaptive model selection (AMSEL)*. Unlike classical debiasing methods, AMSEL preserves the high accuracy of the original ERM model in bias-dominated regimes while matching debiased models under balanced conditions by adaptively selecting its classifier from a family of candidate heads trained on simulated test distributions. This added flexibility requires minimal extra overhead, as the additional forward passes (two for estimating the test distribution characteristics) only involve the lightweight classifier head. Our experiments on both CelebA [20] and ChestX-ray14 [33] demonstrate that AMSEL outperforms ERM [31] and other debiasing approaches [12,18] when evaluated across subsets with varying levels of spurious correlation. For example, on ChestX-ray14, AMSEL achieves an area under the accuracy curve (AuC) of 0.771 ± 0.007, outperforming the strongest baseline by a margin of 0.025. Furthermore, we believe that the core idea of selecting from a set of candidate models, each optimized for a specific strength of spurious correlation, can extend to other debiasing strategies such as reweighting [16], resampling [16], or Group-DRO [28]. Overall, AMSEL provides an efficient solution for robust performance in environments where the degree of spurious correlation may vary significantly.

Disclosure of Interests. The authors have no competing interests to declare that are relevant to the content of this article.

References

1. Almeida, P.R., Oliveira, L.S., Britto, A.S., Sabourin, R.: Adapting dynamic classifier selection for concept drift. Expert Syst. Appl. (2018)
2. Asaad, I., Shadaydeh, M., Denzler, J.: Gradient Extrapolation for Debiased Representation Learning (2025)
3. Beery, S., van Horn, G., Perona, P.: Recognition in terra incognita. In: ECCV (2018)
4. Bishop, C.M.: Pattern Recognition and Machine Learning (2006)
5. Britto, A.S., Sabourin, R., Oliveira, L.E.: Dynamic selection of classifiers—a comprehensive review. Pattern Recogn. (2014)
6. Cruz, R.M., Sabourin, R., Cavalcanti, G.D.: Dynamic classifier selection: Recent advances and perspectives. Inf. Fusion (2018)
7. Dehkharghanian, T., et al.: Biased data, biased AI: deep networks predict the acquisition site of TCGA images. Diagn. Pathol. (2023)
8. Deng, J., Dong, W., Socher, R., Li, L.J., Li, K., Fei-Fei, L.: ImageNet: a large-scale hierarchical image database. In: CVPR (2009)
9. Fan, R.E., Chang, K.W., Hsieh, C.J., Wang, X.R., Lin, C.J.: LIBLINEAR: a library for large linear classification. JMLR (2008)
10. Geirhos, R., et al.: Shortcut learning in deep neural networks. Nat. Mach. Intell. (2020)
11. He, K., Zhang, X., Ren, S., Sun, J.: Deep residual learning for image recognition. In: CVPR (2016)
12. He, Q., Xu, K., Yao, A.: EvA: erasing spurious correlations with activations. In: ICLR (2025)
13. Ho, T.K., Hull, J.J., Srihari, S.N.: Decision combination in multiple classifier systems. PAMI (1994)
14. Howard, F.M., et al.: The impact of site-specific digital histology signatures on deep learning model accuracy and bias. Nat. Commun. (2021)
15. Huang, G., Liu, Z., van der Maaten, L., Weinberger, K.Q.: Densely connected convolutional networks. In: CVPR (2017)
16. Idrissi, B.Y., Arjovsky, M., Pezeshki, M., Lopez-Paz, D.: Simple data balancing achieves competitive worst-group-accuracy. In: Proceedings of the First Conference on Causal Learning and Reasoning. PMLR (2022)
17. Izmailov, P., Kirichenko, P., Gruver, N., Wilson, A.G.: On feature learning in the presence of spurious correlations. In: NeurIPS (2022)
18. Kirichenko, P., Izmailov, P., Wilson, A.G.: Last layer re-training is sufficient for robustness to spurious correlations. In: ICLR (2023)
19. Lapuschkin, S., Wäldchen, S., Binder, A., Montavon, G., Samek, W., Müller, K.R.: Unmasking Clever Hans predictors and assessing what machines really learn. Nat. Commun. (2019)
20. Liu, Z., Luo, P., Wang, X., Tang, X.: Deep learning face attributes in the wild. In: ICCV (2015)
21. Lövdal, S., Biehl, M.: Iterated relevance matrix analysis (IRMA) for the identification of class-discriminative subspaces. Neurocomputing (2024)
22. Murali, N., Puli, A., Yu, K., Ranganath, R., Batmanghelich, K.: Beyond distribution shift: spurious features through the lens of training dynamics. TMLR (2023)
23. Oakden-Rayner, L., Dunnmon, J., Carneiro, G., Ré, C.: Hidden stratification causes clinically meaningful failures in machine learning for medical imaging. CHIL (2020)

24. Pedregosa, F., et al.: Scikit-learn: machine learning in python. JMLR (2011)
25. Penzel, N., Stein, G., Denzler, J.: Reducing bias in pre-trained models by tuning while penalizing change. In: VISIGRAPP (2024)
26. Rosenfeld, E., Ravikumar, P.K., Risteski, A.: Domain-adjusted regression or: ERM may already learn features sufficient for out-of-distribution generalization. In: NeurIPS 2022 Workshop on Distribution Shifts: Connecting Methods and Applications (2022)
27. Rueckel, J., et al.: Impact of confounding thoracic tubes and pleural dehiscence extent on artificial intelligence pneumothorax detection in chest radiographs. Invest. Radiol. (2020)
28. Sagawa, S., Koh, P.W., Hashimoto, T.B., Liang, P.: Distributionally robust neural networks for group shifts: on the importance of regularization for worst-case generalization. In: ICLR (2020)
29. Shah, H., Tamuly, K., Raghunathan, A., Jain, P., Netrapalli, P.: The pitfalls of simplicity bias in neural networks. In: NeurIPS (2020)
30. Teney, D., Abbasnejad, E., Lucey, S., van den Hengel, A.: Evading the simplicity bias: training a diverse set of models discovers solutions with superior OOD generalization. In: CVPR (2022)
31. Vapnik, V.N.: An overview of statistical learning theory. IEEE Trans. Neural Netw. (1999)
32. Villani, C.: Optimal Transport, Old and New (2009)
33. Wang, X., Peng, Y., Le Lu, Lu, Z., Bagheri, M., Summers, R.M.: ChestX-ray8: hospital-scale chest X-ray database and benchmarks on weakly-supervised classification and localization of common thorax diseases. In: CVPR (2017)
34. Xiao, K.Y., Engstrom, L., Ilyas, A., Madry, A.: Noise or signal: the role of image backgrounds in object recognition. In: ICLR (2021)
35. Yang, Y., Zhang, H., Katabi, D., Ghassemi, M.: Change is hard: a closer look at subpopulation shift. In: ICML (2023)
36. Ye, W., Zheng, G., Cao, X., Ma, Y., Zhang, A.: Spurious correlations in machine learning: a survey (2024)
37. Zech, J.R., Badgeley, M.A., Liu, M., Costa, A.B., Titano, J.J., Oermann, E.K.: Variable generalization performance of a deep learning model to detect pneumonia in chest radiographs: a cross-sectional study. PLoS Med. (2018)

SelfExplaNETory: Improving Classification Accuracy with Local Post Hoc Interpretation

Suraja Poštić[(✉)] and Marko Subašić

Faculty of Electrical Engineering and Computing, University of Zagreb, Unska 3,
10000 Zagreb, Croatia
{suraja.postic,fer,marko.subasic}@fer.unizg.hr
https://www.fer.unizg.hr/en

Abstract. Deep classification models have been criticized for showing noise sensitivity, bias toward background, and lack of interpretability. In this paper, we expand the idea of using local post hoc interpretation to improve classification accuracy under the assumption of existing object localization masks. Our classifiers are trained to align their interpretation maps with these objects, and we refer to such models as SelfExplaNETory (SE). The idea of SE is to force the model to focus on important information despite being presented with copious background distractions. We first identify the best interpretation loss function suited for bounding box-type object localization among multiple proposed variations. Then we train a CNN to incorporate this loss in its parameter optimization process using five different input resolution interpretation methods based on both class scores and probabilities. We show that SE improves the classification accuracy of the baseline model, with probability-based MaxPIn as the most successful interpretation method. Next, we design SE MINI as a smaller and faster version of SE based only on the correct class' interpretation. We find SE MINI to be more stable than SE for probability interpretations at the expense of a bit smaller accuracy gain in general. SE MINI proves to be most effective with Expected Gradients probability interpretation. We also compare SE with other interpretation-aided training procedures and find that it mostly achieves better accuracy or gives a comparable result. Finally, we argue that the accuracy gain achieved by SE can serve as an objective evaluation metric for interpretation quality.

Keywords: XAI · model enhancement · image classification

1 Introduction

Understanding the behavior of DNNs has become increasingly important, especially in delicate fields [8], leading to the development of eXplainable Artificial Intelligence (XAI) [12]. XAI research took two main directions: training models interpretable by design [7,19] and explaining black-box models with post hoc

© The Author(s), under exclusive license to Springer Nature Switzerland AG 2026
M. Castrillón-Santana et al. (Eds.): CAIP 2025, LNCS 15622, pp. 112–122, 2026.
https://doi.org/10.1007/978-3-032-05060-1_10

interpretation methods [6, 8, 10, 15, 20–22, 24]. Recently, researchers started incorporating interpretation into parameter optimization directly [1, 3, 11, 14, 16, 17]. In this work, we expand the idea of using local post hoc interpretation to improve the accuracy of image classifiers. Under the assumption of existing class-discriminative object localization masks, we train our models to focus their interpretation maps on these objects. By diverting the attention away from background noise, we aim to prevent the model from making spurious correlations. We first conduct extensive research regarding the definition of the interpretation loss function for bounding box-type object localization. Then we train a classifier, referred to as SelfExplaNETory (SE), based on five different local post hoc interpretation methods [3, 15, 21–23] for both class score and probability interpretations in input image resolution. The schematics of the SE training procedure are shown in Fig. 1. Our experiments show that SE achieves higher accuracy than the model trained without the aid of interpretation. Furthermore, the accuracy gain varies depending on the interpretation method. Next, we design SE MINI as a smaller and computationally more friendly version of SE, which uses only the correct class' interpretation. We demonstrate that SE MINI can still improve the accuracy, though normally not on the same scale as SE. Both SE and SE MINI mostly outperform other interpretation-aided training methods or at least show comparable results. Finally, we discuss how the accuracy gain of SE can serve as an objective interpretation evaluation metric.

2 Related Work

Here we give a brief overview of the papers that used interpretation in the parameter optimization process. In [17], the authors define Right for the Right Reasons (RRR) loss, which forces the model's log-output gradients with respect to input image pixels to be zero in locations that are a priori known to contain irrelevant information. The contextual decomposition [13] interpretation is pushed towards zero in [16] for pixels segmented as spurious patterns. In [3], the authors propose interpretation method called Expected Gradients (EG) and optimize its smoothness using Pixel Attribution Prior (PAP) loss that penalizes differences between neighboring pixels' interpretations. The authors of Guided Attention Inference Network (GAIN) [11] require the model to have small predictions for images where regions associated with the strongest interpretation activations have been obscured. The authors of [14] focus on preserving the consistency of interpretations across spatial input transformations. Region of Interest Activation loss employed in [1] consists of modified Intersection over Union between the thresholded interpretation map and unsupervised object detector output as one component, while the other is simply the interpretation bounding box diagonal.

3 Methods

In this section, we first explain the concept of optimizing model parameters to align its reasoning with human logic using local post hoc interpretation and

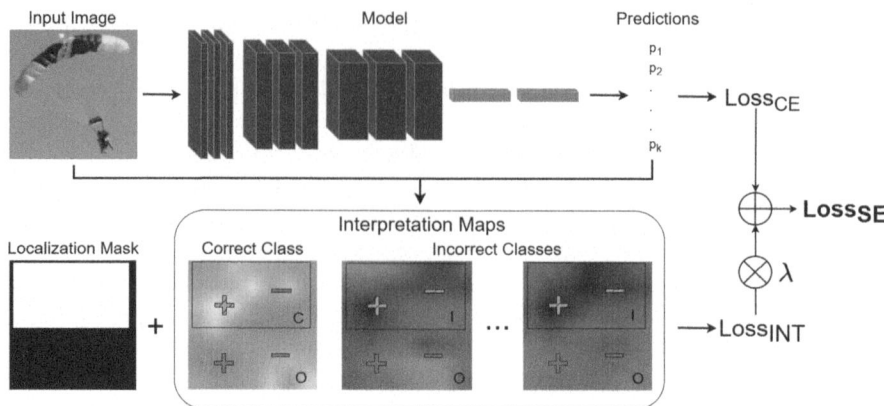

Fig. 1. Schematics of a SelfExplaNETory model training procedure. An input image is fed to a model to produce class predictions, which are used to calculate cross-entropy loss. The same image and model are then used to obtain interpretation maps for all the classes from which the interpretation loss is calculated based on the corresponding localization mask. The interpretation maps are divided into three areas: the correct class' (C) and incorrect classes' (I) area inside the mask and all classes' area outside the mask (O). Positive (P) and negative (N) contributions are either desirable (green hatched interior) or undesirable (red interior) based on which area they belong to. The AO-NC loss shown here penalizes **A**ll contributions in the **O** area and **N**egative contributions in the **C** area. The final SelfExplaNETory loss is calculated as a sum of cross-entropy loss and interpretation loss multiplied by a scaling factor λ.

object localization masks. We refer to this concept as SelfExplaNETory (SE). Then we introduce a set of loss functions appropriate for SE training. Finally, we propose a computationally more friendly SE variant that uses only the correct class' interpretation.

3.1 SelfExplaNETory: Interpretation-Aided Model Optmization

Let $f(\theta)$ denote a classifier model function, and let $\mathcal{I} = \{\mathbf{I}_i\}_i$ be the set of input images with correct class labels $\mathbf{y} = \{y_i\}_i$. We presume to have a priori knowledge about the locations of objects within these images given as a set of corresponding binary masks $\mathcal{M} = \{\mathbf{M}_i\}_i$. These masks can be given in the form of bounding boxes or precise segmentation masks. Apart from correct classification, SE aims to optimize parameters θ such that features responsible for the model's decision, as purported by interpretation method ϕ, align with object localization masks given in $\mathcal{M}$. The resulting loss function is then given by:

$$\mathcal{L}_{SE}(f, \theta, \mathcal{I}, \mathbf{y}, \phi, \mathcal{M}) = \mathcal{L}_{CE}(f(\theta, \mathcal{I}), \mathbf{y}) + \lambda \mathcal{L}_{INT}(\phi(f(\theta, \mathcal{I})), \mathcal{M}) \qquad (1)$$

where $\mathcal{L}_{CE}$ denotes standard cross-entropy loss and hyperparameter λ regulates the strength of interpretation loss $\mathcal{L}_{INT}$.

3.2   Interpretation Loss

Let $\mathbf{I} \in \mathbb{R}^{w \times h}$ be an input image belonging to class c_i and $\mathbf{S} = \phi(f(\theta, \mathbf{I})), \mathbf{S} \in \mathbb{R}^{n_c \times w \times h}$ the corresponding interpretation map, where n_c stands for the number of classes in the dataset. We will use the term $\mathbf{S}^{c_i} \in \mathbb{R}^{w \times h}$ for the correct class' interpretation map and $\mathbf{S}^{-c_i} \in \mathbb{R}^{(n_c-1) \times w \times h}$ for the interpretation maps of all the other classes. Let $\mathbf{M} \in \{0, 1\}^{n_c \times w \times h}$ denote the associated localization mask stacked n_c times to match the dimension of $\mathbf{S}$, with ones in locations where the object of interest is present (object interior) and zeros elsewhere (object exterior). We distinguish between three main interpretation map regions. Pixels outside (O) the object form one region, while the other two are made of object pixels belonging to either the correct (C) class c_i or any of the incorrect (I) classes. Ideally, we would expect $\mathbf{S}^{c_i}$ to have high positive values inside the object, indicating that the model used information present there to increase its confidence in c_i. The area outside the object contains no relevant information for any of the classes, so we expect no contributions there. Regarding the $\mathbf{S}^{-c_i}$ object pixels, we could assume to find negative or no contributions there, but positive contributions might also be logical for datasets where different classes share some features. We discern between the interpretation loss functions according to the type of contributions they penalize, which can be positive (P), negative (N), or any (A), and the region where they penalize them. Once we decide which contributions to penalize and where, we need to choose the type of metric. We can take the total sum (T) or mean (M) over the target region either as absolute (A) values or relative (R) to the sum or mean of absolute $\mathbf{S}$ values, respectively. The general formula for AO-NC-PI interpretation loss calculated using metric $m_1 m_2 \in \{A, R\} \times \{T, M\}$ is given by:

$$
\mathcal{L}_{INT} = \frac{\overbrace{\dfrac{|\mathbf{S}| \odot \neg \mathbf{M}}{\mathbbm{1}\{\sum_{i,j,k}(\neg \mathbf{M}_{ijk})\}_{m_2=M}}}^{\mathcal{L}_{INT}^{AO}} + \overbrace{\dfrac{max(0, -\mathbf{S}^{c_i}) \odot \mathbf{M}^{c_i}}{\mathbbm{1}\{\sum_{i,j} \mathbf{M}_{ij}^{c_i}\}_{m_2=M}}}^{\mathcal{L}_{INT}^{NC}} + \overbrace{\dfrac{max(0, \mathbf{S}^{-c_i}) \odot \mathbf{M}^{-c_i}}{\mathbbm{1}\{\sum_{i,j,k} \mathbf{M}_{ijk}^{-c_i}\}_{m_2=M}}}^{\mathcal{L}_{INT}^{PI}}}{\mathbbm{1}\left\{\dfrac{\sum_{i,j,k}|\mathbf{S}_{ijk}|}{\mathbbm{1}\{n_c wh\}_{m_2=M}} + \epsilon\right\}_{m_1=R}}
$$

$$(2)$$

where $\odot$ denotes the scalar product, $\neg \mathbf{M}$ is $\mathbf{M}$ with inverted values, and $\mathbbm{1}\{value\}_{cond}$ equals the value inside the braces if condition $cond$ is satisfied and 1 otherwise. Terms $\mathbf{M}^{c_i}$ and $\mathbf{M}^{-c_i}$ are defined analogously to $\mathbf{S}^{c_i}$ and $\mathbf{S}^{-c_i}$. Small positive value ϵ serves to avoid division with zero and improve stability. $\mathcal{L}_{INT}$ can include all three components listed in Eq. 2, namely $\mathcal{L}_{INT}^{AO}$, $\mathcal{L}_{INT}^{NC}$, and $\mathcal{L}_{INT}^{PI}$, or any subset of them. Naturally, if a certain area does not exist, e.g., in a case where the localization map spreads across the entire image, that area will be excluded from the loss function. We can also decide to relax the constraint on pixels outside the object and allow negative contributions there since these regions failed to detect any important information. To do so, we would replace the $\mathcal{L}_{INT}^{AO}$ component with the $\mathcal{L}_{INT}^{PO}$ given by $max(0, \mathbf{S}) \odot \neg \mathbf{M}$.

Note how different interpretation loss configurations may be suitable for different interpretation methods and datasets. For example, output-conserving interpretations could benefit from the total (T) metric, while mean (M) can be used to diminish the importance of region size. The relative (R) metric could partially reduce the need for interpreting the softmax layer, as penalizing one type of contribution in a certain region indirectly increases other types of contributions in that region, as well as all contributions in other regions. Absolute (A) loss could be a good fit for probability interpretations where we expect positive contributions in a certain region for one class to imply negative contributions in the same region for all the other classes as described in [15]. For fine-grained classification, we might want to exclude the $\mathcal{L}_{INT}^{PI}$ component.

3.3   SE MINI

Calculating the gradients of interpretations for all the classes can be very computationally expensive, especially for a large number of classes. To mitigate this issue, SE can be restrained to work only with the correct class' interpretation, and we refer to this variant as SE MINI. Any constraints on the object region for incorrect classes no longer apply here, and the area outside the object is reduced to the correct class only. Absolute interpretation map sums and means used for relative metrics are taken only across the correct class as well.

4   Results

In this section, we list dataset and training procedure details, followed by the optimization of the interpretation loss function and multiplier. Then we present the results of interpretation-enhanced image classification with SE, SE MINI, and three state-of-the-art methods.

4.1   Dataset and Implementation Details

We conducted our experiments on the ImageNet [18] localization dataset restricted to the same ten classes present in Imagenette from fastai. Localization bounding boxes were provided for approximately 5.5k images, which we used for training. The rest of the images were divided between validation and test sets.

The implementation was done in the Jax [2] framework with an NVIDIA RTX A6000 GPU. We used the ResNet18 [5] network trained from scratch with the ADAM [9] optimizer for 500 epochs with a batch size of 32. The starting learning rate was set to 0.001 and decreased to 0.0001 after 300 epochs. Interpretation loss hyperparameters were set to $\lambda = 0.001$ and $\epsilon = 10^{-15}$. The central square patch resized to 256×256 pixels was taken prior to the 224×224 random crop and horizontal flip augmentations. The model with maximum validation accuracy was taken as final. We used one seed for the interpretation loss function and multiplier experiments and two different seeds for classification, where we reported a better result.

The interpretation methods we combined with SE were Gradients × Inputs (GI) [21], Integrated Gradients (IG) [23], FullGrad (FG) [22], MaxPIn (MP)

[15], and Expected Gradients (EG) [3]. We chose to work only with methods that support positive and negative contributions since both are required by the tested loss functions. Thus, the methods from the Class Activation Mapping (CAM) [24] family were not considered. Note that for ReLU activation-based networks, GI become very similar to Layer-wise Relevance Propagation (LRP) [10] and DeepLIFT [20], while FG and MP have the same class score interpretation maps. Probability interpretations of FG and MP were calculated as suggested in [15]; only the softmax layer was treated as a single linear function given by $p_i = \sum_{j=1}^{n_c} [(p_i - p_j)/(x_j - x_i)](x_j - x_i) + [1 - p_i(n_c - 1)]$, where $\mathbf{x} = \{x_i\}_i$ and $\mathbf{p} = \{p_i\}_i$ represent class score and probability vectors, respectively. The number of bins for IG was 3 for SE and 5 for SE MINI. EG were based on 1 sample from the current batch.

We compared SE and SE MINI with RRR [17], PAP [3], and GAIN [11]. For RRR we set $\lambda_1 = 1^{-10}$ and omitted the regularization term for fair comparison. For PAP we also used $\lambda = 1^{-10}$, while for GAIN we set $\alpha = 0.00001$, $\omega = 100$, and $\sigma = 0.6$. All three interpretation loss multipliers were selected via small grid search by increasing and decreasing the initial value by a factor of 0.01 until no accuracy gain was achieved in either direction. GAIN was paired with MP and EG probability interpretations for a better comparison with SE.

4.2   Interpretation Loss

We experimented with six contribution constraints combined with all four metric combinations as listed in Table 1. Testing all these losses with multiple interpretation methods would be overly time-consuming, so we used only GI based on class scores (GI_{cs}) and probabilities (GI_{prob}). We also froze λ to 0.001. Classification accuracy of SE trained with listed losses and interpretations is shown in Table 1. All combinations perform rather well, except for AT losses with GI_{cs}, which have significantly worse results. Taking GI_{cs} and GI_{prob} into account, the RT metric most often had the highest accuracy across different constraints, while the AO-NC constraint performed better than others across different metrics. All together, AO-NC-RT loss had the highest average accuracy between GI_{cs} and GI_{prob} based SEs.

Once we identified the best loss function, we tried decreasing and increasing λ by a factor of 0.1. The results are shown in Table 2. The initial λ value of 0.001 performed better than the other two options for GI_{cs} and GI_{prob}, so we decided to use this value in further experiments.

4.3   Interpretation-Enhanced Image Classification

Table 3 compares the accuracy of SE and SE MINI with a baseline model trained only with cross-entropy loss as well as RRR, PAP, and GAIN. All SE and SE MINI models had at least marginally higher accuracy than the baseline. The best-performing SE model with an accuracy gain of 1.6% is based on MP_{prob} interpretation. GI_{cs} and GI_{prob} follow after, while FG_{prob} had the worst result. IG_{prob} and EG_{prob} did not have stable training with SE. SE MINI normally

Table 1. Interpretation Losses: Test set accuracy (%) on Imagenette for GI_{cs} and GI_{prob}-based SE classifiers trained with different loss functions. $SE + GI_{cs}$ achieved the best accuracy with AO-NC-RT loss, while GI_{prob} worked the best with PO-NC-PI-RT loss. Considering both methods, AO-NC-RT loss performed the best on average.

INT Method	GI_{cs}				GI_{prob}			
Loss/Metric	AT	RT	AM	RM	AT	RT	AM	RM
AO	64.14	85.01	85.12	85.64	83.54	85.61	84.82	85.41
AO-NC	57.38	**85.76**	85.44	85.09	83.76	85.47	85.30	82.49
AO-NC-PI	51.35	85.30	84.28	85.49	84.66	84.93	84.82	83.66
PO	66.69	85.41	84.87	84.96	85.33	85.25	85.11	84.50
PO-NC	70.78	84.55	85.26	84.76	84.31	85.42	84.99	85.01
PO-NC-PI	67.07	85.23	85.01	84.87	84.61	**85.77**	84.92	84.49

Table 2. Interpretation Loss Multiplier (λ): Test set accuracy (%) on Imagenette for GI_{cs} and GI_{prob}-based SE classifiers trained with AO-NC-RT loss and different λ values. Both interpretation methods achieved the highest accuracy for $\lambda = 0.001$.

Model	$\lambda = 0.01$	$\lambda = 0.001$	$\lambda = 0.0001$
$SE + GI_{cs}$	85.09	**85.76**	84.09
$SE + GI_{prob}$	81.64	**85.47**	84.74

performed worse than SE but proved more stable for probability interpretations. The best SE MINI accuracy gain, which is only marginally smaller than the best SE result, was achieved with EG_{prob}. RRR and GAIN with both MP_{prob} and EG_{prob} performed worse than almost all other interpretation loss-method combinations but still remained at least marginally above the baseline. PAP with EG_{cs} and EG_{prob} based on all classes' interpretations also exceeded baseline accuracy but did not outperform SE with EG_{cs}. Regarding the correct class-based PAP, its combination with EG_{cs} was only marginally more accurate than SE MINI, while SE MINI outperformed PAP with EG_{prob}.

Figure 2 shows how SE interpretation maps have better focus on target objects than baseline interpretation maps. A single epoch of baseline model training took 53 s, while $SE + GI_{cs}$ took 127 s, which is roughly 2.5x slower. SE MINI $+ GI_{cs}$ took 77 s, which is ~50% more than the baseline.

5   Discussion

As the model learns important features, the absolute values of pixel contributions tend to grow. Thus, it is not surprising that it was easier to minimize the RT metric than the AT. Furthermore, probability contributions are much smaller than class score contributions, which can explain why GI_{prob} performed much better with the AT metric than GI_{cs}.

Table 3. Interpretation-Enhanced Image Classification: Test set accuracy (%) on Imagenette for the baseline model trained only with cross entropy (CE) loss compared to SE, SE MINI, RRR, PAP, and GAIN. All SE and SE MINI models surpassed the performance of the baseline model. SE worked the best with the MaxPIn_{prob} interpretation, with 1.6% higher accuracy than the baseline. Expected Gradients_{prob} was the most successful method for SE MINI, with an accuracy gain of 1.45%. RRR, PAP, and GAIN showed at least marginally higher accuracy than baseline but mostly performed worse than SE models.

Interpretation Loss & Method	SE	SE MINI
None (Baseline)	84.46	84.46
AO-NC-RT + Gradients × Inputs_{cs}	85.76	84.99
AO-NC-RT + Gradients × Inputs_{prob}	85.47	84.99
AO-NC-RT + Integrated Gradients_{cs}	85.68	85.09
AO-NC-RT + Integrated Gradients_{prob}	—	85.20
AO-NC-RT + $\text{FullGrad/MaxPIn}_{cs}$	85.34	85.30
AO-NC-RT + FullGrad_{prob}	84.80	85.41
AO-NC-RT + MaxPIn_{prob}	**86.06**	85.03
AO-NC-RT + Expected Gradients_{cs}	85.50	85.14
AO-NC-RT + Expected Gradients_{prob}	—	**85.91**
RRR + $\text{Log-Gradients}_{prob}$ [17]	84.99	84.65
PAP + Expected Gradients_{cs} [3]	85.11	85.31
PAP + Expected Gradients_{prob} [3]	85.45	85.31
GAIN + MaxPIn_{prob} [11]	—	84.93
GAIN + Expected Gradients_{prob} [11]	—	84.92

IMAGE	Baseline	SE+MaxPIn_{prob}	IMAGE	Baseline	SE+MaxPIn_{prob}

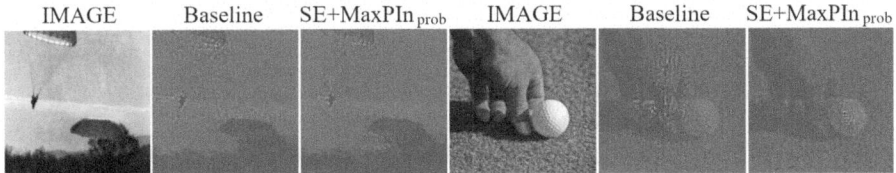

Fig. 2. Baseline and SE correct class' interpretation maps. Original images taken from the ImageNet dataset [18] are followed by the baseline model and SE+MaxPIn_{prob} MaxPIn_{prob} interpretation maps for classes "parachute" (left) and "golf ball" (right). Yellow and purple regions correspond to positive and negative contributions, respectively. It can be seen that SE interpretations have a better focus on class-discriminative objects.

Regarding the classification enhancement, we should keep in mind that the loss function was chosen according to GI-based SE, which gives this method a slight advantage. However, MaxPIn and Expected Gradients still had the highest accuracy for SE and SE MINI. Note that using bounding boxes, which often contain some background features, aggravates SE training. Having precise segmentation masks would guarantee that SE draws the model's attention to the right places and allow SE to reach its full potential.

Finally, since SE improved classification accuracy and we designed the interpretation loss to guide the model's attention to correct places, we can argue that a higher accuracy gain indicates better interpretation quality. Thus, SE can be used to objectively evaluate interpretation methods, which is still an open question [4].

6   Conclusion

In this work, we built upon the idea of incorporating local post hoc interpretation maps into the parameter optimization process of an image classifier. We first performed an extensive interpretation loss function optimization, which pointed out that any contributions outside the object region should be minimized, as well as negative contributions inside the object for the correct class. Furthermore, it was best to penalize the sum of undesirable contributions' absolute values relative to the sum of absolute interpretation map values. We trained our models to include interpretation loss for five different interpretation methods based on class scores and probabilities. These models consistently outperformed the model trained without the aid of interpretation, as well as most of the tested state-of-the-art interpretation-aided training procedures, while remaining comparable to others. The SE accuracy gain depended on the interpretation method, where probability-based MaxPIn performed best when interpretations for all classes were used, while probability-based Expected Gradients worked best with the correct class' interpretation alone. Thus, we believe that SE accuracy gain can serve as an objective evaluation metric for the quality of interpretation methods.

7   Limitations of the Study

The interpretation loss function was chosen based on a single interpretation method and bounding box localization masks. Different losses might be more suitable for other methods or mask types. SE could further be tested on various networks, datasets, and problems for more information about its generalizability.

Acknowledgments. This work was supported by the European union, Interreg VI-A IPA Programme Croatia-Bosnia and Herzegovina-Montenegro 2021–2027. [HR-BA-ME00030 - LADY].

Disclosure of Interests. The authors have no competing interests to declare that are relevant to the content of this article.

References

1. Ahmadi, R., Rajabi, M.J., Khalooie, M., Sabokrou, M.: Mitigating bias: enhancing image classification by improving model explanations (2023). https://arxiv.org/abs/2307.01473
2. Bradbury, J., et al.: JAX: composable transformations of Python+NumPy programs (2018). http://github.com/jax-ml/jax
3. Erion, G., Janizek, J., Sturmfels, P., Lundberg, S., Lee, S.I.: Learning explainable models using attribution priors (June 2019). https://doi.org/10.48550/arXiv.1906.10670
4. Halliwell, N., Gandon, F., Lecue, F., Villata, S.: The need for empirical evaluation of explanation quality. In: AAAI 2022 - Workshop on Explainable Agency in Artificial Intelligence. Vancouver, Canada (February 2022). https://hal.science/hal-03591012
5. He, K., Zhang, X., Ren, S., Sun, J.: Deep residual learning for image recognition. In: 2016 IEEE Conference on Computer Vision and Pattern Recognition (CVPR), pp. 770–778 (2016). https://doi.org/10.1109/CVPR.2016.90
6. Holzinger, A., Saranti, A., Molnar, C., Biecek, P., Samek, W.: Explainable AI Methods - A Brief Overview, pp. 13–38. Springer, Cham (2022). https://doi.org/10.1007/978-3-031-04083-2_2
7. Huang, Z., et al.: Optimizing model performance and interpretability: application to biological data classification. Genes **16**(3) (2025). https://doi.org/10.3390/genes16030297, https://www.mdpi.com/2073-4425/16/3/297
8. Kamble, A., et al.: Enhanced multi-class classification of gastrointestinal endoscopic images with interpretable deep learning model (2025). https://arxiv.org/abs/2503.00780
9. Kingma, D., Ba, J.: Adam: a method for stochastic optimization. In: International Conference on Learning Representations (ICLR) (December 2014). https://doi.org/10.48550/arXiv.1412.6980, https://arxiv.org/abs/1412.6980
10. Lapuschkin, S., Binder, A., Montavon, G., Klauschen, F., Müller, K.R., Samek, W.: On pixel-wise explanations for non-linear classifier decisions by layer-wise relevance propagation. PLoS ONE **10**, e0130140 (2015). https://doi.org/10.1371/journal.pone.0130140
11. Li, K., Wu, Z., Peng, K.C., Ernst, J., Fu, Y.: Guided attention inference network. IEEE Trans. Pattern Anal. Mach. Intell. **42**(12), 2996–3010 (2020). https://doi.org/10.1109/TPAMI.2019.2921543
12. Linardatos, P., Papastefanopoulos, V., Kotsiantis, S.: Explainable AI: a review of machine learning interpretability methods. Entropy **23**(1) (2021). https://doi.org/10.3390/e23010018, https://www.mdpi.com/1099-4300/23/1/18
13. Murdoch, W.J., Liu, P.J., Yu, B.: Beyond word importance: contextual decomposition to extract interactions from lstms (2018). https://arxiv.org/abs/1801.05453
14. Pillai, V., Koohpayegani, S.A., Ouligian, A., Fong, D., Pirsiavash, H.: Consistent explanations by contrastive learning. In: 2022 IEEE/CVF Conference on Computer Vision and Pattern Recognition (CVPR), pp. 10203–10212 (2022). https://doi.org/10.1109/CVPR52688.2022.00997
15. Poštić, S., Subašić, M.: Single-input interpretation of a deep classification neural network. Knowl.-Based Syst. **309**, 112903 (2025). https://doi.org/10.1016/j.knosys.2024.112903, https://www.sciencedirect.com/science/article/pii/S0950705124015375

16. Rieger, L., Singh, C., Murdoch, W.J., Yu, B.: Interpretations are useful: penalizing explanations to align neural networks with prior knowledge. In: Proceedings of the 37th International Conference on Machine Learning. ICML'20, JMLR.org (2020)

17. Ross, A.S., Hughes, M.C., Doshi-Velez, F.: Right for the right reasons: training differentiable models by constraining their explanations. In: Proceedings of the 26th International Joint Conference on Artificial Intelligence, pp. 2662–2670. IJCAI'17, AAAI Press (2017)

18. Russakovsky, O., et al.: ImageNet large scale visual recognition challenge. Int. J. Comput. Vis. (IJCV) **115**(3), 211–252 (2015). https://doi.org/10.1007/s11263-015-0816-y

19. Shah, C., Du, Q., Xu, Y.: Enhanced tabnet: attentive interpretable tabular learning for hyperspectral image classification. Remote Sensing **14**(3) (2022). https://doi.org/10.3390/rs14030716, https://www.mdpi.com/2072-4292/14/3/716

20. Shrikumar, A., Greenside, P., Kundaje, A.: Learning important features through propagating activation differences (2019)

21. Simonyan, K., Vedaldi, A., Zisserman, A.: Deep inside convolutional networks: visualising image classification models and saliency maps (2014)

22. Srinivas, S., Fleuret, F.: Full-gradient representation for neural network visualization (2019). https://doi.org/10.48550/ARXIV.1905.00780

23. Sundararajan, M., Taly, A., Yan, Q.: Axiomatic attribution for deep networks. In: Precup, D., Teh, Y.W. (eds.) Proceedings of the 34th International Conference on Machine Learning. Proceedings of Machine Learning Research, vol. 70, pp. 3319–3328. PMLR, 06–11 August 2017. https://proceedings.mlr.press/v70/sundararajan17a.html

24. Zhou, B., Khosla, A., Lapedriza, A., Oliva, A., Torralba, A.: Learning deep features for discriminative localization (2015)

SimEx-ViT: Explainable Vision Transformer with Similarity-Based Attention Modulation

R. Selventhiran[1,2]([envelope]) [ORCID], Vish Rajalingam[2] [ORCID], and Satyajit Das[1] [ORCID]

[1] Indian Institute of Technology Palakkad, Palakkad, India
112403001@smail.iitpkd.ac.in, satyajitdas@iitpkd.ac.in
[2] Multicoreware Inc., Coimbatore, India
{selventhiran,vish}@multicorewareinc.com

Abstract. Vision Transformers (ViTs) present state-of-the-art performance on various tasks in modern computer vision. However, the black-box nature of ViTs limits their adoption in safety-critical applications where interpretability and explainability are crucial. Although several post hoc explainability methods exist, they often struggle with generalizability across different architectures and input modalities. Inherently explainable methods, which directly link model predictions to interpretable explanations, remain less explored. In this work, we propose SimEx-ViT, a novel framework for enhancing the explainability of attention mechanisms in Vision Transformers. SimEx-ViT introduces a novel learnable explainability matrix that integrates directly into the attention mechanism, enabling interpretable behavior in transformer models. This matrix is composed of trainable parameters that dynamically influence attention weights across heads. To guide the emergence of interpretable patterns, we introduce auxiliary loss functions: an entropy-based loss that encourages confident and focused attention distributions, and a diversity-promoting loss based on Jensen-Shannon divergence to ensure varied representational learning across different attention heads. We also propose a loss to guide the attention weights based on segmentation masks for semantic segmentation. By jointly optimizing these components, SimEx-ViT ensures that the learned representations are both interpretable and semantically rich without compromising model performance. Experimental results demonstrate that our method achieves a mean Intersection over Union (mIoU) of 53.29%, significantly outperforming ViT (41.42%) and eX-ViT (42.03%) when evaluating attention maps against ground truth segmentation masks. SimEx-ViT represents a step forward toward transparent and trustworthy Vision Transformers, enabling their safer deployment in critical applications such as autonomous driving, healthcare, and defense.

Keywords: Vision Transformers · Explainable AI · Self-attention

Supplementary Information The online version contains supplementary material available at https://doi.org/10.1007/978-3-032-05060-1_11.

1   Introduction

Recent advances in machine learning, especially in natural language processing and computer vision, have highlighted the effectiveness of attention mechanisms [1], particularly multi-head attention. Architectures like Transformers [2] have shown exceptional performance in tasks such as image classification, object detection, and segmentation.

However, despite their success, attention-based models face a major challenge [3]: a lack of explainability. Although these models achieve high accuracy, their decision-making process remains opaque, raising concerns in sensitive areas such as healthcare, autonomous driving, and finance. As a result, efforts in explainable AI (XAI) [4] have emerged to address this issue through techniques such as attention visualization [5], feature importance and saliency maps. However, these post hoc methods are limited and often detached from the actual decision-making pipeline of the model.

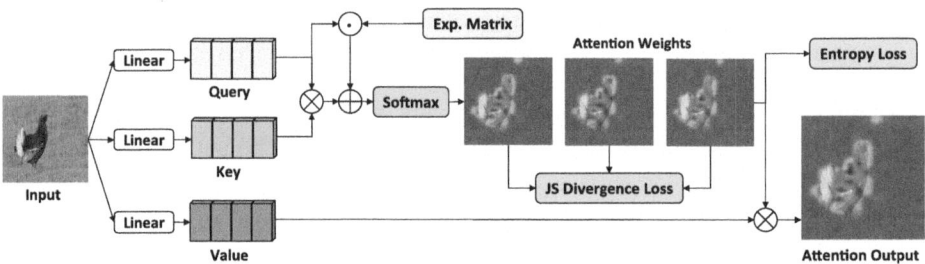

Fig. 1. SimEx-ViT - Illustration of the proposed inherently explainable attention module. Here, $\otimes$ denotes matrix-multiplication, $\odot$ and $\oplus$ denotes element-wise multiplication and addition respectively

To address these challenges, we propose a framework that integrates explainability into the attention-based model training process. Our method introduces a learnable explainability matrix to modulate attention scores and uses an entropy-based loss to encourage meaningful attention distributions. We also employ a diversity loss to promote distinct behaviors across attention heads. Although our model achieves nearly the same accuracy as existing methods, it significantly improves explainability, as shown in experiments for image classification and semantic segmentation.

The key contributions of this work are as follows.

1. We propose a framework that integrates explainability directly into Vision Transformers via an explainability matrix and similarity-based attention modulation.
2. We propose a novel loss function that enhances attention interpretability by combining the standard classification loss with an entropy-based term, Jensen-Shannon divergence, and a ground-truth-guided explainability loss for semantic segmentation.

3. We provide experimental validation showing that SimEx-ViT improves explainability with minimal accuracy trade-offs.

The paper is structured as follows. Section 2 covers related work, Sect. 3 details the proposed method, Sect. 4 presents the experimental setup and results, and Sect. 5 concludes with future research directions.

2   Related Work and Motivation

Explainability in deep learning has been a critical area of research, especially for models deployed in safety-critical environments. Traditional post hoc interpretability methods such as Grad-CAM [6], Integrated Gradients [7] aim to interpret model decisions after training, but these explanations are external and are not directly related to the internal reasoning of the model.

Commonly used attention-specific methods such as Attention Rollout [8], AttnLRP [9], TiBA [10] and GAE [11] have attempted to improve interpretability for Transformer-based models by tracing attention flows or backpropagating relevance scores through attention layers. Although these approaches provide more attention-aware explanations, they are still post hoc in nature and do not influence the training process, making them vulnerable to model biases and adversarial perturbations.

To address these limitations, recent efforts have shifted towards *inherently explainable models*, where interpretability is integrated into the architecture or training pipeline. For example, eX-ViT [12] introduces an L2-norm-based explainable multi-head attention module, which dynamically adjusts attention weights to reduce redundancy and highlight task-relevant regions. Similarly, works such as IA-RED^2 [13] focus on pruning redundant attention heads and patches to enhance interpretability with minimal accuracy loss. ICEv2 [14] promotes comprehensiveness and low attention dispersion using class-specific attention heads. ViT-Net [15] adopts a neural tree decoder to generate hierarchical human-understandable explanations. B-cos-ViT [16] replaces transformer components with dynamic linear mappings (B-cos transforms), enabling model-wide linear summaries that yield more transparent and faithful explanations.

Despite these advancements, existing inherently explainable methods often involve substantial architectural changes or additional computational overhead, limiting their practicality. Our proposed framework, SimEx-ViT, is motivated by the need to strike a balance between architectural simplicity, performance, and inherent explainability. Unlike previous work, SimEx-ViT introduces a lightweight explainability matrix that modulates attention without altering the core structure of ViT. Furthermore, we incorporate entropy regularization [17] and Jensen-Shannon divergence [18] to promote confident and diverse attention distributions across heads, as illustrated in Fig. 1. These components collectively improve the semantic alignment of attention with task-relevant features, as validated through both quantitative metrics and visual heatmap analyses.

In contrast to post hoc methods and inherently explainable architectures, SimEx-ViT offers a training-time explainability mechanism with minimal

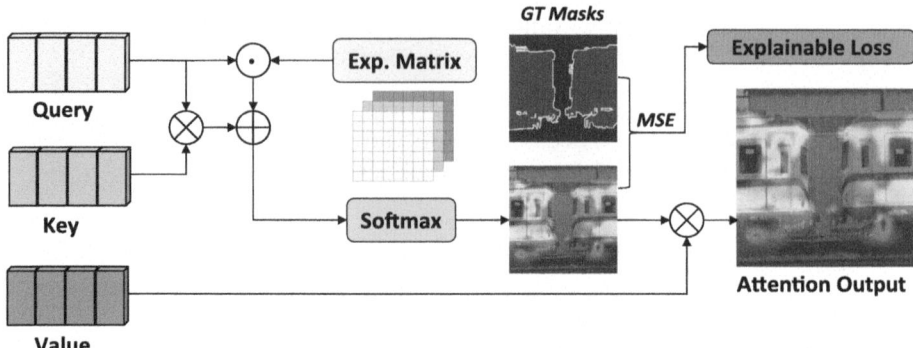

Fig. 2. Proposed Explainable Attention Module with Explainable Matrix and Explainable Loss. We use $\odot$ to denote element-wise multiplication, $\oplus$ for addition, and $\otimes$ for matrix-multiplication and MSE stands for Mean Squared Error.

additional parameters and computation, making it suitable for deployment in resource-constrained environments such as embedded systems and edge AI.

3 Proposed Method

In this section, we present our approach to integrating multihead attention mechanisms with explainability-driven enhancements to achieve improved interpretability in Vision Transformers while maintaining high model performance. Our proposed framework introduces two key components: the **Explainability Matrix** and a **Composite Explainable Loss Function** based on entropy-maximization and diversity of attention (reducing redundancy between heads).

3.1 Explainability Matrix

The *explainability matrix* is a set of learnable parameters introduced to enhance the interpretability of the multi-head attention mechanism. For each attention head, we define an explainability matrix that acts as a scaling factor to modulate the query representations. Formally, let $Q \in \mathbb{R}^{B \times N \times d_k}$ represent the query matrix, where B is the batch size, N is the sequence length and d_k is the head dimension. The explainability matrix $E \in \mathbb{R}^{H \times d_k}$ is initialized as a learnable parameter sampled from a standard normal distribution $\mathcal{N}(0, 1)$, where H is the number of attention heads and it is applied to the query tensor as:

$$\text{Explain Scores} = \sum_{i=1}^{d_k} Q \odot E \tag{1}$$

where $\odot$ denotes element-wise multiplication. These explainability scores are then added to the standard scaled dot product attention scores before the softmax step, acting as a learnable bias that promotes more interpretable attention distributions.

3.2   Loss Functions for Classification

To enhance the explainability and diversity of the attention mechanisms, we introduce two auxiliary losses: **Entropy Loss (EL)** and **Jensen-Shannon Divergence Loss (DL)**, combined with the standard cross-entropy loss.

Entropy Loss (EL): The entropy loss serves as a regularization term that encourages the model to produce more uniformly distributed attention weights, thus promoting a broader contextualization and reducing the dominance of a few tokens in downstream decisions.

Formally, given attention weights (output of Softmax in the attention module) $A \in \mathbb{R}^{H \times N}$, where H is the number of attention heads and N is the sequence length, we interpret each attention vector over tokens as a probability mass function. The entropy for the attention head h is defined as:

$$\text{Entropy}(A_h) = -\sum_{i=1}^{N} A_i \log(A_i) \tag{2}$$

where A_i is the attention probability assigned to token i. To normalize, we compute the maximum entropy corresponding to a uniform distribution, $\log N$. The entropy loss across all heads for an attention layer is computed as follows:

$$\mathcal{L}_{\text{Entropy}} = \log(N) - \frac{1}{H} \sum_{h=1}^{H} \text{Entropy}(A_h) \tag{3}$$

Minimizing $\mathcal{L}_{\text{Entropy}}$ encourages higher attention entropy, promoting spatially diverse feature aggregation across image patches, and mitigating overreliance on localized or isolated regions. This serves as a form of *contextualization regularization*, reducing the sensitivity to spurious visual patterns and aligning with the hypothesis - originally proposed in the NLP domain [17] - that higher entropy fosters more balanced and unbiased information integration.

Jensen-Shannon Divergence Loss (DL): The Jensen-Shannon Divergence (JSD) loss encourages diversity between attention heads by penalizing the similarity of their average attention distributions. This loss function promotes distinct attention patterns for each head, ensuring that different heads learn to focus on different aspects of the input. This form of regularization helps improve the model's robustness by preventing the model from relying on redundant features across heads and promoting more informative and diverse feature representations.

Formally, the JSD between two attention distributions P and Q is defined as:

$$\text{JSD}(P \parallel Q) = \frac{1}{2} D(P \parallel M) + \frac{1}{2} D(Q \parallel M) \quad \text{where} \quad M = \frac{1}{2}(P + Q) \tag{4}$$

and $D(P \parallel M)$ denotes the Kullback-Leibler (KL) divergence between P and M. The diversity loss is computed as the average pairwise JSD across all unique pairs of attention heads:

$$\mathcal{L}_{\text{JSD}} = -\frac{2}{H(H-1)} \sum_{i=1}^{H-1} \sum_{j=i+1}^{H} \text{JSD}(A_i \parallel A_j) \tag{5}$$

where A_i and A_j represent the attention distributions of the heads i and j, respectively. This approach is inspired by regularization techniques such as cosine similarity regularization and orthogonal regularization [?], which encourage distinctness between attention heads to ensure more diversified feature extraction. This diversity in attention distributions makes it easier to interpret the reasoning behind the predictions of the model since each head can be associated with a distinct subset of characteristics, improving the general explainability of the model.

Total Loss Function: The total loss combines the standard cross-entropy loss with the auxiliary losses driven by explainability:

$$\mathcal{L}_{\text{total}} = \mathcal{L}_{\text{CE}} + \lambda_1 \mathcal{L}_{\text{JSD}} + \lambda_2 \mathcal{L}_{\text{Entropy}} \tag{6}$$

where $\mathcal{L}_{\text{CE}}$ is the cross-entropy loss, $\mathcal{L}_{\text{JSD}}$ is the Jensen-Shannon divergence loss (DL), and $\mathcal{L}_{\text{Entropy}}$ is the entropy loss (EL). The hyperparameters λ_1 and λ_2 control the relative contributions of diversity (JSD) and entropy terms to the total loss, respectively. In our experiments, we set λ_1 in the range of 0.1 to 0.3 and λ_2 in the range of 0.2 to 0.6 to balance the trade-offs between the different components of the loss.

3.3   Explainable Loss for Semantic Segmentation

For semantic segmentation, we design an auxiliary explainability loss to align the model's attention with the ground-truth segmentation masks as illustrated in Fig. 2. We first aggregate attention maps across all transformer layers and heads by averaging the attention weights of the transformer based encoder. The resulting global attention map is normalized and reshaped into a 2D spatial map.

Given aggregated attention maps $A \in \mathbb{R}^{B \times N}$, we normalize:

$$A_{\text{norm}} = \frac{A - \min(A)}{\max(A) - \min(A)} \tag{7}$$

The ground truth segmentation masks M are resized to the same spatial resolution as the attention maps. And to reduce sensitivity to background pixels, we apply a background weight $w_{\text{bg}} \in [0, 1]$.

The final explainability loss is defined as a weighted mean squared error:

$$\mathcal{L}_{\text{expl}} = \frac{1}{B} \sum_{b=1}^{B} \frac{1}{N} \sum_{i=1}^{N} W_{b,i} \left(A_{\text{norm},b,i} - M_{b,i}\right)^2 \tag{8}$$

where W is a weighting mask that gives less importance to background regions. Although W is designed primarily to reduce background pixels, it can also be adapted to suppress attention toward specific classes of interest, enabling flexible control over class-level interpretability.

This loss encourages attention to focus more on meaningful foreground objects rather than background regions, improving the interpretability of model predictions. We use this explainability loss as an auxiliary objective alongside the standard cross-entropy loss for semantic segmentation. The total loss is a weighted sum of both, allowing the model to maintain segmentation accuracy while enhancing attention-based interpretability.

4   Experimental Results

We perform experiments on ImageNet-1K image classification, PASCAL VOC semantic segmentation, and Indian Driving Dataset (IDD) semantic segmentation. In the following, we compare the results of image classification, semantic segmentation, and evaluate the attention outputs for the proposed SimEx-ViT with ViT and eX-ViT.

4.1   Image Classification on ImageNet-1K

For image classification, we use ImageNet-1K, which contains 1.28M training images and 50 K validation images from 1,000 classes. We report top-1 and top-5 accuracies for the Transformer architecture using the proposed attention modules, trained and validated on 100 random classes from the ImageNet-1K dataset.

Training Settings: We use the AdamW optimizer with a weight decay of 0.05, a cosine learning rate scheduler over 300 epochs, and 30 epochs of linear warm-up. The initial learning rate is set to 3×10^{-3}, decreasing to 1×10^{-5}.

Model Settings: We use the ViT-Tiny configuration for all the experiments, which is used for its lightweight nature, with a 224×224 input image divided into 16×16 patches. The model has 12 transformer layers, 3 attention heads per layer, and a hidden representation size of 192. The feed-forward network has an intermediate size of 768. ViT, eX-ViT and SimEx-ViT all follow the same configurations as ViT-Tiny. While ViT employs conventional softmax-based attention, eX-ViT [12] replaces this with L2 norm-based attention. On the other hand, SimEx-ViT utilizes the ViT architecture with the proposed loss functions and Explain. Matrix.

Results: Figure 3 shows that compared to ViT and eX-ViT, our model shows significant improvements in the explainability of the model with fewer parameters compared to eX-ViT. Table 1 presents a comparison of the baseline ViT, eX-ViT, and our proposed SimEx-ViT for 100 classes of the ImageNet-1K dataset. Our model achieves a top-1 accuracy of (80.14%) which is slightly better compared to ViT and eX-ViT, but also provides better explanations compared to both. Based on our empirical experiments with 100 classes of ImageNet-1K, we find that setting $\lambda_1 = 0.3$ and $\lambda_2 = 0.6$ gives the best results in terms of both accuracy and

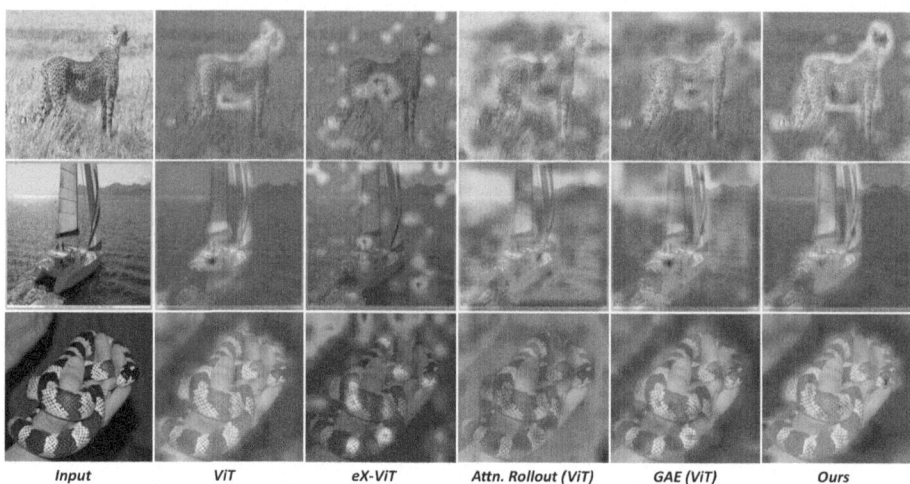

Input   ViT   eX-ViT   Attn. Rollout (ViT)   GAE (ViT)   Ours

Fig. 3. Heat-maps of the attention outputs for ViT, eX-ViT [12], Attention Rollout [8] of ViT, GAE [11] of ViT and SimEx-ViT (ViT + Explain. Matrix + EL + DL Loss). We consider the attention maps of mean values of all the attention layers and attention heads.

explainability. Note that we stopped hyperparameter tuning for Explain. matrix and loss functions, once the model achieved a Top-1 accuracy comparable to ViT. Further tuning could potentially lead to even better performance.

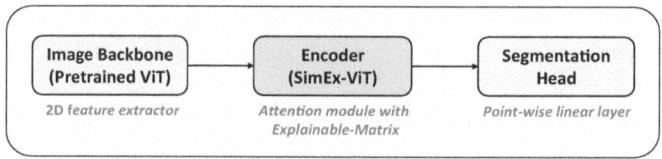

Fig. 4. Schematic of segmentation-model settings.

We also evaluate the attention output, similar to [10], by perturbing the input image by pixel masking based on attention weights. For positive perturbation, pixels with the highest relevance scores are masked first. This is expected to cause a steep drop in model accuracy if the attention weights correctly highlight important regions. For negative perturbation, pixels with the lowest relevance scores are masked first, and ideally this should result in minimal accuracy degradation. We compute the area under the curve (AUC) for the accuracy drop across masking ratios from 10% to 90%. For positive perturbations, a lower AUC indicates better attention quality, while for negative perturbations, a higher AUC suggests better identification of unimportant regions. Table 1 shows that our

Table 1. Evaluation of classification on 100 random classes of ImageNet-1K

Model	No. of Params	Accuracy(%)		AUC(%)	
		top-1	top-5	+ve	-ve
ViT	5.543 M	80.10	94.68	31.73	56.86
eX-ViT [12]	6.94 M	79.62	94.44	33.88	55.04
ViT + Explain. matrix (Ours)	5.545 M	79.88	94.66	**30.48**	**57.48**
ViT + EL + DL Loss (Ours)	5.543 M	**80.14**	94.54	31.31	**57.27**
ViT + Explain. Matrix + EL + DL (Ours)	5.545 M	**80.12**	94.62	**30.54**	**57.39**
Attention Rollout (ViT) [8]	–	–	–	34.05	54.64
GAE (ViT) [11]	–	–	–	30.57	56.89

proposed model outperforms baselines in both perturbation settings (+ve and -ve), reflecting more faithful and informative attention maps.

Table 2. Evaluation of semantic segmentation on PASCAL VOC and IDD

Backbone	Encoder	No. of Params	Seg. mIoU(%)		Exp. mIoU(%)	
			VOC	IDD	VOC	IDD
ViT-Base [2]	ViT	171.03 M	63.13	41.20	41.42	31.19
	eX-ViT	172.87 M	60.03	40.91	42.03	30.02
	ViT + Exp. Matrix Exp. Matrix (Ours)	171.04 M	**63.50**	41.29	50.12	36.27
	ViT + Exp. Loss Exp. Loss (Ours)	171.03 M	**63.57**	41.43	53.29	36.88

4.2 Semantic Segmentation on PASCAL VOC and IDD

For semantic segmentation, we benchmark the results on PASCAL VOC 2012 and the IDD. The PASCAL VOC 2012 dataset consists of 20 object classes and 1 background class, with 1,464 training images and 1,449 validation images. The IDD dataset contains 34 object classes with 10,000 images collected from 182 drive sequences on Indian roads. We consider the training settings and the model settings as follows:

Training Settings: We use the SGD optimizer with an initial learning rate of 0.001. A polynomial learning rate scheduler reduces the learning rate from 0.001 to 1×10^{-5} with a power factor of 0.9. The input image size is 224×224, and Cross-Entropy Loss is used with an ignore index of 255 for background pixels along with the proposed loss function $\mathcal{L}_{exp}$.

Model Settings: The segmentation model setting is briefly described in Fig. 4. We use a pre-trained ViT-Base as the backbone and employ the ViT-Tiny-based Encoder, following the same model configurations used in image classification.

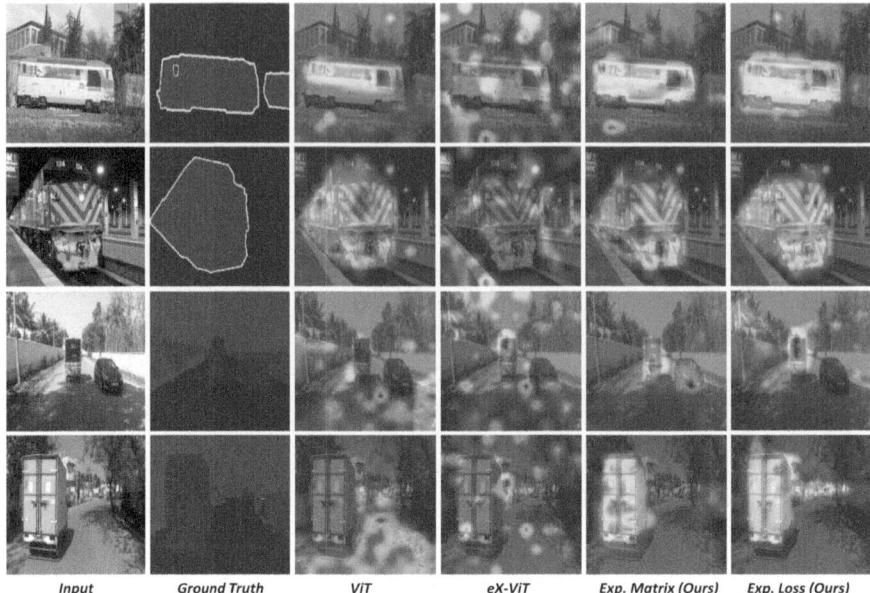

| Input | Ground Truth | ViT | eX-ViT | Exp. Matrix (Ours) | Exp. Loss (Ours) |

Fig. 5. Heatmaps of attention weights for the segmentation task. The top two rows show examples from the PASCAL VOC dataset, while the bottom two rows correspond to images from the IDD dataset.

The segmentation head utilizes a simple point-wise linear decoder, the same as [19], to predict the class labels. We compare results using the same backbone, replacing the encoder with eX-ViT and SimEx-ViT.

Results: Figure 5 compares the attention maps of the baseline ViT, eX-ViT, and the proposed SimEx-ViT. The visualization indicates that SimEx-ViT yields more human-interpretable attention patterns than ViT and eX-ViT. In particular, SimEx-ViT offers clearer and more focused attention even in complex scenarios, such as road scenes from the IDD dataset, illustrating its improved interpretability and robustness.

Table 2 reports the segmentation performance (Seg. mIoU (%)) on the validation sets of PASCAL VOC and IDD. All models are trained only for 20 epochs constrained by available GPU resources. To quantitatively evaluate the attention output, we interpret the attention maps as soft segmentations and compare it with ground-truth masks using the mIoU, reported as Exp. mIoU (%) in Table 2. A higher Exp. mIoU indicates that the model's attention aligns more closely with relevant regions in the input, demonstrating the effectiveness of SimEx-ViT in focusing on semantically meaningful features.

5   Conclusion

In this paper, we proposed a novel framework to enhance the interpretability of Vision Transformers by integrating a learnable explainability matrix and losses based on entropy and Jensen-Shannon divergence into the training process. This approach aligns attention weights with input features, ensuring that the model's decisions are accurate and inherently explainable without significant architectural changes or computational overhead. For semantic segmentation, we introduced an auxiliary explainability loss that aligns attention maps with ground-truth masks, improving spatial interpretability in dense prediction tasks. Our experimental results show that the proposed framework maintains strong performance while significantly improving transparency, outperforming existing inherently interpretable approaches. Future work can extend this framework to other Transformer variants, such as multimodal models, explore scalability on larger datasets, and investigate adaptive weighting or diversity-based techniques to further balance interpretability and accuracy.

References

1. Vaswani, A., et al.: Attention is all you need. In: Proc. of the Int. Conf. on Neural Information Processing Systems, Long Beach, CA, USA, pp. 6000–6010 (2017)
2. Dosovitskiy, A., et al.: An image is worth 16x16 words: transformers for image recognition at scale. In: Proc. of the Int. Conf. on Learning Representations, pp. 1–12 (2021)
3. Jain, S., Wallace, B.C.: Attention is not explanation. In: Proc. of the 2019 Conference of the North American Chapter of the Association for Computational Linguistics: Human Language Technologies, vol. 1 (Long and Short Papers), Minneapolis, MN, USA, pp. 3543–3556 (2019)
4. Barredo Arrieta, A., et al.: Explainable artificial intelligence (XAI): concepts, taxonomies, opportunities and challenges toward responsible AI. Inf. Fusion **58**, 82–115 (2020)
5. Hao, Y., Dong, L., Wei, F., Xu, K.: Self-attention attribution: interpreting information interactions inside transformer. In: Proc. of the AAAI Conference on Artificial Intelligence, vol. 35, no. 14, pp. 12963–12971 (2021)
6. Selvaraju, R.R., Cogswell, M., Das, A., Vedantam, R., Parikh, D., Batra, D.: Grad-CAM: visual explanations from deep networks via gradient-based localization. In: Proc. of the IEEE International Conference on Computer Vision, pp. 618–626 (2017)
7. Sundararajan, M., Taly, A., Yan, Q.: Axiomatic attribution for deep networks. arXiv (2017)
8. Abnar, S., Zuidema, W.: Quantifying attention flow in transformers. In: Proc. of the 58th Annual Meeting of the Association for Computational Linguistics (2020)
9. Achtibat, R., et al.: AttnLRP: attention-aware layer-wise relevance propagation for transformers. arXiv (2024)
10. Chefer, H., Gur, S., Wolf, L.: Transformer interpretability beyond attention visualization. In: Proc. of the Computer Vision and Pattern Recognition Conference, Nashville, TN, USA, pp. 782–791 (2021)

11. Chefer, H., Gur, S., Wolf, L.: Generic attention-model explainability for interpreting bi-modal and encoder-decoder transformers (2021)
12. Yu, L., Xiang, W., Fang, J., Chen, Y.-P.P., Chi, L.: eX-ViT: a novel explainable vision transformer for weakly supervised semantic segmentation. Pattern Recogn. **142**, 109666 (2023)
13. Pan, B., Panda, R., Jiang, Y., Wang, Z., Feris, R., Oliva, A.: IA-RED^2: interpretability-aware redundancy reduction for vision transformers. In: Advances in Neural Information Processing Systems, vol. 34, pp. 24898–24911 (2021)
14. Choi, H., Jin, S., Han, K.: ICEv2: interpretability, comprehensiveness, and explainability in vision transformer. Int. J. Comput. Vision **132**, 1–18 (2024)
15. Kim, S., Nam, J., Ko, B.C.: ViT-Net: interpretable vision transformers with neural tree decoder. In: Proceedings of the 39th International Conference on Machine Learning (ICML), pp. 11162–11172. PMLR (2022)
16. Böhle, M., Singh, N., Fritz, M., Schiele, B.: B-cos alignment for inherently interpretable CNNs and vision transformers (2024)
17. Bartunov, S., et al.: Mitigating unintended bias with contextualization: entropy-aware regularization for transformers. In: Findings of the Association for Computational Linguistics: EMNLP 2020, pp. 1101–1111 (2020)
18. Lin, J.: Divergence measures based on the Shannon entropy. IEEE Trans. Inf. Theory **37**(1), 145–151 (1991)
19. Strudel, R., Garcia, R., Laptev, I., Schmid, C.: Segmenter: transformer for semantic segmentation. In: Proc. of the International Conference on Computer Vision, pp. 510–520 (2021)

Robust Logit to Enhance Stochastic Neural Network Adversarial Robustness

Omar Dardour[1]([✉])[ID], Eduardo Aguilar[2,3][ID], Mourad Zaied[1][ID], and Petia Radeva[2][ID]

[1] Research Team in Intelligent Machines, National Engineering School of Gabes, University of Gabes, Gabes, Tunisia
omar.dardour@isimg.tn, mourad.zaied@univgb.tn
[2] AIBA, Departament de Matemàtiques i Informàtica, Universitat de Barcelona, Barcelona, Spain
{eduardo.aguilar,petia.ivanova}@ub.edu
[3] Department of Computing and Systems Engineering, Catholic University of the North, Antofagasta, Chile

Abstract. It is well-known that Deep Neural Networks can be fooled by adding invisible noise called adversarial examples. Recent work on stochastic neural networks (SNN) shows resistance against adversarial attacks, but these methods are still vulnerable to logits-based attacks. In this paper, we propose a simple but efficient defense method called Robust Logit Stochastic Neural Networks (RL-SNN) that adds a robust block (RL) to a pre-trained SNN model at the inference stage without any additional training cost. The RL block scales and normalizes the logits to prevent attacks to employ large logits. Experiments with different methods show that the proposed RL block can be applied directly to any pre-trained SNN. Tests on two public datasets demonstrate that RL-SNN outperforms previous SNNs methods by a large margin under different attacks. RL-SNN becomes the state-of-the-art under most of attacks evaluated.

Keywords: Deep neural network · Adversarial robustness · SNN · Robust logit

1 Introduction

Deep neural network models (DNN) have shown great success in different fields. However, in image classification, DNNs are still sensitive to adversarial attacks [19]. Such attacks add invisible noise to the input image in order to confuse the model and cause image misclassification, posing an important risk to the safety of the DNNs. Therefore, enhancing adversarial robustness becomes crucial.

In recent years, several strategies for defending against adversarial attacks have been developed, including input reconstruction with a generative adversarial network or variational autoencoder [6] and adversarial training (AT) [17].

M. Castrillón-Santana et al. (Eds.): CAIP 2025, LNCS 15622, pp. 135–145, 2026.
https://doi.org/10.1007/978-3-032-05060-1_12

Adversarial training is commonly used to develop defensive strategies. This technique incorporates created adversarial cases during training, boosting robustness, but resulting in substantially higher computational costs and training time.

Works using stochastic neural networks (SNN) [5, 7, 8, 10, 11, 14–16, 20–22] has shown that SNN improve adversarial robustness. SNN learn stochastic noise information and introduce uncertainty into model predictions.

Based on the variational information bottleneck, Simple and Effective SNN (SE-SNN) [23] increased the robustness of the model in a more efficient training using stochastic latent re-parameterization. Margin-SNN [20] learns stochastic latent information and a margin loss to defend against adversarial attacks. Margin-SNN [20] uses label embedding to discover the semantic representation of class centers and measure the similarity between label embedding and stochastic features to create logits for classification. Margin-SNN [20] optimizes a novel loss function composed of classification loss, stochastic and margin loss to prevent attacks with the final goal of exploring confused classes. Inter-Separability and Intra-Concentration Stochastic Neural Network (ISIC-SNN) [5] uses the same architecture of Margin-SNN [20] and adds Intra-Concentration loss to the learned loss of Margin-SNN. ISIC-SNN [5] showed higher robustness against most attacks. In SNNs, the variability introduced by sampling from latent distributions can help resist some gradient-based attacks. However, they can produce large logits. Attacks such as the Carlini & Wagner (CW) FAB [3], and Square [1] operate by minimizing or manipulating the difference in logits or the decision-boundary and thus can still craft successful adversarial perturbations. This highlights the need to incorporate logit transformations to make the model more robust.

To defend against these advanced logit-targeting attacks and reduce the sensitivity of logits to adversarial perturbations, we propose a novel Robust Logit (RL) transformation that can be added to any SNN model (RL-SNN).

The main contributions of this paper are as follows:

1) First, we address the problem of logits attack of SNN vulnerability by proposing RL-SNN that add a robust logit block to a pre-trained SNN model to enhance robustness against attacks and thus explore logits or boundaries spaces. The RL block focus on transforming logits on the inference stage without any additional training.
2) Extensive tests are done under various attacks, showing that the proposed RL block enhances robustness against adversarial attacks and maintains accuracy in clean data.
3) As a post-training block, RL can enhance a SNN by simply adding it at the inference stage.

The rest of the paper is organized as follows. Section 2 presents background knowledge on adversarial attacks and adversarial defenses of SNNs. Section 3 proposes the Robust Logit method for better adversarial robustness. Section 4 details the experimental settings and the extensive experiments conducted to show the superiority of the proposed method. Finally, Sect. 5 concludes the paper and provides future research lines.

2   Related Works

2.1   Adversarial Attacks

White-box adversarial attacks assume full access to the model's architecture and gradients. Fast Gradient Sign Method (FGSM) [9] perturbs the input in the direction of the loss gradient to quickly induce misclassification. Projected Gradient Descent (PGD) [17] improves upon FGSM by applying multiple iterative updates, each followed by a projection step, making it one of the most effective first-order attacks. The Carlini & Wagner (CW) [2] attack leverages logit-based optimization to minimize perturbation while ensuring high-confidence misclassification, making it particularly robust against defenses that rely on output confidence. Fast Adaptive Boundary (FAB) attack [3] focuses on finding minimal-norm perturbations that cross the decision boundary by iteratively estimating and projecting onto linear approximations of the boundary.

In contrast, black-box attacks do not rely on gradient information. The n-pixel attacks [18] exploit the model's vulnerability to minimal local changes by modifying one or several pixels in the image using optimization techniques like Differential Evolution. These attacks highlight how even small localized changes can drastically alter model predictions. Square Attack [1] operates by applying random square-shaped perturbations in a score-based manner, making it effective even under gradient masking conditions.

AutoAttack method [4] combines white-box and black-box strategies into a robust, parameter-free evaluation framework. It includes four different attacks, offering a comprehensive and reproducible benchmark for evaluating model robustness under diverse adversarial conditions.

2.2   Stochastic Neural Networks for Adversarial Defense

Recent studies have highlighted the effectiveness of SNNs in defending against adversarial attacks.

The popular SNN approach involves injecting noise into the model architecture or parameters. Adv-BNN [15] applies Gaussian noise to network weights, integrating Bayesian principles with adversarial training. Further methods refine how noise is introduced and learned. Parameter Noise Injection (PNI) [10] used learnable noise sensitivity parameters. Learn2Perturb (L2P) [11] enhances PNI using an alternating backpropagation scheme to train both the noise and the neural network. Other works, such as WCA-Net [8] and MFDV-SNN [22], explore the geometry of noise learning anisotropic and isotropic noise distributions, respectively, both with the goal of improving robustness. Several novel techniques take more unconventional approaches. RPF [7] replaces standard convolutional filters with random projection filters. CTRW [16] uses randomly initialized weights during optimization to inject stochasticity directly into the learning process. NINE-SNN [21] relaxes the assumption of Gaussian noise, allowing the model to learn from more general, non-informative distributions. AdaNI [13] adapts the noise injection process based on feature representations, allowing for more context-aware stochastic defenses. Among recent methods, Margin-SNN [20] showed that the low distance between distinct class samples in the latent space of SNN provides class confusion, leading the SNN to adversarial examples. To solve this problem, Margin-SNN [20] and ISIC-SNN [5] add discriminative constraints to increase the robustness of SNNs. Margin-SNN [20] defends adversarial attacks by modeling feature uncertainty, incorporating label embeddings, and applying a margin-enlarging penalty to separate classes. ISIC-SNN [5] enhances adversarial robustness by applying intra-compactness and inter-separability losses.

3   Proposed Method

3.1   Rationale

Margin-SNN [20] and ISIC-SNN [5] offer inherent robustness against gradient-based attacks due to their randomization and uncertainty modeling. They remain vulnerable to logit-based and decision-boundary attacks such as Carlini & Wagner (CW) [2], Fast Adaptive Boundary (FAB) [3], and Square attack. These attacks exploit the model's output confidence or decision surface geometry. CW [2] minimizes perturbation while increasing the logit score of an incorrect class beyond the correct one, leveraging overly confident predictions. FAB [3] iteratively estimates and crosses decision boundaries using linear approximations, targeting points close to the classification surface. Square attack [1], a black-box method, perturbs localized regions and monitors score-based feedback to infer vulnerabilities. To defend against such attacks, it is essential to suppress overconfident outputs through logit smoothing or transformation, and incorporate stochasticity during inference to disrupt deterministic behavior. Enhancing boundary smoothness makes the attack optimization harder or less reliable. This motivation inspires our use of robust logit transformation as a post-hoc defense for pretrained SNNs.

3.2   Robust Logit Block

We introduce a *Robust Logit Transformation* block designed to improve adversarial robustness during inference by applying stochastic perturbations to the output logits of a pre-trained SNN. This technique does not require additional training and is used exclusively during the inference stage.

$$Robustlogits = \exp\left(logits - \max(logits)\right) \tag{1}$$

This transformation enhances the robustness of SNN models at test time by introducing controlled stochasticity into the logit space.

RL transformation showed in Fig. 1 consists of two main steps: (1) centering the logits by subtracting the maximum value $\max(logits)$ across classes to reduce numerical instability and eliminate dominant logits, and (2) exponential scaling of logits to suppress the influence of extreme logit differences. This operation compresses the logit range and makes the model's decision surface less linear in the logit space, a key vulnerability exploited by gradient-based adversarial attacks. By introducing non-linearity and compressing less confident logits, the model becomes less sensitive to small, adversarial changes in the input, contributing to more stable and reliable predictions under attack.

Its key advantages include:

- *Disruption of Logit-Level Exploits:* Since many adversarial attacks rely on precise manipulation of logits, RL block interferes with adversarial optimization paths, reducing the success rate of such attacks without degrading performance on clean data.
- *Decision Boundary Smoothing:* As shown in Fig. 1 by injecting noise using SNN into the logits and rescaling them using the RL block, the resulting predictions exhibit more distributed confidence, effectively smoothing decision boundaries and making them more resistant to small adversarial perturbations.
- *No Retraining Required:* Unlike adversarial training or architecture modifications, this method is lightweight and can be applied directly to any pre-trained SNN, offering a practical and deployment-ready defense.

By converting deterministic logits into a randomized logit using the SNN model and scaled representation during inference using the RL block, the model achieves greater robustness against gradient-based and logits and boundary-sensitive adversarial attacks, while maintaining high classification accuracy on clean inputs.

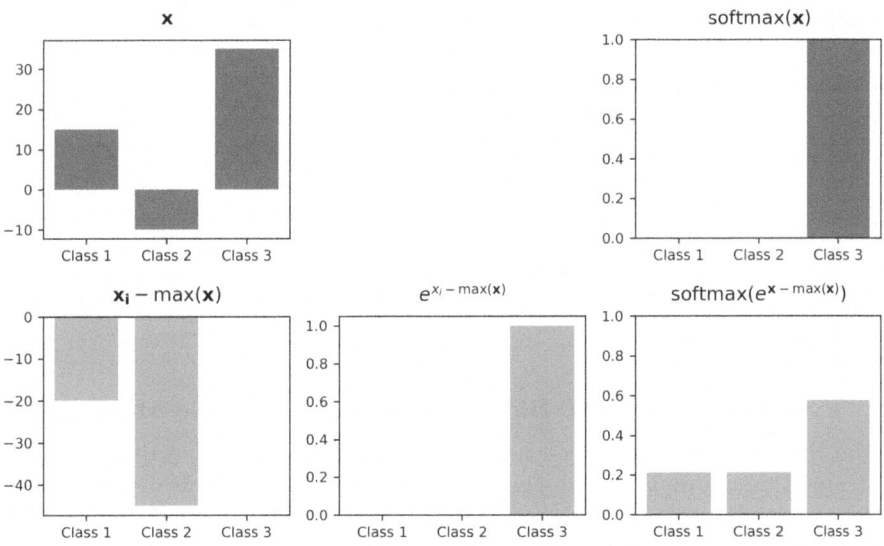

Fig. 1. Result of the original Logits and the robust logit transformation steps.

4   Validation

In this section, we will examine how effective the proposed RL block is on a variety of benchmarks, including datasets and adversarial attacks. First, we explain the datasets used, the implementation setting and the adversarial attacks details. Then, we provide an experimental comparison to the state-of-the-art and a qualitative analysis to show the robustness of the proposed method under different white-box and black-box adversarial attacks.

4.1   Datasets

In this paper, we use two benchmark datasets for comparison against the current state-of-the-art: CIFAR-10 [12] and CIFAR-100 [12] contain 60K 32×32 color images, 50K for training and 10K for testing, evenly spread across 10 and 100 classes, respectively.

4.2   Implementation

In all experiments, for a fair comparison, we follow the validation setting of Margin-SNN [20] and ISIC-SNN [5]. We use a pre-trained SNN model based on ResNet-18 as the backbone for the feature extractor architecture. We conducted all experiments using Pytorch and Advertorch libraries on a computer with one NVIDIA GeForce RTX 2080 Ti GPU and a maximum of 12 GB of memory.

4.3   Adversarial Attacks Aspects

We follow the validation methodology from [5,20] to verify the robustness of the proposed RL-SNN method on different white-box and black-box attacks. For white box attacks, we tested its performance against FGSM [9], PGD [17], CW [2], FAB [3] and AutoAttack [4]. We use L_{inf} norm constraint for FGSM, PGD and FAB, and we adopt L_2 norm for CW and AutoAttack. For the gradient-based attack, we set the attack strength ϵ as $8/255$, the step size as $\epsilon/10$ and the number of iterations as 10. Regarding CW attack, the learning rate is set as 0.0005 with the maximum optimization iterations of 1000. For the AutoAttack, we use the implementation setting [4]. For black-box attacks, we tested performance against n-pixel attacks [18] ($n = 1, 2, 3, 5$) using L_0 norm constraint and against square attack following the implementation in [4].

4.4   Evaluation of the RL Block on SoA SNN Defense Methods

As the RL block does not need any training step and is added directly to the pre-trained model at the inference step, we investigate the effectiveness of the RL block with the Margin-SNN and ISIC-SNN pre-trained models. Table 1 shows that our RL block can be used with the two methods and enhance the robustness with a large margin with both methods under FAB attack.

From Table 1, we can see that the RL block performs better with the ISIC-SNN method. In the rest of the paper, we adopt ISIC-SNN + RL and refer to it as RL-SNN.

Table 1. Accuracy comparison of SNNs on CIFAR-100 under CW and FAB attacks with ResNet-18 backbone.

Method	Clean	CW	FAB
Margin-SNN	70.1	69.6	38.9
Margin-SNN + RL	71.2	71.1	71.0
ISIC-SNN	73.7	67.8	32.0
ISIC-SNN + RL	73.8	73.7	73.1

4.5 Comparison with SNN Defense Methods

We compare the proposed method with recent SNN-based defense methods Adv-BNN [15], PNI [10], L2P [11], WCA-Net [8], SE-SNN [23], MFDV-SNN [22], RPF [7], CTRW [16], NINE-SNN [21], AdaNI [13], Margin-SNN [20] and ISIC-SNN [5]. For a fair comparison, we follow the previous work on the SNNs defense methods.

Under White Box Attacks. Table 2 provides experiment comparisons on CIFAR-10 and CIFAR-100 datasets. Table 2 demonstrates that the proposed RL-SNN provides higher clean data accuracy on CIFAR-10 and CIFAR-100 datasets. It can be seen from Table 2 that RL-SNN produces competitive defense against FGSM and PGD attacks and outperforms all SNNs adversarial defensive methods against the CW attack with large values. On CIFAR-10 and under CW attack, RL-SNN enhances the base method ISC-SNN by 11.7%, outperforming the best competitor (MFDV-SNN) by 4.5%. For CIFAR-100, Table 2 demonstrates that our defense enhances the results under CW attack by 4.1% compared to the prior best defensive technique. Table 3 shows that our RL-SNN outperforms all SNNs defense methods under different strengths of the CW attack. Table 4 resumes experiments under FAB, Square and AutoAttack. The results show that our RL-SNN outperforms recent SNN adversarial defense methods, namely Margin-SNN and ISIC-SNN by 43.4% and 50% respectively under FAB attack, 3.5%, and 15.2% under Square attack and 11.5% under AA for both of the two methods.

Under Black-Box Attacks. In this part, we evaluate the proposed RL-SNN using the black-box n-Pixel attack and Square attack. For n-Pixel attack, the number of pixels that the attack changes determines the strength of the attack. Table 5 shows that the proposed RL-SNN outperforms most of SNNs approaches and shows concurrent performance compared with ISIC-SNN for 1 and 5 pixels attacks. For the black box Square attack, Table 4 demonstrates that our RL-SNN outperforms Margin-SNN and ISIC-SNN and maintains 74% of robustness.

Under Stronger Attacks. In this part, we conduct the test under different strengths of PGD_{100}, Square and AutoAttack attacks. Table 6 shows the

Table 2. Comparison of state-of-the-art SNNs for FGSM and PGD attacks on CIFAR-10 and CIFAR-100 datasets with a ResNet-18 backbone.

Methods	CIFAR-10				CIFAR-100			
	Clean	FGSM	PGD	CW	Clean	FGSM	PGD	CW
Adv-BNN	82.2	60.0	53.6	78.9	50.0	30.0	27.0	–
PNI	87.2	58.1	49.4	66.9	61.0	27.0	22.0	–
L2P	85.3	62.4	56.1	83.6	58.0	30.0	26.0	–
WCA-Net	93.2	77.6	71.4	–	70.1	51.5	42.7	–
MFDV-SNN	**93.7**	85.7	79.6	88.8	69.4	47.1	37.3	–
AdaNI	93.5	79.2	76.0	82.0	–	–	–	–
RPF	83.8	62.7	61.3	–	56.9	37.7	37.4	–
CTRW	83.7	66.5	69.5	–	–	–	–	–
NINE-SNN	93.7	80.4	76.2	86.7	69.4	39.3	31.2	–
Margin-SNN	93.7	92.8	89.8	87.7	70.1	67.5	53.7	69.6
ISIC-SNN	93.6	**93.4**	**91.0**	81.8	**73.7**	**73.4**	**66.5**	67.8
RL-SNN(ours)	**93.7**	93.2	**91.0**	**93.5**	73.8	73.3	65.0	**73.7**

Table 3. Accuracy comparison of SNNs on CIFAR-10 under CW attack with ResNet-18 backbone.

Strength k	Clean	0	0.1	1	2	5
L2P	85.3	83.6	84.0	76.4	66.5	34.8
WCA-Net	93.2	-	89.4	78.4	71.9	55.0
MFDV-SNN	**93.7**	88.8	87.9	87.2	86.6	83.5
NINE-SNN	**93.7**	86.7	86.1	83.1	75.8	57.3
Margin-SNN	**93.7**	87.7	87.0	86.7	85.2	84.2
ISIC-SNN	93.6	81.8	82.4	82.2	82.2	82.1
RL-SNN (ours)	**93.7**	**93.5**	**93.5**	**93.5**	**93.5**	**93.4**

Table 4. Accuracy comparison of SNNs on CIFAR-10 under FAB, Square and AutoAttack attacks with ResNet-18 backbone.

Methods	Clean	FAB	Square	AA
Margin-SNN	**93.7**	50.1	71.5	88.0
ISIC-SNN	93.6	43.5.	58.8	88.0
RL-SNN (ours)	**93.7**	**93.5**	**74.0**	**91.5**

performance of RL-SNN in CIFAR-10 under PGD_{100} and Square attacks while increasing the value of the epsilon. It can also be seen that as the attack strength increases, the model's accuracy decreases, yet it still provides some protection. Table 7 provides the performance of RL-SNN compared to SNNs methods under

Table 5. Accuracy comparison of SNNs on CIFAR-10 under black-box n-pixel attack with ResNet-18 backbone.

Strength (pixels)	Clean	1	2	3	5
WCA-Net	93.2	90.8	85.5	81.2	64.3
MFDV-SNN	**93.7**	85.4	80.4	76.0	68.0
NINE-SNN	**93.7**	84.1	78.9	74.3	66.1
Margin-SNN	**93.7**	90.2	83.9	80.3	75.7
ISIC-SNN	93.6	**93.2**	92.8	92.6	**92.3**
RL-SNN (ours)	**93.7**	**93.2**	92.9	92.7	92.2

Table 6. The effectiveness of the ISIC-SNN method on CIFAR-10 under stronger attacks with ResNet-18.

Attack	$\epsilon/255$	Clean	1	2	4	8	16	32	64	128
PGD_{100}	No defense	92.2	53.1	22.3	11.6	11.1	10.6	7.8	1.3	0
	RL-SNN (ours)	93.7	91.7	91.7	91.6	89.9	80.5	48.2	16.2	10.0
Square	No defense	92.2	66.2	37.2	5.9	0	0	0	0	0
	RL-SNN (ours)	93.7	93.4	92.9	89.7	74.0	32.2	5.6	3.0	2.2

Table 7. Accuracy comparison of SNNs on CIFAR-10 under the AutoAttack attack with ResNet-18 backbone.

$\epsilon/255$	Clean	1	2	4	8	16	32	64	128
NINE-SNN	93.2	80.9	78.1	74.6	60.6	27.8	10.3	4.2	0.2
Margin-SNN	**93.7**	91.8	91.4	90.4	88.0	80.5	59.7	20.6	1.9
ISIC-SNN	93.6	92.0	91.7	90.7	88.0	80.0	61.3	23.5	**2.0**
RL-SNN (ours)	93.6	**92.2**	**92.1**	**91.8**	**91.5**	**91.3**	**90.9**	**90.3**	**88.4**

AutoAttack. The results show that RL-SNN outperforms all previous SNNs defense methods in all values of attack strength. The RL-SNN maintains the strongest robustness while increasing the value of attack strength ϵ. When the attack strength ϵ increases, RL-SNN outperforms all SNN methods by a large margin. The test with the highest attack strength ϵ equal to 128 shows that our RL-SNN outperforms ISIC-SNN by 86.4%.

4.6 Features Visualization Analysis

The visualization for the logits of CIFAR-10 is demonstrated in Fig. 2. Clearly, the suggested RL-SNN produces large separation between different classes and more compactness for the logits from the same class.

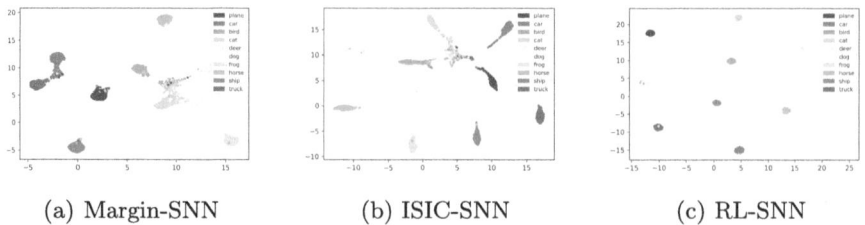

(a) Margin-SNN   (b) ISIC-SNN   (c) RL-SNN

Fig. 2. UMAP distribution logits of Margin-SNN, ISIC-SNN and RL-SNN on CIFAR-10

5   Conclusions

This work proposes a new method to increase robustness against adversarial attacks. Our proposed method uses an added Robust Logit (RL) transformation block to a pre-trained SNN model at inference time without any training effort. Recent work (ISIC-SNN) shows the ability to introduce stochasticity to the latent space in order to enlarge the distance between different classes and make features from the same class compact. RL-SNN is based on the pre-trained ISIC-SNN model and focuses on making logits more robust using logit transformation. Qualitative analysis of results shows that the proposed method achieves better intra-class compactness and inter-class separability in logit space. RL-SNN outperformed several state-of-the-art SNN defense methods under different white and black box attacks and enhanced accuracy in clean images.

Acknowledgments. This work is funded by the Government of Tunisia through its General Direction of Scientific Research (No. UR17ES46), Beatriu de Pinós Programme (2022 BP 00257), the Horizon EU project MUSAE (No. 01070421), 2021-SGR-01094 (AGAUR), Icrea Academia'2022 (Generalitat de Catalunya), Robo STEAM (2022-1-BG01-KA220-VET-000089434, Erasmus+ EU), DeepSense (ACE053/22/000029, ACCIÓ), DeepFoodVol (AEI-MICINN, PDC2022-133642-I00) and PID2022-141566NB-I00 (AEI-MICINN).

Disclosure of Interests. The authors have no competing interests to declare that are relevant to the content of this article.

References

1. Andriushchenko, M., Croce, F., Flammarion, N., Hein, M.: Square attack: a query-efficient black-box adversarial attack via random search (2020)
2. Carlini, N., Wagner, D.: Towards evaluating the robustness of neural networks. In: 2017 IEEE Symposium on Security and Privacy (SP) (2017)
3. Croce, F., Hein, M.: Minimally distorted adversarial examples with a fast adaptive boundary attack. In: ICML (2020)
4. Croce, F., Hein, M.: Reliable evaluation of adversarial robustness with an ensemble of diverse parameter-free attacks. In: ICML (2020)

5. Dardour, O., Aguilar, E., Radeva, P., Zaied, M.: Inter-separability and intra-concentration to enhance stochastic neural network adversarial robustness. Pattern Recognit. Lett. (2025)

6. Dardour, O., Zaied, M., Radeva, P.: DVAE-SR: denoiser variational auto-encoder and super-resolution to counter adversarial attacks. In: 13th International Conference on Machine Vision, vol. 11605. SPIE (2021)

7. Dong, M., Xu, C.: Adversarial robustness via random projection filters. In: CVPR, pp. 4077–4086 (2023)

8. Eustratiadis, P., Gouk, H., Li, D., Hospedales, T.: Weight-covariance alignment for adversarially robust neural networks. In: ICML. Proceedings of Machine Learning Research (2021)

9. Goodfellow, I., Shlens, J., Szegedy, C.: Explaining and harnessing adversarial examples. In: ICLR (2015)

10. He, Z., Rakin, A., Fan, D.: Parametric noise injection: trainable randomness to improve deep neural network robustness against adversarial attack. In: CVPR, pp. 588–597 (2019)

11. Jeddi, A., Shafiee, M., Karg, M., Scharfenberger, C., Wong, A.: Learn2perturb: an end-to-end feature perturbation learning to improve adversarial robustness. In: CVPR (2020)

12. Krizhevsky, A., Hinton, G.: Learning multiple layers of features from tiny images. Technical Report 0, University of Toronto (2009)

13. Li, Y., Zhang, C., Qi, H., Lyu, S.: Adani: adaptive noise injection to improve adversarial robustness. Comput. Vis. and Image Underst. **238**, 103855 (2024)

14. Liu, X., Cheng, M., Zhang, H., Hsieh, C.J.: Towards robust neural networks via random self-ensemble. In: ECCV, pp. 381–397. Springer (2018)

15. Liu, X., Li, Y., Wu, C., Hsieh, C.J.: Adv-BNN: improved adversarial defense through robust Bayesian neural network. In: ICLR (2019)

16. Ma, Y., Dong, M., Xu, C.: Adversarial robustness through random weight sampling. In: NeurIPS (2023)

17. Madry, A., Makelov, A., Schmidt, L., Tsipras, D., Vladu, A.: Towards deep learning models resistant to adversarial attacks. In: ICLR (2018)

18. Su, J., Vargas, D.V., Sakurai, K.: One pixel attack for fooling deep neural networks. IEEE Trans. Evol. Comput. **23**(5), 828–841 (2019)

19. Szegedy, C., et al.: Intriguing properties of neural networks. In: ICLR (2014)

20. Wang, R., Ke, H., Hu, M., Wu, W.: Adversarially robust neural networks with feature uncertainty learning and label embedding. Neural Netw. (2024)

21. Yang, H., et al.: Non-informative noise-enhanced stochastic neural networks for improving adversarial robustness. Inf. Fusion **108**, 102397 (2024)

22. Yang, H., Wang, M., Yu, Z., Zhou, Y.: Rethinking feature uncertainty in stochastic neural networks for adversarial robustness. arXiv abs/2201.00148 (2022)

23. Yu, T., Yang, Y., Li, D., Hospedales, T., Xiang, T.: Simple and effective stochastic neural networks. In: Proceedings of the AAAI Conference on Artificial Intelligence, vol. 35, pp. 3252–3260 (2021)

Corruption Aware Fusion for LiDAR Camera Based 3D Object Detection

Ron Alfia[1,2]([✉]) [iD] and Avi Mendelson[1] [iD]

[1] Technion – Israel Institute of Technology, 3200003 Haifa, Israel
ronalfia@campus.technion.ac.il, mendelson@technion.ac.il
[2] Riskified Ltd., 6492806 Tel Aviv, Israel
ron.alfia@riskified.com
https://www.riskified.com, https://www.technion.ac.il

Abstract. This paper investigates the challenges and solutions associated with modality bias in LiDAR-Camera-based 3D Object Detection (LC-3DOD) systems. Modality bias, where models disproportionately rely on the dominant modality, poses significant risks, particularly in safety-critical applications like autonomous driving. Our research aims to enhance the robustness of multimodal systems by addressing this bias and the associated robustness to sensor failures.

We begin by defining and quantifying modality bias within LC-3DOD systems, demonstrating its impact on system robustness under corrupted conditions. We identify a bias towards the LiDAR signal, which has a stronger correlation to the predictions compared to the camera inputs. We show how a well known modality dropout technique is useful in mitigating this bias, however, we encounter the phenomena of accuracy-robustness trade off for robustness to sensor failures as a limiting practical constraint on the possible robustness enhancement.

We propose an end-to-end adaptive inference architecture that leverages multiple model heads, each optimized for a specific regime of LiDAR input quality called Corruption-Aware-Fusion, which incorporates a LiDAR Corruption Estimation Module to dynamically assess the LiDAR signal quality and select the optimal head variant based on those conditions. CAF robustly balances clean accuracy with enhanced performance under severe corruptions.

Our CAF framework achieves state of the art accuracy on nuScenes-C LiDAR failures—with performance gains increasing as corruption severity rises—thereby safeguarding overall accuracy by mitigating the collapse that typically occurs under severe LiDAR degradations. Central to CAF is a LiDAR Corruption Estimation Module, a critical component that dynamically assesses corruption type and severity to drive adaptive model selection.

Keywords: 3D Object Detection · LiDAR Camera Fusion · Multimodal Robustness

Supplementary Information The online version contains supplementary material available at https://doi.org/10.1007/978-3-032-05060-1_13.

M. Castrillón-Santana et al. (Eds.): CAIP 2025, LNCS 15622, pp. 146–157, 2026.
https://doi.org/10.1007/978-3-032-05060-1_13

1   Introduction

Human perception of the environment is enabled by vision, sound, and other senses, each contributing uniquely to our understanding of the world. Imagine waking to complete darkness while hearing the sound of birds. You might conclude that the curtains are tightly sealed rather than that the birds have changed their habits. This ability to infer meaning from incomplete or conflicting sensory information demonstrates the robustness of human perception to multimodal inputs. Now consider driving in low light conditions. While your car's camera struggles to capture clear visuals, the LiDAR sensor detects the spatial layout of the road and nearby objects with remarkable precision. This interplay of modalities provides the redundancy necessary for safety, showing how combining complementary inputs can mitigate failures.

Deep learning has driven multimodal progress, achieving superhuman results in tasks such as image classification, object detection, and natural language processing [15]. However, neural networks often latch onto spurious correlations rather than true causal patterns [11]. In multimodal systems, this leads to modality bias—where the most informative sensor dominates predictions—and creates a serious vulnerability in safety-critical settings like autonomous driving, since failure in the dominant modality can have catastrophic consequences.

To address these challenges, we focus on LiDAR-Camera-based 3D Object Detection (LC-3DOD), a critical testbed for exploring multimodal robustness. Motivated by the large body of work on LiDAR-and-camera fusion for 3D detection [1,5,16], the availability of public benchmarks such as nuScenes [4], and the immediate applications in autonomous vehicles, robotics, and advanced driving assistance systems, we chose LC-3DOD as the domain for this study on multimodality robustness.

Previous studies in LC-3DOD systems have highlighted a significant lack of robustness to sensor failures, particularly those involving LiDAR, while demonstrating comparatively better resilience to image corruptions [8]. Despite advancements in multimodal integration, the underlying causes of these robustness issues remain insufficiently explored, especially concerning the potential role of modality bias [12]. Current research has so far been concentrated about missing modalities cases e.g. no LiDAR signal [9,18,19]. We aim to design a novel architecture that preserves state-of-the-art accuracy under normal conditions while remaining resilient to the full spectrum of LiDAR failures—from minor to severe sensor degradations.

Our central contribution is the development of a Corruption-Aware Fusion (CAF), an inference-only architecture. CAF leverages a LiDAR Corruption Estimation Module to asset the quality of the LiDAR signal, which is inserted into a Head Selector module (none differentiable) that picks a head with an appropriate bias level, matching high quality LiDARs with lidar-biased head and degraded LiDAR signal with an image biased head. In doing so, our approach not only outperforms state-of-the-art models but also establishes a new paradigm for robust LC-3DOD in challenging, real-world scenarios.

Our key contributions are as follows:

- **Corruption-Aware Fusion (CAF) Architecture**: Achieving state-of-the-art accuracy on numerous robustness cases including sensor failure class of nuScenes-C.
- **Quantitative Modality Bias:** We introduce a novel, normalized metric to quantify the modality bias in LC-3DOD systems and propose methods to control it's level.
- **Point-Cloud Corruption Estimation:** We define novel point-cloud corruption estimation task and solve it to an accuracy level that allows valuable realtime decisions.

Our approach depends on accurately assessing LiDAR signal quality, so it requires a labeled training set for each type of degradation and a model capable of detecting those corruptions. As new LiDAR hardware and techniques emerge, previously unseen failure modes may occur that our estimator cannot recognize. In practice, however, we find that any degradations that escape detection tend to have only a minor impact on 3D object detection performance.

In the next section, we will review related works including robust methods and key architectures for LC-3DOD. Following, we will present our methodologies including the CAF architecture and it's key components. Section 4 will present the results, comparison of CAF to other architectures in terms of LiDAR failures of nuScenes-C and our novel failure classes, followed by an ablation study. Next we will discuss the implications of this study, suggest that CAF can be applied as a framework to other base models. Last sections is a brief conclusion.

2   Related Works

The research community has produced a vast amount of multimodal models, such as BEVFusion [16], FUTR3D [5] and TransFusion [1]. Each of those models is capable of providing top accuracy on nuScenes [4] dataset while fusing LiDAR and camera signals as inputs. However, those models were not inherently designed to be robust into sensor failures, particularly LiDAR failures, and they experience accuracy collapse for major LiDAR signal degradations [8]. Our architecture is set to excitedly tackle those degradations while preserving clean signal accuracy, mitigating the accuracy collapse.

Recently, a concurrent branch of works, namely MetaBEV [9], UniBEV [18] and the Cross-Modal Transformer [19] have leveraged Modality Dropout and Masked Training to enhance robustness against missing LiDAR or camera signal. Through a supportive architectural component namely Cross Modal Attention, which allows processing an input with missing modality, those methods apply a modality drop throughout the training regime thus yielding vast improvements for accuracy of inference on missing modalities. We propose a refined training regime, that doesn't require Cross Modal Attention, by applying DropOut as a percentage on the LiDAR features. Acknowledging the accuracy-robustness trade-off, we train multiple heads with different levels of bias to support robustness for medium and severe LiDAR degradations in contrast to entirely missing modality.

A popular family of methods for quantifying dependence between high-dimensional inputs and outputs are neural mutual information estimators, such as MINE [2], InfoNCE [17], and CLUB [6]. These approaches formulate a variational lower bound on the true mutual information and parameterize the critic (or discriminator) network via deep architectures. During training, they require two data streams—positive samples drawn from the joint distribution and negative samples from the product of marginals—and optimize the critic by gradient ascent. While flexible and theoretically well-founded, they incur substantial computational and implementation overhead: each new model or dataset demands training a separate estimator, tuning its architecture and learning dynamics, and ensuring numerical stability. By contrast, our AUC-based modality bias metric is entirely nonparametric and thus requires no additional training beyond the usual evaluation runs.

With large labelled datasets, convolutional and attention-based networks learn degradation features directly. Kang et al. [14] first regress patch-level quality using a shallow CNN. Bosse et al. [3] enhance this via dual-stream networks that jointly learn spatial attention and pooling for global quality. These end-to-end models achieve high accuracy on image benchmarks. In contrast, our LiDAR corruption estimator is the first to target point-cloud quality: it reuses a voxel-based detection backbone and trains via MSE regression on synthetically corrupted point clouds.

ImageNet-C [13] was the first large-scale benchmark to systematically evaluate model robustness by applying eight common corruptions at five severity levels to the ImageNet validation set and measuring the resulting classification error. Dong et al. introduced both nuScenes-C and KITTI-C [8], defining corruptions for point-clouds and multi-sensor calibrations to the nuScenes [4] and KITTI datasets [10]. We further expand this suite by adding two novel LiDAR failure modes—Accuracy Failure and Mutual Interference—which capture sensor misalignment and crosstalk effects beyond the original corruption types.

3 Methodologies

3.1 Corruption-Aware Fusion

CAF architecture is composed of two feature extractors, multiple decoder heads (in our implementation - 3 such heads), corruptions estimator module for key corruptions (in our implementation - 4 such corruptions), and a head selector which is a non differentiable component. The inference pass is conducted as follows - inputs are processed to produce an image features vector and a pointcloud features vector, the pointcloud features are inserted into the corruptions estimator module producing an output vector of length 4 indicating the predictions for severity level of each of those corruptions. Those predictions are inserted into the head selector. The selected head receives the image and lidar features and produces the final prediction. Figure 1 provides an overview of inference architecture. Details for the algorithm of head selection are in the Supplementary.

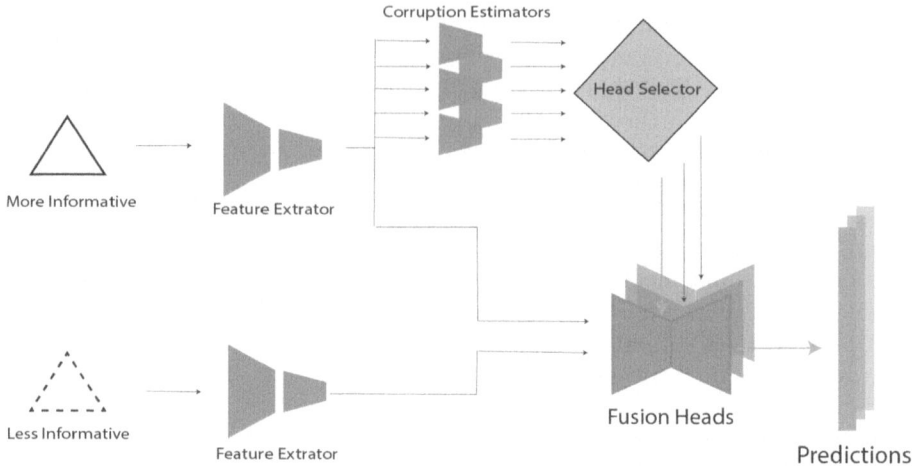

Fig. 1. Overview of the proposed Corruption-Aware Fusion (CAF) architecture. More/Less Informative refer to LiDAR/Camera inputs. Both streams are processed to produce features. The informative features are inserted into a corruption estimation followed by head selection. Each head was pretrained with a certain bias level. The selected head receives the features and produces the prediction.

The training part of the CAF components involves two main categories, corruptions estimator training and the decoder heads training. The decoder heads of CAF are differing by their modality bias towards the LiDAR signal. In Sect. 3.2 we define the Modality Bias and present methods to decrease it. Those methods are the ones we used to generate the multiple decoder heads. In Sect. 3.3 we show the corruptions estimator module and it's training schemes. Together those categories compose the entire algorithm for building the trained CAF architecture.

3.2   Modality Bias

Intuitively, if a multimodal model is heavily biased towards a specific modality e.g. the LiDAR then upon a complete degradation of the LiDAR signal the accuracy of the overall multimodal model will completely collapse. With PixelsErase and PointsDrop as continuous synthetic degradations for the camera, LiDAR sensors respectively, Fig. 2 illustrates this collapse, which indicates a strong bias towards the LiDAR modality.

We defined two synthetic degradation methods, namely PixelErase and PointsDrop. Both methods receive as input a clean signal (images, pointcloud respectively) and a percentage i.e. float between 0 to 1 where 0 keeps the signal as-is and 1 refers to complete degradation of the signal. Exact details for PixelErase and PointsDrop are provided in the Supplementary. To quantitatively assess modality bias, we define a metric based on the Area Under the Curve (AUC) of the accuracy versus corruption-level plots for those two degradations.

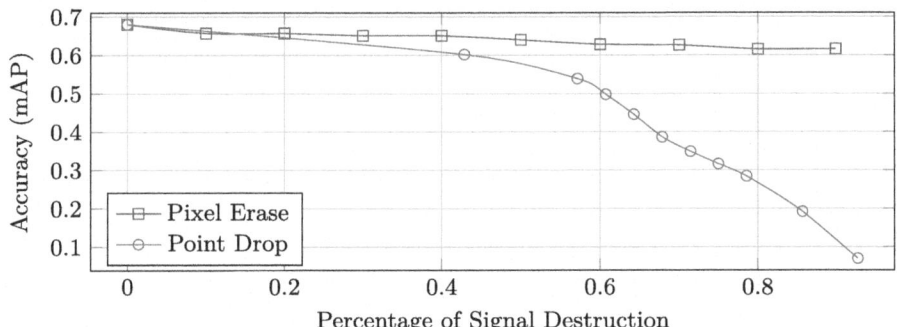

Fig. 2. Accuracy degradation for Pixel Erase (blue) and Point Drop (red) as the respective image or LiDAR signals are progressively destroyed. (Color figure online)

We then define the normalized modality bias metric as

$$B = \frac{\text{AUC}_{\text{PointDrop}} - \text{AUC}_{\text{PixelErase}}}{A_{\text{clean}}},$$

where A_{clean} is the accuracy under clean conditions. This normalization ensures that B is bounded between -1 and 1, facilitating direct comparisons across different models and datasets.

Debiasing Techniques. Based on our observations of modality bias, we propose debiasing methods aimed at reducing the model's over-reliance on the LiDAR input by reducing the correlation between the output and the LiDAR. In the Supplementary we provide theoretical backing to this technique and additional details for the debiasing methods.

3.3   LiDAR Corruption Estimation

Task Definition. We formalize the corruption estimation task as follows. Given a point cloud

$$\mathbf{X} \in \mathbb{R}^{N \times D},$$

where N is the number of points and D the number of features (e.g., spatial coordinates, intensity, reflectivity), and considering C corruption types with severity levels $s \in \{0, 1, \ldots, S\}$ (with $s = 0$ indicating no corruption), our objective is to learn a function

$$f_\theta : \mathbb{R}^{N \times D} \to \{0, 1, \ldots, S\}^C,$$

that predicts a severity vector $\hat{\mathbf{Y}} = f_\theta(\mathbf{X})$ approximating the ground truth $\mathbf{Y}$. We treat this as a regression task and optimize the model using the mean squared error (MSE) loss, thereby leveraging the ordinal structure of the severity levels.

In practice, we leveraged pre-trained LiDAR features, and added a combination of pooling and fully connected layers to reduce the features into the number of corruptions to predict, where prediction value corresponds to a severity level. More details are provided in the Supplementary.

Table 1. nuScenes-C [8] LiDAR sensor failures comparison between CAF and the other models.

Corruption	FUTR3D	TransFusion	BEVFusion	CAF (Ours)
Density	63.72	65.77	**67.79**	67.68
Cutout	62.25	63.66	66.18	**66.48**
Crosstalk	62.66	64.67	67.32	**67.35**
FOV Lost	26.32	24.63	27.17	**34.46**
Gaussian (L)	58.94	55.10	60.64	**62.68**
Uniform (L)	63.21	64.72	66.81	**67.11**
Impulse (L)	63.43	65.51	**67.54**	67.12
Average	57.22	57.72	60.49	**61.84**

3.4   Expanded LiDAR Failures Taxonomy

While prior work has largely focused on weather conditions and sensor noise, we argue that not all LiDAR failure phenomena are adequately modeled. To bridge this gap, we introduce two additional corruption models:

Accuracy Failure (AF): Accuracy Failure represents calibration errors in LiDAR systems. In our model, we simulate this by rotating the point cloud around a specified axis (e.g., the Z-axis or a rotated variant for more severe cases). This deterministic transformation consistently shifts point positions, reflecting a calibration bias rather than random noise.

Mutual Interference (MI): Mutual Interference arises when multiple LiDAR sensors operate in proximity, leading to cross-talk between their signals. In our model, we (i) scale the number of affected points based on the number of neighboring LiDARs (ranging from 1 to 16, as per [7]), and (ii) adjust the affected points by moving them closer to the sensor while retaining their original angular position. This design better mimics real-world behavior where fixed-angle LiDARs experience systematic interference.

4   Results

4.1   Benchmarks on LiDAR Failures

We present how our proposed method defined in Sect. 3.1 performs across multiple LiDAR failures, including sensor failures defined in nuScenes-C [8] and our two additional failures Accuracy Failure and Mutual Interference defined in Sect. 3.4 in comparison with three key multimodal models namely FUTR3D [5], TransFusion [1] and BEVFusion [16] collectively will be denoted as other models.

In Table 1, CAF is compared to other models on nuScenes-C LiDAR failures, acheiving the best score in 5 out of the 7 corruptions. We note that the benefits are more distinguishable on corruptions with high impact on the multimodal models for example FOV Lost is causing a degradation of more than 35 pts on the other models. This degradation is being mitigated by CAF performing 7 pts higher than the next model. This is algo distinguishable on Gaussian - the second highest impact corruption. On the 2 out of 7 where CAF is not the top model, it is being second by a small margin of less than a 1 pt.

The mitigation effect for high impact corruption is also visible on Accuracy Failure and Mutual Interference, presented in Table 2. On Accuracy Failure, which is a high impact corruption, CAF achieved the best score by a margin of 6 pts from the next model. For Mutual Interference, we report that CAF is the top model by a small margin.

Overall, CAF consistently outperforms existing fusion methods on both nuScenes-C and our new LiDAR failures with large gains on the most challenging failure cases. We attribute this behavior to the architectural design which detects LiDAR failures in realtime and selects an appropriate head. In the case of severe corruption this is both easier to detect and important to mitigate.

4.2 Ablation Studies

In this section, we explore two aspects of CAF namely the mitigation property i.e. the ability of CAF to mitigate severe corruption cases, and the selection algorithm which is compared to an oracle selection and a random selection.

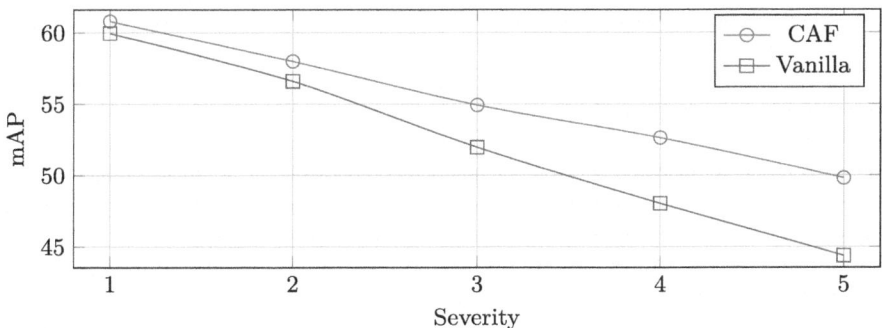

Fig. 3. Mean mAP trend for Vanilla (FUTR3D) and CAF models across severity levels aggregated over 6 key corruptions - FOV Lost, Accuracy Failure, Mutual Inference, Guassian Noise, Cutout and Density.

Figure 3 visualizes the mean mAP trend for both the vanilla FUTR3D and our CAF-enhanced model as the severity of LiDAR corruption increases. While the vanilla model exhibits a marked performance decline at higher severity levels, CAF maintains a shallower drop, consistently outperforming the baseline. This

Table 2. Performance on novel LiDAR-failure corruptions - Mutual Interference and Accuracy Failure, for CAF and the other models.

Corruption	FUTR3D	TransFusion	BEVFusion	CAF (Ours)
Mutual Interference	59.30	58.50	52.00	**60.00**
Accuracy Failure	33.35	26.12	27.69	**39.10**

Table 3. Comparison of mAP for three selection algorithms: **ORACLE** (leverages exact knowledge of the corruption case—unrealistic in real time), **OURS** (ours selection algorithm), and **RANDOM** (randomly selects a head).

Corruption	ORACLE	OURS	RANDOM
FOV-Lost	37.4	34.5	33.5
Gaussian	62.2	62.7	52.1
Mutual Interference	59.8	60.0	50.6
Accuracy Failure	39.5	39.1	30.1
Density	66.8	67.7	55.0
Cutout	65.8	66.5	54.4
Average	55.2	55.1	46.0

performance gap widens particularly in the mid-to-high severity range, underscoring CAF's ability to dynamically adapt to severe LiDAR degradations and sustain detection accuracy in challenging conditions.

Table 3 compares three model-selection strategies—**ORACLE** (best head on average for every corruption type and level), **OURS** (our corruption-aware selector algorithm), and **RANDOM** (randomly select the decoder header). We note that the Oracle selects the best head on average across the validation set, not the best head per sample, as such it could have inferior performance on some cases and we do see a small margin of upside for OURS on Mutual Interference, Density and Cutout.

Comparing Random to Ours we see that the overall selection algorithm is performing marginally better across Gaussian, Mutual Interference, Accuracy Failure, Density and Cutout while across all of those corruptions, the selection algorithms reaches the level of oracle prediction. However, on FOV-Lost the selection algorithm have not made most of the mitigation possible by the additional heads.

Table 4. Parameter breakdown of CAF on a FUTR3D backbone. At inference only *one* head is resident on the GPU; the others can be paged to CPU if memory is constrained.

Module	Params (M)	% of Total
Image & LiDAR backbones (shared)	73.4	72.1
Neck (shared)	9.2	9.0
Decoder Head 0 (clean)	5.3	5.2
Decoder Head 1 (mild)	5.3	5.2
Decoder Head 2 (debias)	5.3	5.2
CAF Total	101.9	100

4.3   Resource Footprint of CAF

CAF keeps *one* feature extractor per modality and adds **two extra decoder heads** (Table 4). Each head composed of a fusion block followed by a three-stage transformer decoder identical to FUTR3D's decoder (5.4 M parameters). Consequently, CAF increases total parameters from 88.1 M to 101.9 M i.e. an increase of +15.7%.

The corruption-estimator MLP and head-selection logic are lightweight, introducing only marginal latency and keeping the overall pipeline within the same runtime requirements of FUTR3D.

5   Discussion

In the methodologies section we demonstrated that modality bias—specifically, the over-reliance on LiDAR inputs—plays a critical role in the robustness of LiDAR-Camera 3D object detection systems. Under clean conditions, LiDAR-based features drive high detection accuracy; however, when the LiDAR signal is degraded by various corruptions, the performance can collapse dramatically.

A key insight is the inherent trade-off between accuracy and robustness. Our experiments indicate that while aggressive debiasing can substantially improve robustness, it may lead to a modest decline in performance on clean data. This trade-off raises the question of whether a single continuous model can handle the vast amount of input quality scenarios. In contrast to most current approaches that seek one deep learning model to handle all cases, we presented here a different approach that finetune different heads for different areas of the trade-off spectrum.

CAF is a specific inference architecture, however we highlight the potential of using it as a framework. In our implementation we have selected FUTR3D to act as the base model, for multiple reasons including the relatively fast training. However, CAF can be implemented on top of every architecture that processes image and LiDAR features independently. Given that, corruption estimators can be trained on top of the LiDAR features, and decoder headers can be tuned with the additional of debaising layers between the LiDAR and the fusion layers.

Our corruption estimators are trained in a fully supervised manner, using synthetic examples generated by each corruption model we wish to recognise. This design has two clear drawbacks. First, it assumes that future LiDAR failure modes can be simulated in advance; truly novel faults (e.g., a new scanning pattern, a firmware bug, or unexpected multi-path reflections) may fall outside the estimator's learned decision boundaries and be mis-classified or left unrecognised. Second, every additional corruption family demands its own labelled dataset, either by corruption modelling or through data acquisition, which introduces manual effort whenever new hardware or operating conditions appear.

Furthermore, as new architectures will emerge, CAF as a framework can be implemented on top to provide for the robustness safety net. As LiDAR is more informative for the 3D object detection task, we predict that model accuracy collapse due to sensor failure will be related to the LiDAR sensor. We note, that as new failure modes for LiDAR will be defined, they can be added on top of existing corruption estimators, provided that they can be detected, by adapting the head selection logic and without retraining former estimators.

Future Work: Two directions merit immediate attention. Light-weight CAF: future iterations should reduce the overhead of maintaining multiple heads—for example, through weight-sharing schemes, compact adapter layers, or post-training distillation—so that corruption awareness comes with only a minimal memory and latency footprint. Enhanced FOV-Lost resilience: our current head-selection strategy still leaves measurable headroom under extreme field-of-view loss; closing this gap remains an open problem for subsequent research.

6   Conclusion

In this work, we have introduced CAF, a corruption-aware fusion framework that achieves new state-of-the-art performance on nuScenes-R LiDAR failure scenarios. Our experiments demonstrate that the mitigation effect of CAF grows stronger as the severity of LiDAR degradation increases, highlighting its robustness under challenging conditions. Furthermore, we show that the underlying LiDAR corruption estimators are accurate enough to guide real-time head selection, enabling valuable, on-the-fly decisions in deployed systems. Overall, CAF offers a practical and effective solution for maintaining high 3D detection performance in the presence of severe LiDAR corruptions.

References

1. Bai, X., et al.: Transfusion: robust lidar-camera fusion for 3D object detection with transformers. In: Proceedings of the IEEE/CVF Conference on Computer Vision and Pattern Recognition, pp. 1090–1099 (2022)
2. Belghazi, M.I., et al.: Mutual information neural estimation. In: International Conference on Machine Learning, pp. 531–540. PMLR (2018)
3. Bosse, S., Manfred, D., Bethge, M., Wichmann, F.A.: Deep neural networks for no-reference and full-reference image quality assessment. IEEE Trans. Image Process. **27**(1), 206–219 (2018)
4. Caesar, H., et al.: nuscenes: a multimodal dataset for autonomous driving (2020). https://arxiv.org/abs/1903.11027
5. Chen, X., Zhang, T., Wang, Y., Wang, Y., Zhao, H.: Futr3d: a unified sensor fusion framework for 3D detection (2023). https://arxiv.org/abs/2203.10642
6. Cheng, P., Hao, W., Dai, S., Liu, J., Gan, Z., Carin, L.: Club: a contrastive log-ratio upper bound of mutual information. In: International Conference on Machine Learning, pp. 1779–1788. PMLR (2020)
7. Diehm, A.L., Hammer, M., Hebel, M., Arens, M.: Mitigation of crosstalk effects in multi-lidar configurations. In: Electro-Optical Remote Sensing XII, vol. 10796, pp. 13–24. SPIE (2018)
8. Dong, Y., et al.: Benchmarking robustness of 3D object detection to common corruptions in autonomous driving (2023). https://arxiv.org/abs/2303.11040

9. Ge, C., et al.: Metabev: solving sensor failures for 3D detection and map segmentation. In: Proceedings of the IEEE/CVF International Conference on Computer Vision, pp. 8721–8731 (2023)

10. Geiger, A., Lenz, P., Stiller, C., Urtasun, R.: Vision meets robotics: the kitti dataset. Int. J. Robot. Res. **32**(11), 1231–1237 (2013)

11. Geirhos, R., et al.: Shortcut learning in deep neural networks. Nat. Mach. Intell. **2**(11), 665–673 (2020)

12. Hazarika, D., Li, Y., Cheng, B., Zhao, S., Zimmermann, R., Poria, S.: Analyzing modality robustness in multimodal sentiment analysis (2022)

13. Hendrycks, D., Dietterich, T.: Benchmarking neural network robustness to common corruptions and perturbations. arXiv preprint arXiv:1903.12261 (2019)

14. Kang, L., Ye, P., Li, Y., Doermann, D.: Convolutional neural networks for no-reference image quality assessment. In: Proceedings of the IEEE Conference on Computer Vision and Pattern Recognition, pp. 1733–1740 (2014)

15. Krizhevsky, A., Sutskever, I., Hinton, G.E.: Imagenet classification with deep convolutional neural networks. In: Advances in Neural Information Processing Systems, vol. 25 (2012)

16. Liu, Z., et al.: Bevfusion: multi-task multi-sensor fusion with unified bird's-eye view representation (2024). https://arxiv.org/abs/2205.13542

17. Oord, A.V.D., Li, Y., Vinyals, O.: Representation learning with contrastive predictive coding. arXiv preprint arXiv:1807.03748 (2018)

18. Wang, S., Caesar, H., Nan, L., Kooij, J.F.: Unibev: multi-modal 3D object detection with uniform BEV encoders for robustness against missing sensor modalities. In: 2024 IEEE Intelligent Vehicles Symposium (IV), pp. 2776–2783. IEEE (2024)

19. Yan, J., et al.: Cross modal transformer: towards fast and robust 3D object detection. In: Proceedings of the IEEE/CVF International Conference on Computer Vision, pp. 18268–18278 (2023)

Effective Relationship Between Characteristics of Training Data and Learning Progress on Knowledge Distillation for Image Recognition

Minoru Mori[✉] [ID] and Yuta Kogawa

Kanagawa Institute of Technology, Atsugi-shi, Kanagawa 243-0292, Japan
mmori@ic.kanagawa-it.ac.jp

Abstract. In image recognition, knowledge distillation is a valuable approach to train a compact model with high accuracy by exploiting outputs of a highly accurate large model as correct labels. In knowledge distillation, studies have shown the usefulness of data with high entropy output generated by image mix data augmentation techniques. Other strategies such as curriculum learning have also been proposed to improve model generalization by the control of the difficulty of training data over the learning process. In this paper, we explore the relationship between the learning process and data characteristics, focusing on the entropy of the output distribution and learning difficulty in knowledge distillation. To validate this relationship, we propose a method to readily generate data that yields high entropy output and adjust the degree of learning difficulty by controlling bounds and a sampling range for mix ratios in mixing images between classes. In evaluation experiments using multiple datasets on several numbers of epochs, our proposed method outperformed conventional approaches in terms of accuracy.

Keywords: Deep neural network · Knowledge distillation · Image mix data augmentation · Curriculum

1 Introduction

In image recognition using Deep Neural Networks (DNNs), larger models often provide higher accuracies [4]. However, such an approach is facing major problems like increased computational resources and longer inference time. Small models can avoid these problems but have limitations in accuracies. Knowledge distillation [3,11] is one of the approaches that allows small models to achieve high accuracies. In knowledge distillation, a small model is generalized by using the outputs of a pre-trained large model with high performance as correct labels. After the work of [11], various improvement methods have been proposed [16]. Function Matching [2] is one of the improved methods by using images with stronger augmentation like Mixup [30] in training and learning models on more

M. Castrillón-Santana et al. (Eds.): CAIP 2025, LNCS 15622, pp. 158–169, 2026.
https://doi.org/10.1007/978-3-032-05060-1_14

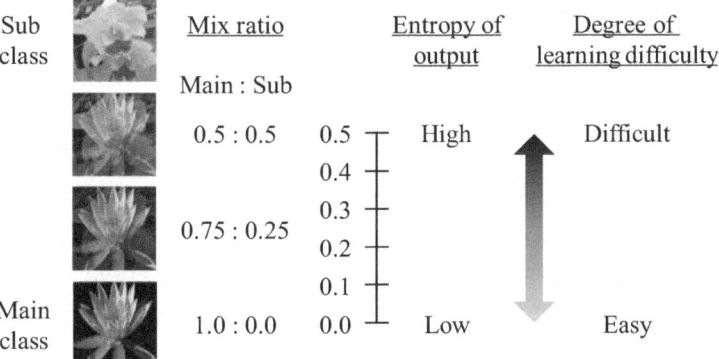

Fig. 1. Our concept. The image mixed with ratio of 0 is same as a main class image. This is an image with the lowest entropy of its output and the easiest to learn. The image mixed with ratio of 0.5 is an equal mixture of two classes. This has high output entropy and is the most difficult to learn.

epochs. Mixup is one of the techniques that generate diverse images by mixing images and labels in random ratios between classes.

To extend Function Matching, we discuss two aspects of the effective relationship between the properties of training data and the learning process on knowledge distillation using image mix data augmentation. First, to achieve better knowledge transitions, we propose a simple and effective method of generating training data that provides better outputs of the teacher model. Recently, Wang et al. [26] have provided that a higher entropy of output distribution of training data is more appropriate for knowledge distillation. In this paper, we propose a method to increase the output entropy of training data by restricting the sampling range of mix ratios instead of random sampling on data augmentation. Although our procedure allows us to frequently generate images with high output entropy, this reduces the diversity of augmented images. In our experiments, we show that restricting the mix ratio range on data augmentation is effective. In particular, when learning on more epochs, narrowing the range aggressively becomes beneficial. Next, we examine the relationship between training progress and the degree of augmentation of data as curriculum. As for the training process involving knowledge distillation with data augmentation, Gontijo-Lobez et al. [9] have proposed an approach that terminates data augmentation in the middle during training. On the basis of [9], we can say that keeping a high degree of augmentation can be applied in the early stage of learning, but decreasing the degree of augmentation toward the end of training is needed for better knowledge distillation. From this thought, we show that in knowledge distillation, gradual reduction of the augmentation degree in the latter half of training is effective in improving generalization performance. We also show that combining this procedure with the increase of output entropy mentioned above dramatically improves classification accuracies. Figure 1 shows our concept of the relationship among mixed images with several mix ratios, their entropies of outputs and their degrees of learning difficulty.

2 Related Works

2.1 Knowledge Distillation

Knowledge distillation [3,11] is one of the model compression techniques such as pruning and quantization. Knowledge distillation transfers knowledge from a large pre-trained model with high accuracy to another small model. This technique provides a compressed model with little loss in accuracy. The pre-trained model is called the teacher model and the smaller one is the student model. Classification tasks using DNNs generally use one-hot vectors as correct labels of training data. On the other hand, knowledge distillation uses the predictions of the teacher model as the correct labels. By training the student model using prediction probabilities of not only the correct class but also other classes, knowledge other than the correct class in each data is reflected in the student model, and the student model achieves higher generalization performance.

Since knowledge distillation has been introduced into DNNs in [11], a wide variety of methods have been proposed [2,5,21,24,26,27,31]. Noisy student [27] inputs a fixed image into the teacher model but inputs an augmented image into the student model. On the other hand, Function Matching [2] proposes that the use of the same images with strong data augmentation for both the teacher and student models prevents saturation of generalization and improves the student one on more epochs. In Function Matching, Mixup [30] is used as strong data augmentation. Moreover, some studies [6,26,28] describe that a high entropy of an output given by mixed images is effective for knowledge distillation.

2.2 Image Mix Data Augmentation

Data augmentation usually transforms images of training data for generalizing models, and many techniques have been proposed [17,22]. Image mix is one of the strategies, including rotation and flipping or image deletion like Cutout [8] and Random Erasing [32]. Image mix mainly mixes two images into one. Mixup [30] is a typical technique and combines not only images but also labels between classes with a ratio sampled from [0, 1]. This processing produces a variety of data and is expected to improve discriminant ability between classes. Since the proposal of Mixup, several image mix data augmentation techniques such as Cutmix [29] and Attentive Cutmix [25] have been proposed [18].

2.3 Curriculum Learning

Techniques that control the degree of learning difficulty of training data according to learning steps have been proposed. Curriculum learning [1,23] is a typical method that aims to generalize a model by gradually increasing the learning

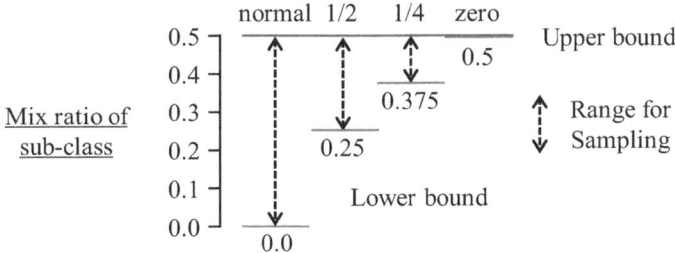

Fig. 2. Examples of sampling ranges of mix ratio for sub-class. More difficult data are often produced using sampling range restriction by lifting lower bound.

difficulty of training data as the learning process progresses. In knowledge distillation, for example, Gontijo-Lobez et al. [9] have reported that generalization can be improved by discontinuing data augmentation in the middle of the learning process. Li et al. [14] also have proposed learnable temperature parameters as a curriculum used in the distillation process.

3   Proposed Method

3.1   Restricting Sampling Range of Mix Ratio

When using data augmentation that mixes images between classes, outputs of the teacher model tend to be multimodal with peaks in multiple classes. As a result, the entropy of the outputs tends to be high. Therefore, the use of training data forcing output to be multimodal and more softly distributed can induce more knowledge transfer between models. In other words, applying a mix ratio that is close to an equal one between classes is suited for knowledge distillation. To accomplish this thought, we propose a strategy that restricts the range for sampling mix ratio by setting the lower bound close to the upper bound. Detailed procedures are as follows: First, we set the lower bound (default = 0) close to the upper bound. Next, we uniformly sample a mix ratio from not [0, 1] but [lower bound, 0.5] for each sub-class. This, therefore, tends to sample a mix ratio near the upper bound from the narrowed range. Restricting the range reduces a diversity of mixed data, but can aggressively generate data with high entropy of output distribution. Figure 1 shows relations between several mixed images using Mixup and their entropy of outputs. We validate the following four kinds of lower bounds shown in Fig. 2; normal (lower bound is 0), 1/2 (= middle between upper and lower bounds), 1/4 (= 1/4 of width of upper and lower bounds), and zero (= same as upper bound).

3.2   Control of Degree of Learning Difficulty According to Learning Progress

As mentioned in Sect. 2.3, the study of [9] provides that interrupting data augmentation before the end of the learning process improves model generalization.

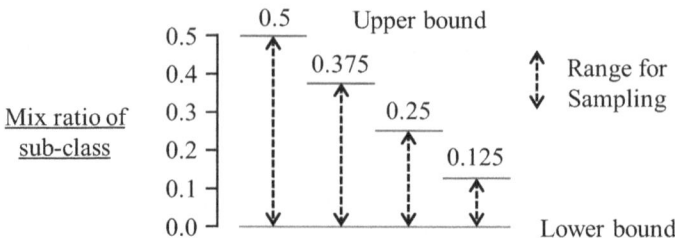

Fig. 3. Examples of upper bounds of sampling ratio for sub-class. Easier data are generated by lowering upper bound.

From this study, we can say that applying data augmentation aggressively in the early stages of the learning process and reducing the degree of data augmentation toward the end of training are important. Contrast this with the process of model generalization; learning using highly augmented data in the initial stage seems to correspond to the acquisition of feature representation capability as a foundation for the model, and learning using data with little augmentation can be regarded as fine-tuning in the final stage. Also, when the degree of data augmentation is high, inferring properly an inherent probability distribution of training data seems to be difficult in insufficient learning conditions. In other words, data with a higher degree of augmentation is higher in the degree of learning difficulty.

In this paper, we delve deeper into the relationship between the degree of data augmentation as the learning difficulty and the progress of learning in knowledge distillation. Specifically, the degree of learning difficulty is controlled by systematically lowering the upper bound of the mix ratio on the sampling range for sub-class in accordance with the learning step. By lowering the upper bound, only low mix ratios can be sampled, which reduces the learning difficulty shown in Fig. 1. An overview of the lower upper bound and each corresponding range for sampling mix ratios for sub-class is also shown in Fig. 3. In the context of this paper, the original Mixup constantly holds 0.5 as the upper bound on every epoch. Interrupting data augmentation used in [9] corresponds to lowering the degree of learning difficulty to the easiest level. Concrete examples of the control of the upper bound are described in the section of validation experiments.

3.3   Combination of Range Restriction and Bound Control

Finally, we combine the sampling range restriction with the control of the degree of learning difficulty according to the learning step for validating synergistic effects of them. Specifically, the sampling range is restricted for aggressively increasing the entropy of the output in the early stages of learning, while the upper bound for ratio sampling is lowered for reducing the degree of learning difficulty in the late stages. The objective is to obtain a discriminant feature representation through effective knowledge transfer between models, while finally

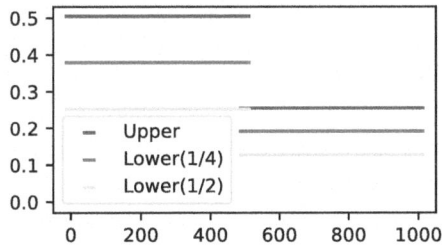

Fig. 4. Combinations of lowering upper bound and narrowing a sampling range for mix ratios on 1,000 epochs. Red line is the lower bound for 1/4 the width and pink one is for 1/2. (Color figure online)

Table 1. Experimental setups for each dataset and results by teacher models [%].

	Flowers102		Pets37		Food101	
Epochs	1,000	10,000	300	1,000	3,000	100	1,000
Learning rate	0.002	0.003	0.01	0.01	0.003	0.003	0.001
Accuracy of teacher	98.02		90.65		87.13	

obtaining a high generalization performance through fine-tuning. Examples of adjusting the sampling range to 1/4 or 1/2 and the decreasing upper bound that is divided into two steps are shown in Fig. 4.

4 Evaluation Experiments

4.1 Experimental Setups

To compare our methods with the conventional Function Matching, we conducted evaluation experiments. Setups are based on [2]. We used the following three datasets for validating our method: Flowers102 [19], Pets37 [20] and Food101 [12]. Their detailed specifications are tabulated in our supplemental material. BiT-ResNet101×3 [13] was used as teacher models and trained by using train images of each dataset. ResNet50V2 [10] was used for student models. The batch size was 64 images due to limitations on our GPU resources. The number of epochs, the initial learning rates, and recognition accuracies by teacher models are shown in Table 1. Each learning rate gives the best accuracies for Function Matching. The learning rate was increased by warm-up in 1,500 steps and then decreased by cosine decay. The optimizer is AdamW [15]. The loss function is KL Divergence. Every result in Sect. 4.2 was given by 3 trials with different seeds.

4.2 Experimental Results

Table 2. Classification accuracies [%] on range restriction for sampling mix ratio. Second row expresses epoch number. **Bold** expresses a better result than that by normal range on each condition.

Range	Flowers102	
	1,000	10,000
normal	72.15±0.63	80.83±0.95
1/2	71.57±0.30	**81.51±0.93**
1/4	**72.49±0.46**	75.70±0.88
zero	67.81±0.46	77.99±0.96

Range	Pets37		
	300	100	3,000
normal	72.55±0.09	78.11±0.18	82.34±0.14
1/2	**73.78±0.17**	**80.15±0.75**	**84.27±0.33**
1/4	68.36±0.96	**78.40±0.57**	**83.88±0.24**
zero	58.29±0.89	73.61±0.67	80.97±0.05

Range	Food101	
	100	1,000
normal	80.21±0.15	83.94±0.11
1/2	79.20±0.19	**84.33±0.16**
1/4	76.90±0.22	83.77±0.08
zero	72.98±0.53	83.05±0.08

Sampling Range Restriction for High Output Entropy/ First, we tested the effect of restricting the sampling range for achieving a high entropy of output distribution from training data in learning. We used the three kinds of range restrictions that are 1/2, 1/4, and zero of the normal sampling range as shown in Fig. 2. Classification rates of them are shown in Table 2. From Table 2, at least one or more among narrow ranges for mix ratios are effective to improve accuracies in many cases without Food101 on 100 epochs. These results indicate that the conventional idea, in which more varieties based on the diversity of mix ratios is more effective, is not necessarily true, but the high entropy of output distribution contributes to the accuracy improvement. On the other hand, no range for sampling, "zero", degraded the performance. Such results are apparently caused by fewer varieties on no randomness of mix ratios. They also indicate that only high entropy of output from training data is not a factor to generalize models on knowledge distillation.

Control of Degree of Learning Difficulty. Next, we provide experiments with smaller epochs for investigating characteristics and effectiveness of the con-

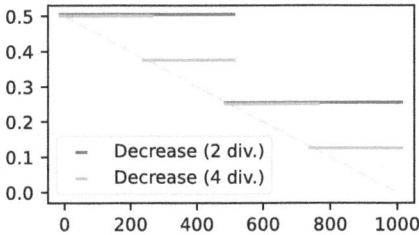 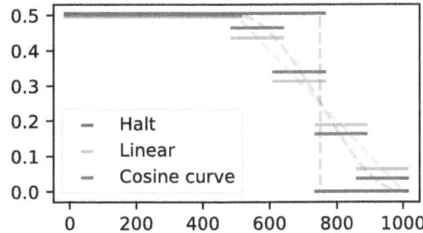

(a) Divisions in linear decrease on upper bound. The blue and orange lines indicate reductions in 2 and 4 steps.

(b) Characteristics of decrease of upper bound on latter half of epochs. Halt, linear decrease and cosine curve.

Fig. 5. Examples of upper trends for mix ratio for sub-class.

Table 3. Classification acuracies [%] on each upper bound change as degree of augmentation with small epochs. Second row is epoch number. **Bold** expresses the best result on each condition. <u>Underline</u> expresses the second best result.

Trend	Flowers102 1,000	Pets37 300	Food101 100
Constant	72.15±0.63	72.55±0.09	80.21±0.15
Dec-2	73.18±0.62	73.09±0.11	79.93±0.04
Dec-4	70.84±0.38	72.24±0.24	79.88±0.10
Halt	<u>73.79±0.58</u>	71.58±0.25	80.28±0.18
Linear	73.42±0.58	<u>73.88±0.26</u>	**80.63±0.13**
Cosine	**74.31±0.65**	**74.27±0.17**	<u>80.60±0.07</u>

trol of the degree of augmentation in the training process. As trends of the upper bound for the control of the degree of augmentation, we validated six trends. As the default trend, "constant" is a normal augmentation for keeping the initial value (=0.5) for the whole training process. "Dec-2" and "Dec-4" mean cases that the upper bound is decreasing in 2 and 4 steps as shown in Fig. 5a. The reason for such a division is that mix ratios are sampled randomly in each range, so changing the upper bound on every epoch seems to be meaningless. Therefore, we divide learning epochs into some sections and each section keeps the upper bound. "Halt" is terminating the augmentation toward the end of the training process described in [9]. In order to identify more effective trends regarding the reduction of augmentation level in the second half of the learning process, we verify two more kinds of trends, monotonically decreasing in linear and cosine curve, shown in Fig. 5b. Each latter half is divided into 4 intervals. Table 3 shows accuracies of controlling the upper bound of possible augmentation degree. From Table 3, "Dec-2" improved accuracies in many cases compared to the default "Constant", but "Dec-4" has just degraded results. These results indicate that reducing the difficulty level can improve performance, but simply reducing it is not always effective. Table 3 also shows that while simply stopping augmentation

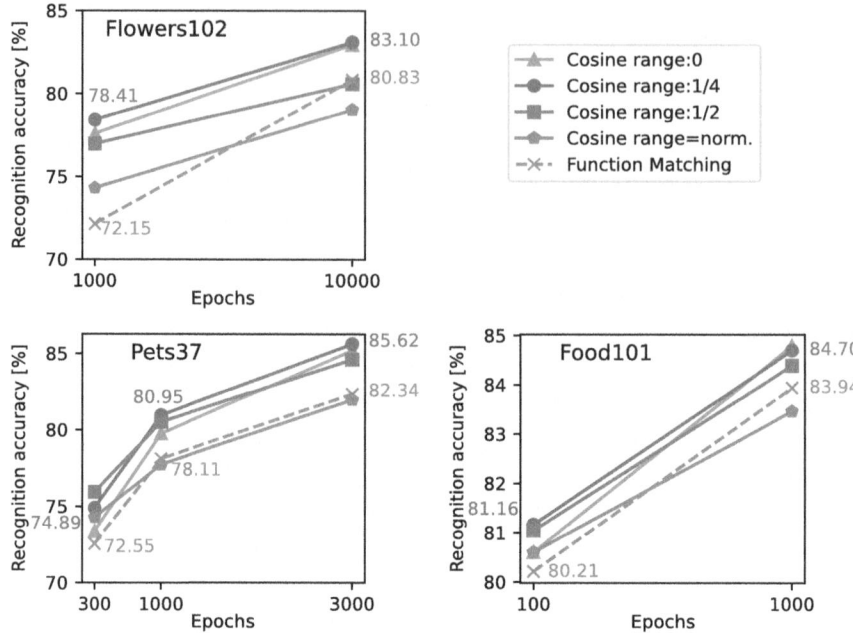

Fig. 6. Classification accuracies [%] using the combination of our proposals with the decrease using cosine curve and each sampling range. "Range: norm, 1/2, 1/4, and zero" express normal (default) sampling range, a half sampling range, a quarter range for sampling, and no range for sampling, respectively. Values with blue and gray color express results by ours (Range:1/4) and the conventional Funciton Matching, respectively. (Color figure online)

can be more effective, the other two kinds of reduction in the latter half can be more appropriate. This is because one step in the latter half is too simple. To summarize Table 3, we can say that keeping the initial upper bound for at least the first half of the training process seems to be needed for model generalization and decreasing it gradually to the end of the learning process seems appropriate.

Combination of Range Restriction and Difficulty Control. Finally, we examine the combined effect of the sampling range restriction and the control of the upper bound for mix ratios. We used the gradual decrease of cosine as upper bound control because this trend often gives better accuracies in Table 3. Results of combining the restriction of each sampling range and the cosine curve in upper bound decreases are figured in Fig. 6. Here, the number of epochs on the horizontal axis is in log scale. The vertical axis expresses the recognition accuracy. Conventional results were obtained by Function Matching. Specific values of Fig. 6 are shown in our supplementary material. Figure 6 shows that better generalization can be achieved by the combination of our proposals. We can also confirm that when the number of epochs is increased, almost the best results are obtained by narrower ranges including "zero". This tendency seems to be caused by the fact that the diversity of combinations of training data becomes more significant than that of mix ratios in the case with a greater number of epochs. In such cases, the effectiveness of higher output entropy becomes obvious. On the other hand, the decrease of the upper bound with the normal sampling range is effective only when the number of epochs is small. And this is a little worse than the conventional method when the number of epochs increases. The reason for such results seems that the increase of sampling low mix ratios is caused by the decrease of the upper bound with more epochs.

5   Conclusion

In this paper, we have proposed a method based on the relationship between the characteristics of training data and the learning progress for improving generalization performances of a student model on knowledge distillation using image mix data augmentation. On the basis of the finding that data with high output entropy is effective for knowledge transfer, our method aggressively and simply produces data with higher entropy of output by restricting the range of sampling for mix ratios of a sub-class in the data augmentation that mixes images between classes. We also have controlled the learning difficulty of training data according to learning progress. Specifically, the upper bound for mix ratios is lowered as the learning progresses, so that difficult data can be used in the early stages of the learning process and easy data toward the end of learning. Evaluation experiments using several datasets have provided that each of our methods has made some improvements, and the combination of them has achieved significantly higher rates than the conventional one. In particular, our results show that in the case with a large number of epochs, high entropy of training data is more important than the diversity of mix ratios.

Validating our concepts using other image mix data augmentation techniques such as CutMix is one of the future works. If we can access high-performance GPUs, experiments using larger datasets like ImageNet [7] should be conducted.

Acknowledgment. This work was supported by the ISM Cooperative Research Program (2024-ISMCRP-0007).

References

1. Bengio, Y., Louradour, J., Collobert, R., Weston, J.: Curriculum learning. In: ICML, pp. 41–48 (2009)
2. Beyer, L., Zhai, X., Royer, A., Markeeva, L., Anil, R., Kolesnikov, A.: Knowledge distillation a good teacher is patient and consistent. In: CVPR (2022)
3. Bucila, C., Caruana, R., Niculescu-Mizil, A.: Model compression. In: KDD, pp. 535—-541 (2006)
4. Cherti, M., et al.: Reproducible scaling laws for contrastive language-image learning. In: CVPR (2023)

5. Cho, J., Hariharan, B.: On the efficacy of knowledge distillation. In: ICCV, pp. 4794–4802 (2019)

6. Choi, H., Jeon, E.S., Shukla, A., Turaga, P.: Understanding the role of mixup in knowledge distillation: an empirical study. In: WACV, pp. 2319–2328 (2023)

7. Deng, J., Dong, W., Socher, R., Li, L.J., Li, K., Fei-Fei, L.: Imagenet: a large-scale hierarchical image database. In: CVPR, pp. 248–255 (2009)

8. Devries, T., Taylor, G.: Improved regularization of convolutional neural networks with cutout. CoRR (2017)

9. Gontijo-Lopez, R., Smullin, S., Cubuk, E., Dyer, E.: Tradeoffs in data augmentation: an empirical study. In: ICLR (2021)

10. He, K., Zhang, X., Ren, S., Sun, J.: Deep residual learning for image recognition. In: CVPR (2016)

11. Hinton, G., Vinyals, O., Dean, J.: Distilling the knowledge in a neural network. In: NIPS Workshop (2015)

12. Kaur, P., Sikka, K., Divakaran, A.: Combining weakly and webly supervised learning for classifying food images (2017). arXiv preprint arXiv:1712.08730

13. Kolesnikov, A., Beyer, L., Zhai, X., Puigcerver, J., J.Y., Gelly, S., Houlsby, N.: Big transfer (bit): general visual representation learning. In: ECCV (2020)

14. Li, Z., et al.: Curriculum temperature for knowledge distillation. In: AAAI (2023)

15. Loshchilov, I., Hutter, F.: Decoupled weight decay regularization. In: ICLR (2019)

16. Moslemi, A., Briskina, A., Dang, Z., Li, J.: A survey on knowledge distillation: recent advancements. Mach. Learn. Appl. **18**, 100605 (2024)

17. Mumuni, A., Mumuni, F.: Data augmentation: a comprehensive survey of modern approaches. Array **16**, 100258 (2022)

18. Naveed, H., Anwar, S., Hayat, M., Javed, K., Mian, A.: Survey: image mixing and deleting for data augmentation. Eng. Appl. Artif. Intell. **131**, 107791 (2024)

19. Nilsback, M.E., Zisserman, A.: Delving deeper into the whorl of flower segmentation. In: BMVC (2007)

20. Parkhi, O., Vedaldi, A., Zisserman, A., Jawahar, C.: Cats and dogs. In: CVPR (2012)

21. Romero, A., Ballas, N., Kahou, S., Chassang, A., Gatta, C., Bengio, Y.: Fitnets: hints for thin deep nets. In: ICLR (2015)

22. Shorten, C., Khoshgoftaar, T.: A survey on image data augmentation for deep learning. J. Big Data **6**(60) (2019)

23. Soviany, P., Ionescu, R., Rota, P., Sebe, N.: Curriculum learning: a survey. IJCV **130**, 1526–1564 (2022)

24. Tarvainen, A., Valpola, H.: Mean teachers are better role models: weight-averaged consistency targets improve semi-supervised deep learning results. In: NeurIPS (2017)

25. Walawalkar, D., Shen, Z., Liu, Z., Savvides, M.: Attentive cutmix: an enhanced data augmentation approach for deep learning based image classification. In: ICASSP (2022)

26. Wang, H., Lohit, S., Jones, M., Fu, Y.: What makes a "good" data augmentation in knowledge distillation – a statistical perspective. In: NeurIPS (2022)

27. Xie, Q., Luong, M.T., Hovy, E., Le, Q.: Self-training with noisy student improves imagenet classification. In: CVPR, pp. 10684–10695 (2020)

28. Yang, C., et al.: Mixskd: self-knowledge distillation from mixup for image recognition. In: ECCV, pp. 534–551 (2022)

29. Yun, S., Han, D., Chun, S., Oh, S., Yoo, Y., Choe, J.: Cutmix: regularization strategy to train strong classifiers with localizable features. In: ICCV (2019)

30. Zhang, H., Cisse, M., Dauphin, Y., Lopez-Paz, D.: mix-up: beyond empirical risk minimization. In: ICLR (2018)
31. Zhang, L., Song, J., Gao, A., Chen, J., Bao, C., Ma, K.: Be your own teacher: improve the performance of convolutional neural networks via self distillation. In: ICCV (2019)
32. Zhong, Z., Zheng, L., Kang, G., Li, S., Yang., Y.: Random erasing data augmentation. In: AAAI, pp. 13001–13008 (2020)

Multimodal and Vision-Language Models

Automatic Audio Description: A Training-Free Approach Using Foundation Models

Ruxandra Tapu[1,2](✉) [iD] and Bogdan Mocanu[1,2] [iD]

[1] SAMOVAR, Télécom SudParis, Institut Polytechnique de Paris, 91120 Palaiseau, France
{ruxandra.tapu,bogdan.mocanu}@upb.ro
[2] Department of Telecommunications, Faculty of ETTI, National University of Science and Technology "Politehnica", Bucharest, Romania

Abstract. In this paper, we propose a training-free framework for generating audio descriptions (ADs) by leveraging large pretrained Video-Language Models (VLMs) and Large Language Models (LLMs) without task-specific fine-tuning. Our method enhances video understanding through a semantic-constrained prompting strategy that incorporates temporally coherent context into VLM inputs, while an adaptive character recognition module ensures consistent identity tracking across frames. By explicitly linking visual character observations to narrative elements, the system produces contextually rich and coherent visual descriptions. Finally, the video captions are then refined into a single, concise audio description sentence through a LLM operating exclusively on text inputs, ensuring clarity, brevity, and narrative cohesion.

The experimental evaluation performed on the MAD-eval-Named and TV-AD benchmarks, validates the approach achieving CIDEr scores of 23.2 and 23.4, respectively. Compared to state-of-the-art training-free baselines, our framework consistently yields relative improvements ranging from 3.6% to 8% across multiple evaluation metrics.

Keywords: Audio description generation · video-language models · training-free methods · semantic-constrained prompting

1 Introduction

Ensuring equitable access to information for individuals with visual impairments has become a critical concern in modern media. In response, broadcasting regulations worldwide mandate the provision of *audio description* (AD) services. AD involves the insertion of narrated explanations of key visual elements, synchronized with the original audio-visual (A/V) content, to convey essential information that would otherwise be inaccessible to blind and visually impaired audiences [1]. The AD aims not only to improve accessibility but also to promote inclusivity and equal participation in cultural, educational, and entertainment experiences.

In contrast to the automated processes used for generating subtitles, the production of AD remains heavily dependent on manual workflows. This reliance results in substantial production costs, estimated at approximately $30 per minute of video content

M. Castrillón-Santana et al. (Eds.): CAIP 2025, LNCS 15622, pp. 173–183, 2026.
https://doi.org/10.1007/978-3-032-05060-1_15

[2]. Furthermore, creating AD for extended visual sequences, such as a 90-min film, can require between 35 and 40 h of work by trained professionals. These resource-intensive demands have contributed to a persistent gap in accessible media, as traditional methods struggle to keep up with the growing volume of audiovisual content [3]. Addressing this bottleneck requires the development of scalable and cost-effective solutions capable of delivering high-quality audio descriptions at scale.

The development of intelligent systems capable of automatically generating high-quality AD offers the potential to transform multimedia accessibility. This paper introduces a novel framework for the automatic generation of audio descriptions, designed to efficiently handle videos of extended duration, including those spanning several hours, without requiring human involvement.

The technical contributions of this work are summarized as follows:

(1) A novel framework for the automatic generation of AD, capable of directly processing continuous video streams to produce structured textual narratives. The architecture builds upon pretrained video-language models, which are leveraged to interpret complex visual scenes while maintaining both temporal and contextual coherence. The system prioritizes narrative fidelity by explicitly modeling character development and scene dynamics, leading to descriptions that are coherent, immersive, and enriched with narrative depth.
(2) We design a character-aware prompting strategy that explicitly grounds visual identity information into the video-language model through temporally consistent character representations. Unlike the prompting approach in [4], which operates on individual frames without persistent identity modeling, our framework integrates a tracking-based character recognition module that aggregates facial features over time using an adaptive temporal attention mechanism. This enables robust identity assignment across shots, even under challenging visual conditions such as occlusions and motion blur. The resulting temporally coherent descriptors are then linked to character names within structured text prompts, providing stable and consistent narrative generation across long-form sequences.
(3) We show that pretrained models can be effectively leveraged through advanced visual and textual prompting mechanisms. The modular design of our framework ensures seamless compatibility with emerging multimodal large language models, enabling straightforward integration and extension as new architectures are developed.

The paper is organized as follows: Sect. 2 reviews the state-of-the-art methodologies dedicated to AD. Section 3 introduces the proposed framework, detailing its architecture and operational pipeline. Section 4 describes the experimental protocol, including dataset specifications and evaluation metrics. While Sect. 5 concludes the paper and discusses prospective research directions.

2   Related Work

Traditional approaches often rely on supervised pipelines requiring annotated video-text pairs or domain-specific rules. Wang *et al.* [5] propose a dense video captioning framework based on a dual-network architecture, combining a sentence localizer and caption generator trained on ActivityNet Captions. Their approach requires task-specific training and annotated temporal segments, which may limit generalization across domains with sparse labeled data. Campos *et al.* [6] introduced CineAD, a rule-based system that generates AD scripts by analyzing screenplays and subtitle timing. The method does not perform any visual analysis and depends on access to subtitles and scripts. Since visually impaired users can access spoken dialogue through the audio track, subtitle-derived information may offer limited additional value for describing visual content.

The rapid advancement of large-scale Video-Language Models (VLMs) has significantly expanded the capabilities of multimodal learning systems, achieving remarkable results across a range of tasks. However, the high computational demands associated with fine-tuning these models have motivated the exploration of training-free adaptation paradigms. Recent strategies focus on enhancing VLMs through visual and textual prompting, external feature conditioning, and minimal optimization techniques, thereby preserving the generalization ability of the original pretrained models while enabling task-specific customization.

In the context of audiovisual description generation, MM-VID [7] proposes a two-stage pipeline that incorporates GPT-4V [8] to produce narrative outputs conditioned on visual inputs. Despite its effectiveness, MM-VID does not explicitly address the challenge of consistent character identification, as it lacks an integrated character recognition mechanism. This limitation reduces its ability to maintain coherent entity references over extended video sequences, particularly in scenarios involving multiple visually similar individuals.

Character-centric understanding is fundamental to effective movie interpretation, particularly in the context of audio description generation, where accurate and consistent entity recognition is critical for maintaining narrative coherence. Existing automated pipelines [9–11] have explored multimodal strategies, integrating facial, vocal, and bodily features to improve person identification across varying visual conditions. Although these methods have demonstrated success, they often rely heavily on static feature extraction and may struggle in dynamic scenes involving occlusions, rapid appearance changes, or character interactions. Complementary approaches [12–14] have utilized actor portraits as initialization priors for recognition tasks, subsequently refining identity predictions through auxiliary visual cues such as pose and clothing attributes. However, these frameworks typically lack mechanisms for maintaining temporal consistency across extended sequences, limiting their robustness in long-form video narratives.

Prior works such as MM-Narrator [14], LLM-AD [4], and AutoAD-Zero [15] utilize video-language models for automated audio description generation, yet they exhibit limitations in modeling temporal continuity and maintaining character-level consistency that are key requirements for producing coherent long-form narratives.

MM-Narrator [14] integrates GPT-4V(ision) [8] within a multimodal in-context learning framework to enable progressive AD generation. While it incorporates memory mechanisms that support high-level contextual flow, it lacks explicit modules for fine-grained character tracking, reducing the specificity and coherence of descriptions in character-centric scenes. LLM-AD [4] adopts a visual prompting strategy by annotating movie frames with character names, guiding GPT-4V(ision) [8] to generate more semantically aligned audio descriptions. However, its frame-level operation does not account for temporal dependencies or character identity consistency over time, resulting in discontinuities in extended video sequences. AutoAD-Zero [15] presents a zero-shot framework that leverages pretrained VLMs without task-specific fine-tuning, offering scalability and efficiency. Nonetheless, the absence of temporal modeling or entity tracking leads to inconsistencies in referencing characters across all video scenes.

Our method addresses these limitations by incorporating explicit character-grounded visual prompting, a tracking-based character recognition module, and an adaptive temporal attention mechanism to ensure temporally coherent and identity-consistent AD generation across long video sequences.

3 Proposed Approach

Our proposed framework (Fig. 1) is structured around three interconnected modules that collectively enable automated video comprehension and audio description generation.

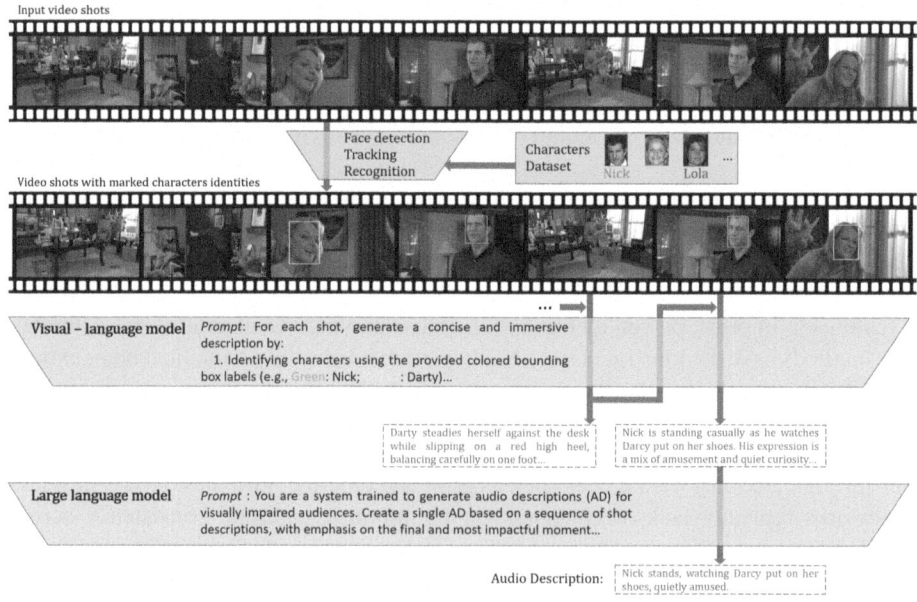

Fig. 1. The proposed system architecture with the main stages involved.

First, a character recognition module is responsible for maintaining consistent character identity tracking across video frames. Second, a video-language prompting mechanism incorporates character-specific information along with short-term temporal memory to enhance the contextual understanding and interactions across adjacent shots. Finally, a LLM-based summarization component synthesizes the extracted information into coherent and narratively rich audio descriptions.

3.1 Character Recognition

Starting from a given video shot S_t extracted using a shot boundary detection method [16], the system performs *face detection* and *feature extraction* across the corresponding set of frames $\{I_1, I_2, \ldots, I_N\}$. Face candidates are localized using RetinaFace [17], followed by the extraction of deep feature embeddings via ArcFace [18]. The resulting set of face detections and associated features is represented as:

$$\Psi(I) = \left\{ \left(\left(x_{ul}^i, y_{ul}^i \right), \left(x_{lr}^i, y_{lr}^i \right), e_i^{face} \right) \right\}_{i=1}^M, \tag{1}$$

where $\left(x_{ul}^i, y_{ul}^i \right)$, and $\left(x_{lr}^i, y_{lr}^i \right)$ define the bounding box coordinates for the ith detected face, $e_i^{face} \in R^D$ is the corresponding feature vector, and M denotes the total number of face instances identified in frame I.

The *face tracking* component receives bounding box outputs from the detection stage and establishes temporal associations between face instances across frames. By linking detections over time, the system promotes identity persistence throughout the video sequence, mitigating fragmentation due to intermittent detection errors and enhancing overall tracking robustness. We extend the ATLAS algorithm [19] to support robust multi-instance face tracking under complex motion patterns and rapidly changing scene contexts. Our adaptation explicitly addresses detection gaps induced by occlusions, motion blur, and variable illumination, maintaining stable identity trajectories over extended temporal windows despite challenging visual conditions. Face trajectories with low confidence scores or insufficient temporal duration are systematically pruned to retain only stable and relevant character representations.

The reliability of *person re-identification* (Re-ID) in video sequences is often compromised by inconsistencies in the discriminative quality of face detections, caused by factors such as occlusions, motion blur, extreme head poses, and resolution degradation. Treating all instances equally amplifies feature noise and weakens identity association across frames. To address this, we introduce a temporal attention mechanism that prioritizes high-quality face representations, improving the robustness and consistency of re-identification over time.

Given a face track with its corresponding embeddings $E = \{e_1, e_2, \ldots, e_L\}$ across L frames, the features are reweighed via a temporal attention module (Fig. 2) based on their discriminative relevance. To ensure uniform input size, each track is temporally sampled to a fixed length $L' = 128$ elements.

The temporal attention mechanism builds upon the scaled dot-product attention formulation, as defined in Eq. (3). Query Q, key K and value V matrices are all in $\mathbb{R}^{L' \times D}$ and computed via independent learned projections applied to the embeddings:

$$Q = E \cdot W_Q; K = E \cdot W_K; V_{VS} = E \cdot W_V; \tag{2}$$

where $W_Q, W_K, W_V \in \mathbb{R}^{D \times D}$ are trainable weight matrices.

The attention output is given by:

$$Attention(Q, K, V) = softmax\left(\frac{Q \cdot K^T}{\sqrt{D}} \right) \cdot V = A \cdot V, \tag{3}$$

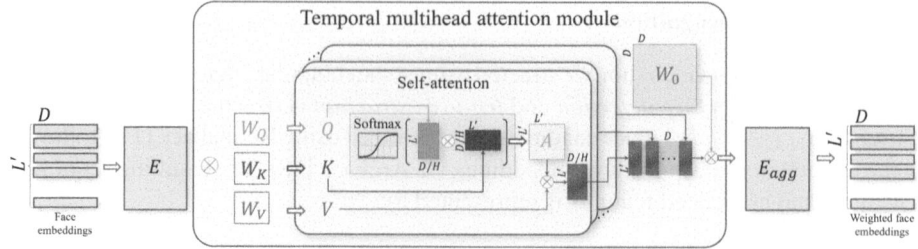

Fig. 2. Adaptive facial feature weighting using temporal attention.

with $A \in \mathbb{R}^{L' \times L'}$ represents the attention scores across temporal instances.

To further enhance representational capacity, the attention computation is performed independently across H heads, each operating on distinct learned subspaces of the input features. The resulting outputs are concatenated and linearly projected through an additional learned matrix $W_O \in \mathbb{R}^{D \times D}$, yielding a relevance vector:

$$E_{agg} = Concat(z_1, \ldots, z_H) \cdot W_O, \tag{4}$$

here, $z_H \in \mathbb{R}^{L' \times d}$ denotes the output of the hth attention head, and d is the dimensionality of each head output ($d = D/H$) and $E_{agg} \in \mathbb{R}^{L' \times D}$ represents the weighted track descriptors associated to the face instances, which can be interpreted as emphasizing temporally discriminative face observations while attenuating less informative instances. Finally, the global face descriptor E_{track} for the entire face track is obtained as the weighted mean across the rows of E_{agg}.

To enable character identity assignment, we construct a character bank Ch_{bank} for each video $\mathcal{V}$ based on cast list information retrieved from an external database $\mathcal{P}$. The character bank is defined as: $Ch_{bank} = \left\{ \left(name_j^{char}, name_j^{actor}, p_j \right) \right\}_{j=1}^{C}$, where C represents the number of characters associated with the video stream, $name_j^{char}$ is the character name, $name_j^{actor}$ is the actor's name, and $p_j \in \mathbb{R}^{W \times H \times 3}$ represents the corresponding profile image. For each profile image p_j, a face embedding $e_j^{prof} \in \mathbb{R}^D$ is extracted using the face encoder (*i.e.*, ArcFace [18]).

Given a track-level descriptor $E_{track} \in \mathbb{R}^D$ obtained via temporal attention-based aggregation of face embeddings across frames, we perform identity assignment by computing cosine similarities between E_{track} and each profile embedding e_j^{prof}. The character identity yielding the highest similarity is selected for assignment.

3.2 Semantically Constrained Prompting for Video-Language Models

Recent advances in VLMs have enabled the generation of semantically rich and temporally coherent descriptions by jointly modeling visual and textual inputs. This makes them particularly suitable for audio description, which requires accurate interpretation of visual scenes, character interactions, and evolving narrative context. However, in complex multi-character settings, unguided VLMs may still produce vague or inconsistent outputs due to a lack of temporal grounding and character disambiguation.

To address these limitations, we propose a semantically constrained prompting strategy aimed at enhancing the quality, coherence, and character specificity of VLM-generated audio descriptions. Rather than relying solely on visual features from isolated frames, our method enriches the input prompts with structured context obtained using a temporal sliding window of three shots. Specifically, we combine information extracted from the current video shot (*i.e.,* background information, character appearances, and object interactions) with semantic predictions accumulated from preceding shots. This integration provides the VLM with temporally coherent knowledge, enabling it to disambiguate entities, maintain character consistency, and adapt its narrative generation to the scene dynamics.

To establish explicit character association during VLM prompting, we assign each detected character a distinct color-coded identifier, systematically embedded within the text prompts (*e.g.,* "Nick" marked with a green bounding box). Character recognition is achieved via our face identification module, ensuring temporally consistent identity association across frames through robust feature aggregation and matching. For each detected video shot, eight representative frames are extracted to capture diverse scene contexts and character poses. These frames are processed by the VLM, conditioned to generate detailed descriptions emphasizing character-specific actions and interactions.

To enable contextual reasoning across consecutive shots, we augment the input to the VLM by appending semantic information extracted from the previous three shots. Specifically, the textual prediction generated for the preceding shots are incorporated into the current prompt, providing temporal continuity and preserving narrative coherence. We employ GPT-4V(ision) [8] as the VLM to produce textual representations.

3.3 LLM-Driven Summarization for Audio Description Generation

In the third stage of the pipeline, the system refines the initially generated video descriptions into a single, concise audio description (AD) sentence through a Large Language Model (LLM) operating exclusively on text inputs. Prior to summarization, character identifiers are replaced with the corresponding character names to ensure linguistic naturalness and narrative coherence. The LLM is prompted to focus on the most salient actions, emphasizing events that span within the temporal analysis window while omitting static visual details and dialogue-related interactions, which are typically inferable by the user from the audio track. Additionally, the prompt specifies the temporal duration of the target AD segment, enabling the LLM to adjust the granularity and verbosity of the summarization accordingly. This strategy ensures that the final AD output remains contextually rich, temporally aligned, and adapted to the dynamic pacing of the video content. We leverage LLaMa3.2-3B-Instruct [20] to refine audio descriptions, improving linguistic coherence, contextual relevance, and narrative consistency.

4   Experimental Evaluation

This section evaluates the proposed system highlighting its performance relative to existing methods and demonstrating the benefits of a training-free audio description approach.

Datasets: We evaluate the proposed audio description generation framework on two benchmark datasets [15]: MAD-eval-Named and TV-AD. MAD-eval-Named consists of 6,520 audio description sentences and 10,602 subtitle segments collected from 10 feature-length films, offering a variety of narrative and visual contexts. TV-AD includes approximately 3,000 audio description segments sourced from 100 television episodes, providing a broader range of episodic structures and character interactions.

Evaluation Metrics: We evaluate the quality of the generated audio descriptions using three complementary metrics, each targeting a distinct dimension of performance [15]. The CIDEr (Consensus-based Image Description Evaluation) score quantifies content similarity between generated and reference descriptions by computing TF–IDF–weighted n-gram overlap. This emphasizes not just surface-level word matching, but the salience of information based on how frequently it appears across human references. The CRITIC metric provides a more semantic evaluation by measuring factual correctness and contextual alignment. It rewards descriptions that accurately reflect visual content, penalizing hallucinations or omissions. This metric is particularly useful for identifying whether the generated sentences preserve the essential visual semantics beyond lexical similarity. In addition, we use LLM-AD-Eval, an evaluation framework based on large language models (LLMs), specifically adapted to assess the quality of audio descriptions. It evaluates three key properties: (1) narrative coherence, which assesses whether the description forms a logically consistent account of events; (2) entity consistency, which measures whether characters and objects are referred to consistently across segments; and (3) overall descriptive quality, which captures fluency, informativeness, and redundancy.

Experimental Results: We evaluate the proposed framework across two benchmarks to comprehensively assess its ability to generate high-quality audio descriptions across diverse audiovisual domains. For broad evaluation, we employ the TV-AD dataset, which covers a wide range of episodic television content characterized by varied narrative structures, character interactions, and visual styles. To ensure direct comparability with existing training-free methods, we also benchmark performance on the MAD-eval-Named dataset, where CIDEr scores are widely reported. All baseline methods (LLM-AD [4], MM-Narrator [14], and AutoAD-Zero [15]) are evaluated under the same protocol, using the official dataset splits, standardized shot segmentation, and metric computation procedures as established in AutoAD-Zero [15].

As shown in Table 1, our approach achieves state-of-the-art performance among training-free methods, surpassing AutoAD-Zero [15], MM-Narrator [14] (both GPT-4 and GPT-4V variants), and LLM-AD [4] on MAD-eval-Named, thereby validating the effectiveness of our semantic-constrained prompting and adaptive character reasoning modules. Table 1 demonstrates the effectiveness of our proposed framework in improving training-free audio description performance. Compared to the strongest baseline, AutoAD-Zero, our method achieves consistent improvements across all metrics, with relative gains ranging from approximately 3.6% on CIDEr (MAD-eval: 23.2 vs. 22.4) to 8.0% on CRITIC (TV-AD: 29.8 vs. 27.6). The performance gains are expressed as relative differences with respect to the baseline metrics. These improvements are primarily attributed to two key innovations introduced in our framework: the semantic-constrained prompting strategy, which provides temporally coherent and entity-aware context to the

Video-Language Model, and the adaptive character recognition module, which ensures consistent identity association across frames. By explicitly modeling scene dynamics, our approach mitigates common issues such as fragmented narrative flow and inconsistent character grounding observed in prior methods, resulting in more accurate, coherent, and natural audio descriptions.

Table 1. Performance comparison between the proposed framework and state-of-the-art training-free methods on MAD-eval-Named and TV-AD datasets.

Method	MAD-eval	TV-AD		
	CIDEr	CIDEr	CRITIC	LLM-AD-eval
LLM-AD [4]	20.5	–	–	–
MM-Narator [14]	13.9	–	–	–
AutoAd-Zero [15]	22.4	22.6	27.6	2.94
Ours	**23.2**	**23.4**	**29.8**	**2.98**

To better understand the contribution of each module, we conducted an ablation study by selectively disabling key components of our character naming pipeline. We evaluated three configurations: (i) without any character information, (ii) with per-frame character recognition without temporal integration, and (iii) with our proposed character recognition module, which leverages a global facial descriptor and temporal attention across frames. All models use the same underlying VLM and LLM modules. Table 2 reports CIDEr and CRITIC scores on both the MAD-eval-Named and TV-AD datasets.

Table 2. Ablation study of the various modules involved

Experiment		MAD-eval	TV-AD	
Character information	With character tracking	CIDEr	CIDEr	CRITIC
–	–	17.8	18.1	24.5
X	–	21.9	22.8	29.3
X	X	**23.2**	**23.4**	**29.8**

The ablation results in Table 2 confirm the importance of both components for overall performance. The baseline, which lacks character naming, yields the lowest CIDEr and CRITIC scores, underscoring the limitations of generic prompting. Incorporating only per-frame character recognition improves fluency but lacks temporal grounding. The full system achieves the highest scores, with the character recognition module enabling identity consistency and referentially coherent descriptions across shots.

While our framework avoids training and fine-tuning, inference with foundation models incurs computational costs. The end-to-end processing pipeline has an average cost of approximately $0.05–$0.09 per minute of video, contingent on parameters such as frame sampling rate (typically 8 frames per shot), prompt length, and video complexity. Based on current API pricing [21], this remains substantially lower than the manual AD production cost of ~ \$30 per minute [2], representing an estimated 300 × cost reduction and enabling scalable deployment.

In addition to monetary cost, the computational latency of large-scale VLMs such as GPT-4V(ision) poses practical challenges for large-scale deployment. To constrain inference time, all processing in our framework is performed offline, enabling us to manage latency without compromising system responsiveness during runtime. Moreover, we adopted a fixed sampling strategy (8 frames per shot) to balance coverage and efficiency. The architecture is inherently modular and model-agnostic, allowing straightforward substitution with more efficient VLMs (*e.g.,* LLaVA, MiniGPT-4) to reduce latency with minimal impact on system design.

5   Conclusions and Perspectives

In this paper, we have introduced a training-free framework for automatic audio description (AD) generation, leveraging large pretrained Video-Language Models (VLMs) and Large Language Models (LLMs) without the need for task-specific fine-tuning operations. Our system incorporates a semantic-constrained prompting strategy, enriching VLM inputs with temporally coherent context, and an adaptive character recognition module that ensures consistent identity association across frames. Extensive experiments on the MAD-eval-Named and TV-AD datasets confirm the effectiveness of the proposed approach, with our model achieving CIDEr scores of 23.2 and 23.4, respectively. Across all evaluation metrics, the system consistently outperforms state-of-the-art training-free methods, achieving relative improvements between 3.6% and 8%. Future directions include extending the framework to egocentric video sequences, integrating multimodal memory for persistent context modeling, and adapting the approach for real-time assistive video understanding applications.

Acknowledgments. This work was supported by a grant of the Ministry of Research, Innovation and Digitization, CCCDI - UEFISCDI, project number PN-IV-P6–6.3-SOL-2024–2-0238 and PN-IV-P6–6.3-SOL-2024–0049, within PNCDI IV.

Disclosure of Interests.   The authors have no competing interests to declare that are relevant to the content of this article.

References

1. Snyder, J.: The Visual Made Verbal: A Comprehensive Training Manual and Guide to the History and Applications of Audio Description. American Council of the Blind (2014)
2. Reviers, N.: Audio description services in Europe: an update. JoSTrans **26**, 232–247 (2016)
3. Fresco, P.R., Fryer, L.: Could audio-described films benefit from audio introductions? J. Vis. Impair. Blind. **107**(4), 287–295 (2013)
4. Chu, P., Wang, J., Abrantes, A.: LLM-AD: large language model based audio description system. arXiv:2405.00983 (2024)

5. Wang, Y., Liang, W., Huang, H., et al.: Toward automatic audio description generation for accessible videos. In: Proceedings of the CHI, pp. 1–12 (2021)
6. Campos, V.P., de Araújo, T.M.U., de Souza Filho, G.L., et al.: CineAD: a system for automated audio description script generation. In: UAIS, vol. 19, pp. 99–111 (2020)
7. Lin, K., Ahmed, F., Li, L., et al.: MM-VID: advancing video understanding with GPT-4V(ision). arXiv:2310.19773 (2023)
8. OpenAI: GPT-4 Technical Report. OpenAI (2023)
9. Brown, A., Kalogeiton, V., Zisserman, A.: Face, body, voice: person-clustering with multiple modalities. In: ICCV Workshops, pp. 1–8 (2021)
10. Tapaswi, M., Bäuml, M., Stiefelhagen, R.: Knock! Knock! Who is it? Person identification in TV series. In: CVPR, pp. 3306–3313 (2012)
11. Tapaswi, M., Law, M.T., Fidler, S.: Video face clustering with an unknown number of clusters. In: ICCV, pp. 5027–5036 (2019)
12. Huang, Q., et al.: Person search in videos with one portrait. In: ECCV, pp. 425–441 (2018)
13. Han, T., Bain, M., Nagrani, A., et al.: AutoAD II: who, when, and what in movie audio description. In: ICCV (2023)
14. Zhang, C., Lin, K., Yang, Z., et al.: MM-narrator: narrating long-form videos with multimodal in-context learning. arXiv:2311.17435 (2023)
15. Xie, J., Han, T., Bain, M., et al.: AutoAD-Zero: a training-free framework for zero-shot audio description. arXiv:2407.15850 (2024)
16. Țapu, R., Mocanu, B., Petrescu, T.: A complete framework for video temporal segmentation,.U.P.B. Sci. Bull., Series C **73**(4), 1–12 (2011). ISSN: 1454–234X
17. Deng, J., Guo, J., Ververas, E., et al.: RetinaFace: single-shot face localisation in the wild. In: CVPR, pp. 5203–5212 (2020)
18. Deng, J., Guo, J., Xue, N., Zafeiriou, S.: ArcFace: additive angular margin loss for deep face recognition. In: CVPR, pp. 4690–4699 (2019)
19. Mocanu, B., Tapu, R., Zaharia, T.: Single object tracking using offline deep regression networks. In: IPTA, pp. 1–6 (2017)
20. Meta AI: LLaMA 3.2: revolutionizing edge AI and vision with open-source models. https://ai.meta.com/blog/llama-3-2-connect-2024-vision-edge-mobile-devices/
21. OpenAI: GPT-4 Pricing. https://openai.com/pricing Accessed June 2025

AMEST: Adaptive Morphologically Enhanced Subword Tokenization for Improved Language Model Perplexity

Labib Asari, Dhruvanshu Joshi, Soham Mulye, Viraj Shah,
Sandeep S. Udmale[✉][iD], and Girish P. Bhole

Department of Computer Engineering and Information Technology, Veermata Jijabai
Technological Institute (VJTI), Mumbai, India
{laasari_b21,dhjoshi_b21,ssmulye_b21,gpbhole}@ce.vjti.ac.in,
{vbshah_b21,ssudmale}@it.vjti.ac.in

Abstract. Tokenization inherently affects the way information is processed within Large Language Models (LLMs), with dramatic implications for performance. Common strategies such as Byte Pair Encoding (BPE) used in the likes of GPT-2 tend to output linguistically inconsequential segments simply because they operate statistically, posing a risk of impeding predictability of models (quantified by perplexity). This work explores tokenization techniques to get lower perplexity. We present AMEST, a new hybrid tokenization that leverages both linguistic insight (data-driven morphological segmentation) and statistical soundness (probabilistic sampling of Unigram), with primary emphasis on favouring morphemes for familiar terms and sampling to handle unknown ones. We measured by training models of GPT-2 Small from scratch across eight varied sets of data to demonstrate that AMEST reliably registers significant perplexity decreases, improvements in convergence and competitive throughput vs. GPT-2 BPE and Baseline Unigram tokenizer. Our analysis, including ablation studies and cross-domain evaluations, confirms that the synergy between morphological awareness and probabilistic fallback is key to these improvements.

Keywords: Language Modeling · Morphology · Perplexity ·
SentencePiece · Subword Segmentation · Tokenization · Unigram

1 Introduction

The advent of Large Language Models (LLMs), like GPT-3 [3], has transformed natural language processing. At the center of such models, the critical task of tokenization converts raw text into a sequence of discrete tokens upon which the model operates. The method used for doing that has a profound influence on the model learning linguistic patterns, contextual relationships, and eventually the model's predictive power. A crucial measure that reflects such predictive power is perplexity (PPL) [5], a measure of model uncertainty in predicting the next token. Lower perplexity indicates higher agreement between the model's prediction and

M. Castrillón-Santana et al. (Eds.): CAIP 2025, LNCS 15622, pp. 184–194, 2026.
https://doi.org/10.1007/978-3-032-05060-1_16

the underlying data distribution and, in most instances, the model's better performance across different downstream tasks [3].

The most powerful LLMs, including the widely studied GPT-2 [11], employed Byte Pair Encoding (BPE) [12] or its variants like Byte-Level BPE. BPE constructs a vocabulary iteratively by combining the frequent paired adjacent bytes or characters. Though effective in managing vocabulary size and handling any input string, this solely statistical, greedy approach at times may produce segmentations without linguistic sense. As per Boström and Partanen [1], BPE may split meaningful word parts or join across morpheme boundaries indiscriminately and, in turn, complicate the language modeling task and generate more perplexity.

Noticing these constraints, alternative approaches such as Unigram language model tokenization, popularized by SentencePiece [7], were created. The Unigram method [6] offers a probabilistic model, beginning with a large set of candidates and removing tokens based on corpus likelihood. Furthermore, approaches such as subword regularization [6] introduce randomness by sampling over a set of possible segmentations at training time, promoting model robustness. While frequently yielding improvements over BPE, these approaches remain largely reliant on statistical co-occurrence and not necessarily linguistic knowledge of word construction.

This gap encourages the pursuit of incorporating morphology - the examination of word structure - into tokenization. Words tend to be made up of semantically meaningful subunits (morphemes: prefixes, stems, suffixes). Segmenting words at these boundaries (e.g., "unpredictable" → "un-", "predict", "-able") may give the LLM more consistent, semantically grounded units. More consistency may make the prediction task easier, resulting in reduced perplexity and perhaps more stable model behavior.

In this paper, we introduce the **Adaptive Morphologically Enhanced Subword Tokenizer (AMEST)**. AMEST is envisioned as an operational hybrid system which takes advantage of the strengths of both linguistic structure and statistical reliability:

1. The approach employs a morphological lexicon based on unsupervised analysis, which employs techniques such as Morfessor [4], to decompose recognized words into their morpheme constituents.
2. For out-of-vocabulary words or other cases not found in the lexicon, the system resorts to the strong probabilistic sampling mechanism of SentencePiece Unigram that includes subword regularization [6].

The goal of the strategy is to maximize linguistic coherence where possible, while at the same time achieving full coverage and flexibility. We systematically compare with both the baseline GPT-2 BPE tokenizer and a typical deterministic Unigram tokenizer. By training similar GPT-2 Small [11] models from scratch on eight diverse datasets with vocabularies drawn from WikiText-2 [10], we measure the impact on validation perplexity, convergence rate, and evaluation throughput. Our main contributions are: (1) Design of AMEST, a new hybrid morphologically-aware tokenizer. (2) Experimental results over eight datasets proving AMEST leads to a highly significant reduction in perplexity. (3) A comprehensive analysis including ablation and cross-domain studies, proving AMEST achieves this with successful convergence and comparable throughput.

2   Background and Related Work

Tokenization fills the gap between raw text and neural network inputs, substituting for the previous word-based approaches [9] that were plagued by huge vocabularies and out-of-vocabulary (OOV) words. Subword tokenization was the paradigm.

Byte Pair Encoding (BPE) [12], having originated in data compression, builds a lexicon by repeatedly merging high-frequency adjacent character or byte pairs. Its byte-pair variant [11], used in GPT-2, operates on any input but its greedy, frequency-driven merging readily overlooks grammatical structure at the cost of creating less than optimal units [1].

Unigram LM Tokenization [6], used in tools such as SentencePiece [7], provides a probabilistic alternative. It trains a subword vocabulary to optimize the likelihood of generating the corpus under the unigram assumption. One of its most notable features is that it can generate alternative segmentations of a piece of text. **Subword Regularization** [6] takes advantage of this by sampling alternative segmentations during training, a type of data augmentation and often leading to more robust and performing than deterministic BPE. Even with this, the segmentations are still mostly statistical patterns and not explicit linguistic knowledge.

Morphologically-Aware Tokenization attempts to merge the internal structure of words. Morphemes are the smallest units of meaning (e.g., "un-", "predict", "-able"). Unsupervised approaches such as Morfessor [4] use principles such as Minimum Description Length (MDL) to learn morpheme-like segmentations from corpora, aiming to produce linguistically plausible units. These segmentations have been shown to be successful on a large variety of NLP tasks [2]. Sole use of morphology can, however, be vulnerable to out-of-vocabulary (OOV) words, noise, or sophisticated linguistic phenomena not covered by the unsupervised analyzer. AMEST is characterized as an effective amalgamation, with the objective of leveraging linguistic knowledge derived from unsupervised morphology [4], while simultaneously preserving the comprehensive applicability and resilience inherent in probabilistic subword sampling techniques [6].

3   The AMEST Approach

We introduce AMEST (Adaptive Morphologically Enhanced Subword Tokenizer) to create more robust input representations for LLMs through the union of linguistic knowledge and statistical subword modeling.

3.1   Core Principles and Motivation

The underlying reason is that standard statistical tokenizers are likely to disregard the internal morphology of words, perhaps dividing words such as

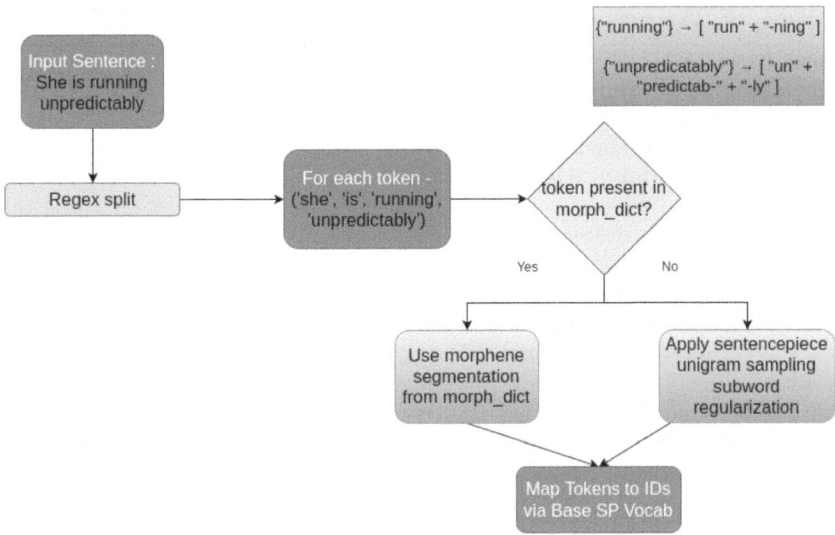

Fig. 1. The AMEST tokenization workflow. Input words are checked against the morphological lexicon (M). If found, morphemes are used; otherwise, SentencePiece Unigram sampling (SP) provides subword tokens.

"unhappily" into less semantically meaningful pieces such as ["unhapp", "ily"]. A morphologically-sensitive strategy looking for ["un-", "happy", "-ly"] may provide tokens with more semantic and structural coherence. We suspect that this coherence facilitates next-token prediction for LLMs, leading to lower perplexity – an important measure of model confidence and predictability. Pure morphological segmentation, however, can go wrong on out-of-vocabulary words or tricky cases. AMEST avoids this by taking a hybrid strategy: use morphology when we can rely on the information, but fall back to a strong statistical backup otherwise.

3.2 AMEST Tokenization Process

AMEST is built from two main components derived from a representative training set (Refer Fig. 1), e.g., WikiText-2:

- **Morphological Lexicon (M):** A lexical resource mapping surface words to their underlying morphemes, learned with an unsupervised morphological analyzer tool called Morfessor [4].
- **Base Statistical Tokenizer (SP):** A SentencePiece Unigram model [7] trained on a corresponding corpus. This model is the base vocabulary and the fallback segmentation strategy that can generate multiple probabilistic segmentations for any word using subword regularization [6].

The tokenization procedure is detailed in Algorithm 1. It processes input text word by word. For each word, it first queries the morphological lexicon M. If the

Algorithm 1. AMEST Tokenization Process

Require: Input text T
Require: Morphological dictionary M (Word → List of Morphemes)
Require: Pre-trained SentencePiece Unigram model SP (with sampling capability)
Require: Sampling parameters: $nbest\_size$, α
Ensure: Sequence of token IDs $ID\_Sequence$
 1: $OutputTokens \leftarrow \emptyset$ ▷ Initialize list for string tokens
 2: $Words \leftarrow$ WhitespaceSplit(T)   ▷ Or other appropriate word segmentation
 3: **for** each $word$ in $Words$ **do**
 4:     **if** $word$ is in M **then**   ▷ Check morphological dictionary
 5:         $morphemes \leftarrow M[word]$
 6:         $OutputTokens \leftarrow OutputTokens \cup morphemes$ ▷ Append morphemes
 7:     **else**   ▷ Word not in dictionary, use fallback
 8:         $subwords \leftarrow$ SP.SampleEncodeAsPieces($word, nbest\_size, \alpha$)
 9:         $OutputTokens \leftarrow OutputTokens \cup subwords$ ▷ Append subwords
10:     **end if**
11: **end for**
12: $ID\_Sequence \leftarrow \emptyset$ ▷ Initialize list for token IDs
13: **for** each $token$ in $OutputTokens$ **do**
14:     $id \leftarrow$ SP.PieceToID($token$)   ▷ Convert token string to ID
15:     $ID\_Sequence \leftarrow ID\_Sequence \cup \{id\}$ ▷ Append ID
16: **end for**
17: **return** $ID\_Sequence$

---

word exists, its pre-computed morpheme list is used. If the word is not found, the base SP model's 'SampleEncodeAsPieces' function is called to produce a likely subword segmentation probabilistically. This hybrid framework generates a token stream which takes advantage of linguistic structure wherever possible, and otherwise resorts to a strong, regularized statistical method.

4   Experimental Setup

We conducted experiments to rigorously evaluate AMEST's impact on language model perplexity and efficiency compared to standard baselines.

4.1   Tokenizers Compared

Three tokenizers were evaluated:

1. **Original GPT-2 (BPE):** The standard Byte-Level BPE tokenizer from Hugging Face's 'gpt2' implementation (50,257 vocab size) [11].
2. **Base Unigram:** A SentencePiece [7] Unigram model trained on WikiText-2 using deterministic encoding (32,000 vocab size).
3. **AMEST (Ours):** Uses a morphological lexicon (from Morfessor [4] trained on WikiText-2 vocabulary) combined with the same Base Unigram model as the fallback mechanism, but utilizing sample-based encoding ('nbest_size=64', 'alpha=0.1') [6]. (32,000 vocab size).

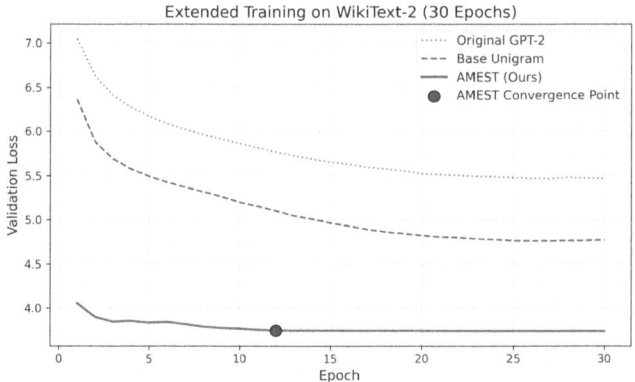

Fig. 2. Extended training on WikiText-2 to 30 epochs. The plot highlights AMEST's rapid convergence, reaching its optimal performance range by epoch 12 and subsequently plateauing. In contrast, the baseline models exhibit a much slower, more prolonged convergence. This demonstrates the superior training efficiency of the AMEST approach.

Table 1. Validation Perplexity (PPL) after 3 epochs. Lower is better. AMEST consistently achieves the lowest values.

Dataset	Original GPT-2 (BPE, 50k)	Base Unigram (Unigram, 32k)	AMEST (Ours) (AMEST, 32k)
WikiText-2	716.33	340.96	**46.93**
Penn Treebank	447.26	235.47	**44.53**
Tiny Shakespeare	2804.91	2959.84	**886.74**
TweetEval	1075.59	289.46	**79.23**
Rotten Tomatoes	8685.00	3464.61	**512.61**
AG News	392.55	132.91	**35.54**
DailyDialog	141.63	118.56	**29.32**
GoEmotions	1733.73	439.21	**108.29**

4.2   Language Model and Training

To account for the tokenizer, we trained individual GPT-2 Small models [11] (12 layers, 12 heads, 768 embedding dim) *from scratch* on each tokenizer-dataset pair. Each model was initialized with a 'GPT2Config' matching the tokenizer's 'vocab_size'. We chose 3 epochs as it proved sufficient to establish a clear and consistent performance hierarchy. While extended training shows continued gradual improvement for the baselines, AMEST achieves the vast majority of its performance gains within the first few epochs and quickly settles into a performance plateau, as shown in Fig. 2. This demonstrates a much more efficient convergence profile. Therefore, the 3-epoch mark serves as a practical and decisive point for

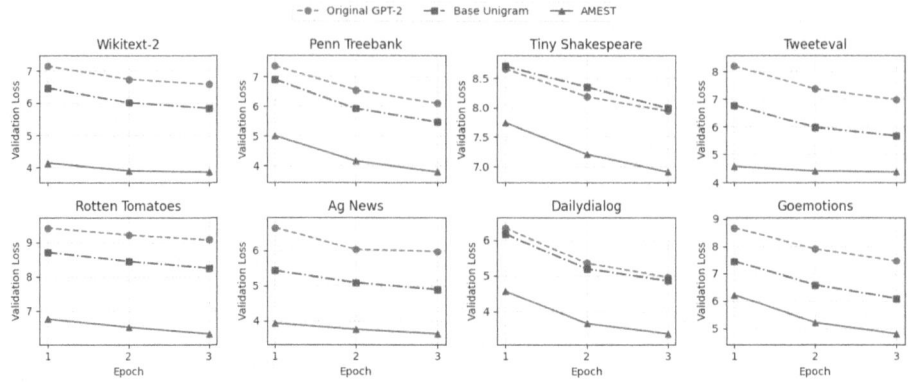

Fig. 3. Validation loss curves over 3 training epochs. AMEST models consistently reach lower validation loss.

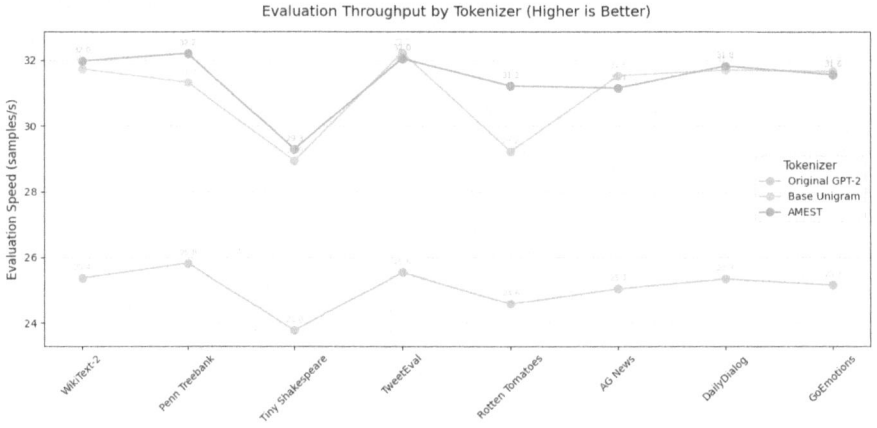

Fig. 4. Evaluation throughput (samples per second). Higher values indicate faster processing.

comparison, capturing the critical performance difference between the methods. Key hyperparameters were AdamW optimizer, learning rate 5e-5, effective batch size 32, and FP16 precision.

4.3   Datasets

The AMEST and Base Unigram custom tokenizers and Morfessor model were trained solely on the WikiText-2 training corpus [10]. The language models were then trained and tested separately on eight varied benchmark datasets from Hugging Face Datasets [8]: WikiText-2, Penn Treebank (PTB), Tiny Shakespeare, TweetEval Sentiment, Rotten Tomatoes, AG News, DailyDialog, and GoEmotions. Standard validation splits were used for evaluation.

Table 2. Average tokens per word by tokenizer.

Dataset	AMEST	Base Unigram	Original GPT-2
wikitext	1.557	1.314	1.156
ptb_text_only	1.540	1.374	1.228
tiny_shakespeare	2.003	1.720	1.775
tweet_eval	1.967	1.744	1.398

Table 3. AMEST fallback rate to Unigram sampling.

Dataset	Fallback Count	Fallback %
wikitext	681	0.04%
ptb_text_only	59,806	6.72%
tiny_shakespeare	9,531	5.08%
tweet_eval	54,360	6.30%

4.4 Evaluation Metrics

Our primary metric is end validation Perplexity (PPL). We also analyze validation loss curves by epoch to track convergence and evaluation throughput (samples/second) to assess processing efficiency.

5 Results and Discussion

This section presents the empirical results comparing AMEST against the baseline tokenizers across the evaluation metrics.

5.1 Quantitative Results

Table 1 presents the end validation perplexity scores. AMEST achieves significantly reducing perplexity across both baselines on all eight datasets, with reductions varies between 2x and 8x over Base Unigram. This is to say that AMEST token sequence modeling is much easier for the GPT-2 Small architecture. The validation loss curves in Fig. 3 show that AMEST models not only reach a lower final loss but often converge faster. Figure 4 demonstrates that this performance gain comes with no efficiency penalty, as AMEST's evaluation throughput is highly competitive with, and often superior to, the baselines.

Table 4. Cross-Domain PPL: Models trained on WikiText-2, evaluated on other datasets.

Evaluation Dataset	Original GPT-2	Base Unigram	AMEST (Ours)
Penn Treebank	4571.01	898.12	**49.48**
Tiny Shakespeare	17677.58	6682.07	**221.72**
TweetEval	9479.86	2629.11	**162.62**

Table 5. Ablation study on WikiText-2. Perplexity (lower is better).

Tokenizer Configuration	Perplexity
AMEST (Morphology + Sampling Fallback)	**46.93**
Ablation: Morphology + Deterministic Fallback	91.02
Base Unigram (Deterministic)	340.96
Ablation: Unigram with Sampling Only	379.37

5.2   Analysis and Discussion

The results of the experiment clearly demonstrate the effectiveness of the AMEST approach.

Tokenization Quality and Features: AMEST's morphological Preference typically produces more detailed tokenizations, as exemplified in Table 2. Though this stretches the sequence length marginally, hence providing the model with more cohesion, key elements. For example, the unsupervised Morfessor properly divides such as 'handheld' → ['hand', 'held'] and 'recasting' → ['re', 'cast', 'ing'], providing unique semantic content. The system is far from perfect; sometimes forming awkward splits like 'mobbed' → ['mob', 'bed'] but the statistical The fallback facility gives essential robustness. The low fallback rates presented in Table 3 show that the morphological lexicon is highly effective on its own, covering the vast majority of words across diverse domains.

Cross-Domain Generalization and Ablation: AMEST's power is further illustrated by its cross-domain generalization (Table 4), where the model trains on WikiText-2 alone handsomely outperforms baselines on other areas. This means AMEST learns more basic, hands-on linguistic representations. Our ablation study (Table 5) verifies the source of this power: applying morphology naively provides a strong gain over the baseline, but the entire AMEST model, combining structure with random sampling for out-of-vocabulary words, is needed to attain the minimum perplexity. The synergy between linguistic structure and statistical robustness is key.

6   Conclusion and Future Work

In this work, we introduced AMEST, a hybrid tokenization method that integrates unsupervised morphological segmentation with probabilistic subword sampling. Our goal was to create a tokenizer with lower perplexity for language models via input representations that are more linguistically informed. By training identical GPT-2 Small models from scratch on eight diverse datasets, we have demonstrated that AMEST consistently and substantially outperforms the baseline GPT-2 BPE as well as a baseline Unigram tokenizer. The significant perplexity reductions, along with successful convergence and competitive throughput, highlight the real-world benefits of morphological awareness in tokenization.

This work emphasizes the significance of tokenization to LLM performance and suggests that a shift from purely statistical methods to linguistically motivated methods like AMEST offers a promising direction for developing more predictable, effective, and perhaps more resilient language models.

Future research directions are: (1) Experimenting with AMEST on higher-scale language models (LLMs) and diverse range of downstream natural language pro- Natural Language Processing tasks. (2) Examining the use of heterogeneous unsupervised or supervised morphological analyzers. (3) Transferring and experimenting with AMEST on languages that possess intricate and diverse morphological patterns. (4) Examining the use of adaptive processes that will alter segmentation strategies dynamically based on model feedback during training. (5) Venturing into an in-depth analysis of the compromises inherent in sequence length, vocabulary size, and processing efficiency.

References

1. Boström, K., Partanen, M.G.: Byte-pair encoding is suboptimal for language model pre-training. In: Proceedings of the 23rd Nordic Conference on Computational Linguistics (NoDaLiDa), pp. 244–254. Linköping University Electronic Press, Reykjavik, Iceland (Online) (2021)
2. Botha, J.A., Blunsom, P.: A compositional-morphological model for learning phrase representations. In: Proceedings of the 2014 Conference on Empirical Methods in Natural Language Processing (EMNLP), pp. 1898–1909. Association for Computational Linguistics, Doha, Qatar (2014)
3. Brown, T.B., et al.: Language models are few-shot learners. In: Advances in Neural Information Processing Systems 33 (NeurIPS 2020), pp. 1877–1901. Curran Associates, Inc. (2020)
4. Creutz, M., Lagus, K.: Unsupervised models for morpheme segmentation and morphology learning, vol. 4, pp. 3:1–3:34 (2007)
5. Jelinek, F., Mercer, R.L., Bahl, L.R., Baker, J.K.: Perplexity—a measure of the difficulty of speech recognition tasks. In: The Journal of the Acoustical Society of America, vol. 62, p. S63. Acoustical Society of America (1977)
6. Kudo, T.: Subword regularization: Improving neural network translation models with multiple subword candidates. In: Proceedings of the 56th Annual Meeting of the Association for Computational Linguistics (Volume 1: Long Papers), pp. 66–75. Association for Computational Linguistics, Melbourne, Australia (2018)

7. Kudo, T., Richardson, J.: Sentencepiece: A simple and language independent sub-word tokenizer and detokenizer for neural text processing. In: Proceedings of the 2018 Conference on Empirical Methods in Natural Language Processing: System Demonstrations, pp. 66–71. Association for Computational Linguistics, Brussels, Belgium (2018)

8. Lhoest, Q., et al.: Datasets: a community library for natural language processing. In: Proceedings of the 2021 Conference on Empirical Methods in Natural Language Processing: System Demonstrations. pp. 175–184. Association for Computational Linguistics, Online and Punta Cana, Dominican Republic (2021). https://aclanthology.org/2021.emnlp-demo.21

9. Manning, C.D., Schütze, H.: Foundations of Statistical Natural Language Processing. MIT Press, Cambridge, MA, USA (1999)

10. Merity, S., Xiong, C., Bradbury, J., Socher, R.: Pointer sentinel mixture models. In: Proceedings of the International Conference on Learning Representations (ICLR) (2017). https://arxiv.org/abs/1609.07843

11. Radford, A., Wu, J., Child, R., Luan, D., Amodei, D., Sutskever, I.: Language models are unsupervised multitask learners. OpenAI Blog 1(8) (2019)

12. Sennrich, R., Haddow, B., Birch, A.: Neural machine translation of rare words with subword units. In: Proceedings of the 54th Annual Meeting of the Association for Computational Linguistics (Volume 1: Long Papers), pp. 1715–1725. Association for Computational Linguistics, Berlin, Germany (2016)

Leveraging Vision-Language Models for Improving Detection of Obstacles on Railway Tracks

Vincenzo Carletti[1], Antonio Greco[1], Alessia Saggese[1], Camilla Spingola[1(✉)], and Bruno Vento[2]

[1] University of Salerno, Fisciano, Italy
{vcarletti,agreco,asaggese,cspingola}@unisa.it
[2] University of Naples Federico II, Naples, Italy
bruno.vento@unina.it

Abstract. Detecting obstacles on railway tracks, such as rocks, is crucial for train safety. In this paper we propose a two-shot architecture for rocks detection: semantic segmentation is used to identify track regions, and a patch extractor is employed to guide a multi-expert system combining the decisions of a convolutional neural network (CNN) classifier and of a Vision Language Model (VLM). The former offers rock detection capability learned from the domain-specific training set, while the latter, pre-trained on millions of general image-text tuples, can recognize rocks and related concepts and distinguish them from similar object categories; these features make them complementary tools that can enhance each other's performance by combining precise expertise with adaptive generalization. As the experiments confirm, the proposed approach achieves 0.897 F1-score, outperforming the CNN classifier of 5 percentage points and the VLM of 7 percentage points, demonstrating a notable reliability in rocks detection on railway tracks.

Keywords: Rock detection · Railway · VLM · CNN · multi-expert

1 Introduction

Maintaining safety and efficiency in rail transportation depends on preserving the integrity of the rail infrastructure, which in turn requires ongoing monitoring and proactive mitigation of potential hazards. Obstacles such as people, vehicles, animals, or rocks on railway tracks pose significant threats, leading to safety concerns and operational challenges. A possible solution could be to install fixed surveillance cameras to monitor the entire track; however, this solution becomes expensive, so it is typically adopted only at the crossing levels [3]. This is why a common alternative solution proposed in the scientific literature involves equipping trains with cameras, to automatically analyze captured image sequences, identifying railway tracks and obstructions. In the last decade, machine learning and deep learning techniques have led to significant advancements. Indeed,

M. Castrillón-Santana et al. (Eds.): CAIP 2025, LNCS 15622, pp. 195–206, 2026.
https://doi.org/10.1007/978-3-032-05060-1_17

employing these methods allows for more than just detecting lines and inter-ruptions. Instead, they enable the analysis of the entire scene, facilitating the identification of specific categories of objects, including railway tracks and various obstacles.

Within this context, two main architectures have emerged: single-shot and two-shot. In single-shot architectures, a single detector is designed for both railway track detection and other types of obstacles, including people, vehicles, or animals. For example, a Feature Fusion Refine Neural Network (FR-Net), built on top of VGG backbone, has been proposed to simultaneously identify both railway tracks and obstacles, including trains, pedestrians and helmets [15]. In [6] the authors propose to use a different detector, namely an extended version of Faster R-CNN [10], called Improved R-CNN, including a cascade method to avoid overfitting during training. Other than the rails, the following obstacles are identified: people, boxes, signs, billboards, power distribution, schoolbags, signals and platforms. One of the main limitations of the aforementioned approaches is the generation of bounding boxes that are larger than the actual railway track regions. This implies that an obstacle, such as a person, within the bounding box may not necessarily be on the track itself, thereby leading to false positives. A potential solution to the aforementioned problems is to adopt a semantic segmentation approach rather than a bounding-box-based one, in order to identify at the pixel level the regions to which railway tracks belong. It is worth noting that the same semantic segmentation network, properly trained, could also be used to identify the presence of obstacles, such as persons or vehicles. Although promising, this approach is impractical due to the need for a large number of pixel-level annotated images for training. Thus, a two-shot architecture [1] can be the solution: semantic segmentation is used to identify the railway track regions, followed by object detection to identify specific obstacles within these regions. As for semantic segmentation, several networks specifically devised for railway tracks have been proposed in the last years, including some inspired by SegNet [13], Bilateral Segmentation Network (BiSeNetV2) [16], and DFA-UNet [19]. Vice-versa, for the obstacle detector stage, widely adopted object detectors properly trained for detecting specific categories of obstacles have been explored, including Residual Neural Networks (RNN) [14] and YOLOv3 [11]. Anyway, among the different categories of obstacles, rocks remains relatively unexplored. Indeed, only a few examples of methods specifically devised for rocks are available in the literature; for instance, in [17] the authors propose to use a camera mounted on an Unmanned Autonomous Vehicle and a single-shot detector (SSD) based on VGG16 to detect persons, trains, and also rocks. The authors in [1] instead use a camera mounted on board of a train, exploiting domain-specific knowledge about intrinsic characteristics of the camera, its positioning with respect to the rails and rail-track characteristics: a deep neural network for semantic segmentation is combined with a mathematical model to identify the railway track in the image. Once the track area has been identified, it is partitioned into patches and a convolutional neural network is applied to analyze these patches and to detect the presence of rocks. While these approaches are very promising, they require a huge amount of annotated data and face challenges in detecting objects at various scales. Distinguishing between large rocks close to the camera and/or small rocks far from the camera within the same scene is particularly challenging due to the difficulties in identifying suitable anchor boxes for region proposals. Although enhancing the representativeness of the training data can mitigate these issues, the problem remains unresolved. This is because it is impractical to create a sufficiently large training dataset that encompasses all possible rock samples, given the critical nature of the application. Similarly, to minimize false alarms, it is crucial to implement a more robust approach that does not depend solely on the representativeness of negative samples in the training set. Relying exclusively on these samples is insufficient, as they are unlikely to cover all possible scenarios. Furthermore, given the specificity of the scenario itself, it is evident that acquiring negative examples also poses severe challenges.

Furthermore, we cannot neglect the recent advances in the fields of computer vision and artificial intelligence: Visual Language Models (VLMs) have indeed emerged, demonstrating impressive capabilities in interpreting visual inputs for tasks such as image captioning [7] and Visual Question-Answering (VQA) [9], making them particularly attractive also in autonomous driving [20]. These models are trained on multimodal large-scale datasets, incorporating both images and text, and comprising a variety of visual and textual information. Consequently, they are capable of extracting information from images based on user-defined queries for specific tasks. This training procedure is the reason for one of their main advantages, which is the capability to perform zero-shot learning, recognizing and interpreting new objects such as rocks, without requiring specific training data for the task of interest. This property renders the system particularly useful for applications such as rock detection on rail tracks, where the creation of exhaustive training datasets is impractical. However, applying VLMs on the entire image would not be suited for this kind of applications, where the rocks of interest that may be the cause of accidents for the moving train only lie on the rail track region, corresponding to a limited part of the image. In order to face this issue but still exploring the benefits deriving from zero-shot learning, in [4] the authors propose to identify the potential regions of interest for the VLM by means of an object detector, devised for the specific task (fire detection in the case of the abovementioned paper).

Starting from the above considerations and in order to face the challenges related to the lack of datasets, but also to exploit the power of zero-shot learning, in this paper we introduce a two-shot architecture in which the first step is a semantic segmentation algorithm to identify the track region, which is then divided into patches and the second step is a multi-expert system combining the decisions of a properly trained convolutional neural network (CNN) classifier and a VLM. Therefore, the proposed solution aims to explore the achievable benefits from the integration of a VLM in a more traditional pipeline, like the one proposed in [2], exploiting the potential complementarity of data-driven classification and visual-language reasoning to achieve higher reliability in the recognition process.

2   Proposed Method

The overall architecture of the proposed method is shown in Fig. 1. It is composed of distinct interconnected modules, each contributing to a specific stage of the prediction pipeline, as detailed in the following subsections.

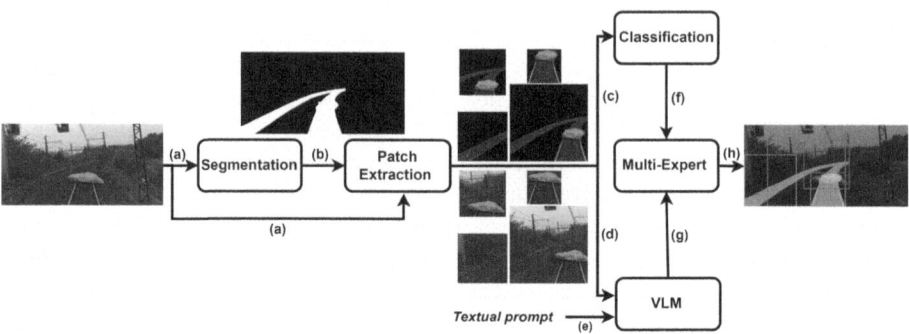

Fig. 1. Architecture of the proposed approach. The source image is resized (a) and given as input to the segmentation module to obtain a binary mask which distinguishes between railway area and background (b). The extracted patches feed a CNN classifier (c) and a VLM (d). The multi-expert system combines the probability outcomes of the CNN classifier (f) and of the VLM (g), obtained for the latter through the use of an appropriate textual prompt (e), to assign a label to each patch (h) from the source image.

2.1   Segmentation

To detect rocks that could disrupt the normal flow of the train, a semantic segmentation module has been integrated to identify the specific track areas where these rocks could be located. In a more formal way, semantic segmentation is a function $L : F \rightarrow C$, which assigns each pixels of the input frame F to a class in $C = \{c_1, c_2\}$, where c_1 represents the railway track and c_2 encompasses everything else. We decided to approximate this function with a modified version of the state-of-the-art BiSeNetV2 neural network [16]. This deep neural network has proved to be the best solution compared to other models trained on the same dataset, RailSem19 [18], and tested on the same problem, such as UNET [12], in obtaining a binary mask that closely represents the railroad region, generalizing well even in the presence of rocks on railway tracks [2].

2.2   Patch Extraction

Rather than working on the entire image, we propose to focus our attention only on the track region area, thus reducing the overall computational and semantic burden on the subsequent components. Inspired by [2], we divide the track region into a set of patches. In more details, the raw image is divided into horizontal slices of decreasing height as the distance from the observation point increases, in order to follow the perspective trend. Using the binary mask produced by the segmentation module, we proceed to extract the connected components within each slice. Starting from the consideration that the patches extracted in this way could be excessively small and lacking useful informational content, the coordinates of the bounding boxes of each connected components were modified by increasing the height or width of the extracted patch to make it square, preserving the spatial characteristics and avoiding any stretching or distortion of the patches before being passed to the downstream modules.

2.3 CNN Classifier

A binary classifier based on a state-of-art CNN, namely ResNet50, was designed to predict the presence or the absence of the rock within each extracted patch. This architecture proved to be particularly effective compared to other models tested with different sizes, namely MobileNetV2 and ResNet101, in terms of the trade-off between performance and computational load [2]. In this model, downstream of ResNet50 backbone, there is a custom classification head, consisting of linear, dropout, batch-normalization layers and ReLU activation functions. The model parameters were optimized using stochastic gradient descent (SGD) with a learning rate of 1×10^{-3}, momentum set to 0.9, and a learning rate scheduler with a step size of 10 and a decay factor (γ) of 1×10^{-4}. The binary cross-entropy loss function was employed as the base criterion and, to prevent overfitting, early stopping was used during training.

In order to overcome the lack of specific data and to take into account the complexity of the application scenario, several augmentation techniques were applied during the training process, both on the source images and on the extracted patches. For the former, zoom-in and zoom-out with different scaling factors (0.6-0.8-1.2-1.4) were applied, while for the latter, horizontal flip, random rotation, and horizontal and vertical shifts were used to make the system robust to variations in the observation point; grayscale conversions and color jittering to simulate illumination and weather variations in the acquisition conditions; and gaussian blur to simulate image noise due to the speed of train movement.

2.4 VLM

This module consists of a specific version of BLIP-2 [8] which employs pre-trained vision and language models with a lightweight Querying Transformer (Q-Former). Q-Former employs learnable query vectors to extract key visual features from the frozen image encoder and transmit them to the frozen language model, acting as a bottleneck. During the initial pre-training, the model is optimized for three objectives: Image-Text Contrastive Learning (ITC), Image-Text Matching (ITM), and Image-grounded Text Generation (ITG). In ITM, which enhances fine-grained alignment via a binary classification task, the model predicts whether an image-text pair is semantically consistent (matched or unmatched), feeding each multimodal query embedding into a two-class classifier to obtain a logit. The output matching score is obtained averaging the logits across all queries. Since the task of detecting the presence of a boulder on tracks in a patch is approached as a binary classification problem, we used only the ITM head of BLIP-2, omitting the generative component for efficiency. The adopted implementation is pre-trained on approximately 130 million images from COCO, Visual Genome, CC12M, SBU and LAION400M and relies on *Blip2ForImageTextRetrieval* available in HuggingFace. The use of this BLIP-2 version is motivated by the inclusion in the pre-training dataset of the concept of "rock", suggesting a good alignment between the visual and textual representations of the objects of interest. Nevertheless, it is still necessary to identify the optimal textual prompt for maximizing the model's ability to recognize this concept. We validated four different prompts, obtaining the best results with *"there is a rock blocking the railway tracks"* (see Sect. 4).

2.5 Multi-expert System

This module is responsible for combining the outputs of the CNN classifier and the VLM, enhancing the reliability of the final prediction by exploiting their complementarity. The former, being a binary classifier, produces a confidence score representing the probability that the rock is present in the input patch, while the latter produces a probability score of how much the input prompt is relevant to the image. Among the various integration strategies that we considered, a probability-based fusion rule was ultimately selected. Specifically, the confidence scores generated by the two experts are summed, and if the total exceeds 1, the system predicts the presence of a rock; otherwise, the prediction is negative. The choice of using a multi-expert system based on the probabilities given in output by the two modules is motivated by two fundamental factors. First, in this way the proposed system can incorporate the notion of confidence from each expert, by relying on the predicted probabilities. This is particularly valuable in scenarios where there is partial or ambiguous evidence. Secondly, the resulting decision function is based on just two input features, rendering the overall system highly interpretable. Explainability is a critical factor in security applications, such as railway monitoring, where decision can directly impact human safety and operational continuity. In such contexts, it is imperative to understand the rationale underlying the predictions of the models in order to ensure the development of trust, facilitate verification and enable real-time interventions.

3 Dataset

Given the difficulty of obtaining a real dataset and the lack of publicly available large datasets for the purpose of our analysis, we decided to use the dataset proposed in [2]. It consists of automatically generated images, exploiting Generative Fill, an artificial intelligence tool built into Adobe Firefly, which, based on deep neural networks, makes it possible to obtain images in which the generated objects turn out to be truthfully integrated into the original image, as opposed to traditional copy-paste techniques. Examples of generated images are reported in Fig. 2. Therefore, images from two available state-of-the-art datasets collected in a railway scenario were used as background images: RailSem19 [18] and FRSign [5]. From the images of the first dataset, collected by vehicle-mounted cameras and highly variable in terms of lighting, weather and landscape conditions, 5500 images were generated and used for the training and validation phase (about 75% and 25%, respectively). From the images of the second dataset, focused on railway traffic light recognition, 550 images were generated, selecting only those whose observation point corresponded to the application scenario, these are used for the test phase. In total, the dataset includes 4096 training images, 1404 validation images, and 550 test images, corresponding to 17399, 5885, and 2434 patches, respectively.

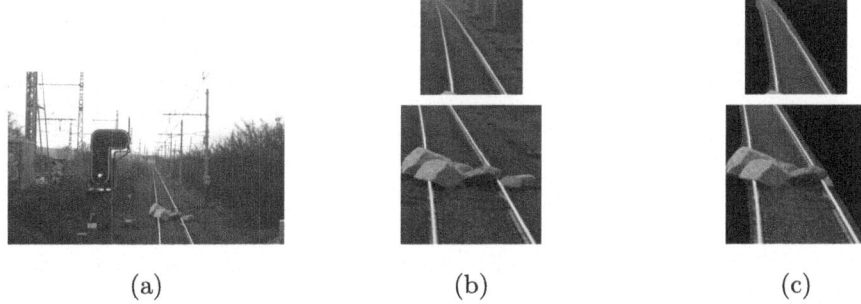

(a)   (b)   (c)

Fig. 2. Example of test images. Source image before resizing (a), extracted patches to be analyzed by the VLM (b) and by the CNN classifier (c).

4 Experimental Analysis

This section describes the experimental evaluation conducted to assess the effectiveness of the proposed approach. Quantitative comparisons among the methods are based on Precision (P), Recall (R) and F1-Score (F). In this context true positives (TP), true negatives (TN), false positives (FP) and false negatives (FN) are defined according to the binary ground-truth (1 if a rock is present in the patch, and 0 otherwise), and based on whether the probability score produced by the model (or by the models combination) exceeds a predefined threshold (indicating a positive prediction, i.e., the detection of a rock) or not.

As a preliminary step, it was necessary to analyze the performance of the CNN classifier and of the VLM to identify their optimal operating conditions, through analysis conducted on the validation set.

- **CNN classifier**: We determined the optimal decision threshold, that is the probability value above which a sample classified as positive (indicating the presence of rocks on the tracks) and below which it is classified as negative. This threshold selection, carried out on the validation set, was guided by a performance analysis across a range of values (from 0 to 1 with a step of 0.05), with the objective of maximizing the F1-Score.
- **VLM**: in addition to the optimal decision threshold, obtained with the same procedure adopted for the CNN classifier, four natural language prompts, exhibiting varied descriptions of the presence of rocks on the tracks, were evaluated on the validation set, in order to choose the most effective.

We verified on the validation set to what extent the combination of the two models could yield performance enhancements. Our analysis revealed that a substantial portion of misclassified samples by one model were correctly predicted by the other. Hypothetically speaking, if all correct predictions were to be accepted, the resulting F1-Score would reach 0.928. Therefore, the multi-expert system has the potential to improve the overall performance of the approach. To this end, two distinct integration strategies, with some variants, were compared with the proposed one.

- **Cascade**: in this setup, the CNN classifier and the VLM are organized in a cascade, in which one of them acts as the upstream module, while the other acts only on a subset of samples, filtered according to the predictions of the former. Specifically, the downstream module only receives those samples predicted as positive/negative by the upstream module. Depending on the adopted solution, we distinguish *Cascade on Positives* (CP) or *Cascade on Negatives* (CN), and two different versions of these schemes were explored. In the *Cascade on Positives with Fixed Thresholds* (CPFT) and in the *Cascade on Negatives with Fixed Thresholds* (CNFT), both modules were utilized with their respective optimal thresholds; in the *Cascade on Positives with Optimized Thresholds* (CPOT) and in the *Cascade on Negatives with Optimized Thresholds* (CNOT), the thresholds of the two modules were selected optimizing the overall F1-Score of the cascade on the validation set.
- **MLP**: this multi-expert system was designed using only two input features, namely the confidence outcomes of the CNN classifier and of the VLM, to determine an optimal nonlinear decision rule. The model, consisting of a hidden layer with 4 neurons with ReLU activation function and a single neuron output layer, was trained with a batch size of 64 and the BCELoss function. The AdamW optimizer was employed with a learning rate of 0.001 and the training was conducted for 50 epochs. The weights were learned using data from the training and validation sets and the MLP was finally evaluated on the test set by using a confidence threshold equal to 0.6.

The quantitative results of our experiments are reported in Table 1. Moreover, we depict in Fig. 3 the decision boundaries of all the multi-expert systems.

Table 1. Results of the considered approaches in terms of Precision (P), Recall (R) and F1-Score (F). The best results for each metric are highlighted in bold.

Model	Prompt	Threshold		P	R	F
		VLM	CNN			
CNN	-	-	0.550	0.946	0.769	0.849
VLM	*a rock on the rails*	0.700	-	0.879	0.725	0.794
	a rock on the railway tracks	0.650	-	0.897	0.729	0.804
	the railway tracks are blocked by a rock	0.900	-	0.953	0.532	0.682
	there is a rock blocking the railway tracks	0.900	-	0.923	0.742	0.823
CPFT		0.900	0.550	**0.996**	0.643	0.781
CNFT				0.894	0.868	0.881
CPOT		0.050	0.550	0.946	0.769	0.849
CNOT		0.950	0.750	0.928	0.804	0.862
MLP		-	-	0.975	0.761	0.855
Proposed		-	-	0.922	**0.873**	**0.897**

We can note that the CNN classifier alone achieves $F = 0.849$, while the VLM with the best prompt obtains $F = 0.823$; both the experts are more precise than sensitive, since P is substantially higher than R. For this reason, the results obtained by the CP approaches do not improve those achieved by standalone models, since R is not increased; however, it is worth noting that CPFT achieves the highest $P = 0.996$, but the increased precision is paid in terms of R. On the contrary, the CN approaches showed to be more effective, since they allow to improve the overall R. The CNFT multi-expert achieved the second best $F = 0.881$, obtaining a good trade-off between P and R; CNOT obtained the third best $F = 0.862$, being more precise and less sensitive. The multi-expert system based on MLP was able to improve the performance of the CNN classifier ($F = 0.855$), but the positive results achieved on the validation set were not confirmed on the test set; this finding suggests a limitation in the MLP's capability to effectively generalize beyond the observed domain during the test phase.

The proposed method, with its $F = 0.897$, demonstrated to be the most effective in exploiting the complementarity between the CNN classifier and the VLM. The performance on the test set show the capability of the proposed multi-expert system to achieve a good tradeoff between sensitivity ($R = 0.873$) and specificity ($P = 0.922$), thereby ensuring a better generalization capability on test samples compared to other more complex strategies.

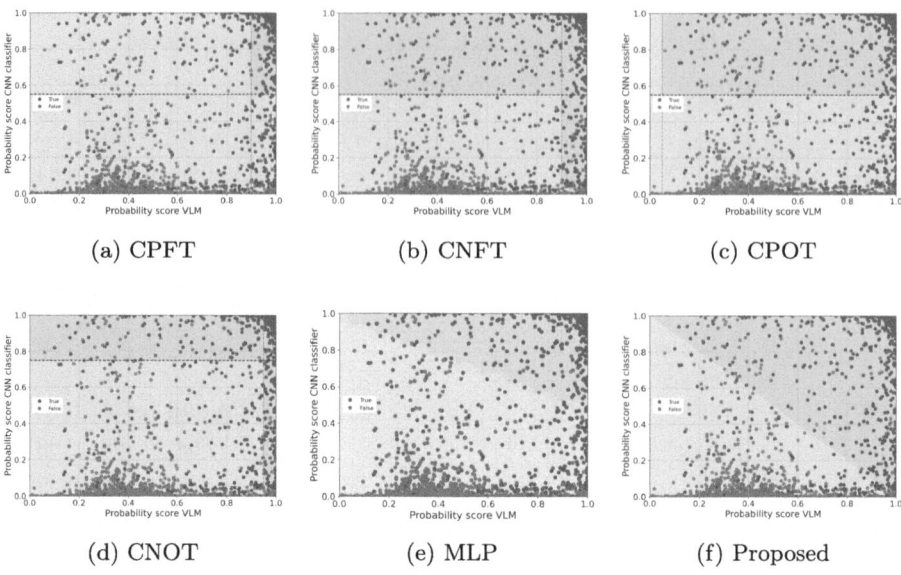

(a) CPFT   (b) CNFT   (c) CPOT

(d) CNOT   (e) MLP   (f) Proposed

Fig. 3. Decision boundaries of the considered methods. The chart shows the probability scores produced by the VLM on the x-axis and by the CNN classifier on the y-axis. Blue points represent positive samples and red points represent the negative ones, while the regions with the corresponding colors indicate positive and negative decision boundaries, respectively. (Color figure online)

5   Conclusions

This work introduced a two-shot architecture for rocks detection on railway tracks, combining semantic segmentation, patch extraction, a CNN classifier, and a VLM. By leveraging the complementary strengths of the CNN's domain-specific expertise and the VLM's generalization capabilities, the system achieved higher detection performance with a multi-expert system based on an effective and efficient combination rule. Experimental results demonstrated that the proposed method outperformed individual models, achieving a F1-score of 0.897, improving upon the standalone CNN and VLM by 5 and 7% points respectively. These results confirm the effectiveness and reliability of the combined multi-expert approach for the targeted application, leaving open the door to future research directions. In particular, the performance of the individual modules may be further improved by expanding the domain-specific training datasets for the CNN classifier and refining the VLM's outputs through more targeted and specific prompts. Additionally, future works may explore novel strategies for combining the decisions of the different experts, aiming for context-aware fusion mechanisms that could further boost the overall rock detection accuracy.

References

1. Carletti, V., et al.: Combining deep networks with model-based scene segmentation for reliably detecting rocks on railway tracks. In: 2024 IEEE 8th Forum on Research and Technologies for Society and Industry Innovation (RTSI), pp. 524–529. IEEE (2024)

2. Carletti, V., et al.: Onboard vision based system for automatic rock detection on rail tracks. In: International Conference on Pattern Recognition, pp. 263–274. Springer (2024)

3. Carletti, V., Greco, A., Saggese, A., Vento, B.: Enhancing safety by obstacle detection at railway level crossings. In: 2024 IEEE 8th Forum on Research and Technologies for Society and Industry Innovation (RTSI), pp. 482–487. IEEE (2024)

4. Gragnaniello, D., Greco, A., Sansone, C., Vento, B.: Video fire recognition using zero-shot vision-language models guided by a task-aware object detector. ACM Trans. Multimedia Comput., Communications and Applications (2025)

5. Harb, J., et al.: FRSign: A large-scale traffic light dataset for autonomous trains. arXiv preprint arXiv:2002.05665 (2020)

6. He, D.: Urban rail transit obstacle detection based on improved R-CNN. Measurement **196**, 111277 (2022)

7. Hossain, M.Z., Sohel, F., Shiratuddin, M.F., Laga, H.: A comprehensive survey of deep learning for image captioning. ACM Comput. Surv. (CsUR) **51**(6), 1–36 (2019)

8. Li, J., Li, D., Savarese, S., Hoi, S.: Blip-2: Bootstrapping language-image pre-training with frozen image encoders and large language models. In: International conference on machine learning, pp. 19730–19742. PMLR (2023)

9. Manmadhan, S., Kovoor, B.C.: Visual question answering: a state-of-the-art review. Artif. Intell. Rev. **53**(8), 5705–5745 (2020). https://doi.org/10.1007/s10462-020-09832-7

10. Ren, S., He, K., Girshick, R., Sun, J.: Faster R-CNN: towards real-time object detection with region proposal networks. IEEE Trans. Pattern Anal. Mach. Intell. **39**(6), 1137–1149 (2017)

11. Ristić-Durrant, D., et al.: Artificial intelligence for obstacle detection in railways: project SMART and beyond. In: Bernardi, S., et al. (eds.) EDCC 2020. CCIS, vol. 1279, pp. 44–55. Springer, Cham (2020). https://doi.org/10.1007/978-3-030-58462-7_4

12. Ronneberger, O., Fischer, P., Brox, T.: U-Net: convolutional networks for biomedical image segmentation. In: Navab, N., Hornegger, J., Wells, W.M., Frangi, A.F. (eds.) MICCAI 2015. LNCS, vol. 9351, pp. 234–241. Springer, Cham (2015). https://doi.org/10.1007/978-3-319-24574-4_28

13. Wang, Z., Wu, X., Yu, G., Li, M.: Efficient rail area detection using convolutional neural network. IEEE Access **6**, 77656–77664 (2018)

14. Xu, Y., Gao, C., Yuan, L., Tang, S., Wei, G.: Real-time obstacle detection over rails using deep convolutional neural network. In: 2019 IEEE Intelligent Transportation Systems Conference (ITSC), pp. 1007–1012. IEEE (2019)

15. Ye, T., Wang, B., Song, P., Li, J.: Automatic railway traffic object detection system using feature fusion refine neural network under shunting mode. Sensors **18**(6), 1916 (2018)

16. Yu, C., et al.: BiSeNet V2: bilateral network with guided aggregation for real-time semantic segmentation. Int. J. Comput. Vision **129**, 3051–3068 (2021)

17. Yundong, L.: Multi-block SSD based on small object detection for UAV railway scene surveillance. Chin. J. Aeronaut. **33**(6), 1747–1755 (2020)
18. Zendel, O., et al.: RailSem19: a dataset for semantic rail scene understanding. In: Proceedings of the IEEE/CVF Conference on Computer Vision and Pattern Recognition Workshops (2019)
19. Zhang, Y., Li, K., Zhang, G., Zhu, Z., Wang, P.: DFA-UNet: efficient railroad image segmentation. Appl. Sci. **13**(1), 662 (2023)
20. Zhou, X., et al.: Vision language models in autonomous driving: a survey and outlook. IEEE Trans. Intell. Veh. (2024)

Generalizable Detection of Student Engagement in Online Learning Environments

Lu Pang[✉], Tony Siu, Anis Alazzawe, Krishna Kant, and Longin Jan Latecki

Temple University, Philadelphia, USA
{lpang,yun.sing.siu,aalazzawe,kkant,latecki}@temple.edu

Abstract. Automated recognition of student engagement in online learning is crucial as it enables teachers to adapt content delivery to improve learning. In this paper, we explore a method that finetunes a pretrained vision language model (VLM) to recognize student engagement markers in still images. Our model learns to avoid incorrect answers during finetuning by using the emerging direct preference optimization techniques on self-generated preference pairs based on the correct and incorrect VLM answers. On publicly available student engagement datasets, our model shows superior performance over other approaches and substantially better generalizability over the traditional vision methods.

Keywords: Direct Preference Optimization · VLM · Engagement

1 Introduction

With the development of network and multimedia technology, online learning environments have become viable in education at all levels. Its adoption has skyrocketed following the COVID-19 pandemic. Online learning is now well-entrenched and increasingly preferred because of its numerous benefits both to the students (e.g., avoidance of physical transportation) and the providers (e.g., lower cost and ability to handle larger classes). However, online learning has some definite disadvantages due to the inability of the instructor to make face-to-face contact and assess comprehension or lack thereof. Research has demonstrated that student engagement provides a positive influence on academic achievement [8]. In the online learning environment, automated monitoring of the student engagement level becomes essential because scanning the camera feeds from a large number of students is distracting and impractical; furthermore, it requires that the video stream from each student be transmitted to the instructor. Ideally, we would like to obtain the same kind of assessment that experienced teachers can easily do in real classrooms without the necessity to transmit student videos.

This research was supported by NSF grants CNS-2333611 and IIS 2331768.

Traditionally, visual behavior recognition has been done via specially designed deep learning models that must be trained on well-crafted labeled datasets. While such methods can do extremely well in recognizing the targeted behaviors, they are limited by the difficulties in model crafting and the creation of good quality and varied training datasets. They also tend to struggle in generalizing well to out-of-distribution samples, as shown in our experiments.

Large Vision-Language Models (VLM) have recently been researched extensively to provide good descriptions (or "captions") of activities in images and short video segments. In this paper, we focus on determining how well the students are engaged in a virtual classroom environment. Video analytics determines this primarily based on the facial expressions and gestures, although the VLM based method explored here considers all visual features implicitly. We explore an open-source Vision Language Model, MiniGPT-4 [16] with Vicuna [3], a multi-modal chatbot based on Llama [15]. We further finetune it for a few student engagement marker tasks.

To enhance the generalizability of the model, we propose a novel variant of Direct Preference Optimization (DPO) [11] method that finetunes the VLM using self-generated preference pairs. In particular, we design a specialized DPO finetuning method that yields a VLM capable of accurately predicting student engagement markers from images (Sect. 2.2). Our design includes a self-generation pipeline to capture the broad distribution of the generated answers so that the preference pairs used to optimize the VLM can capture the actual distribution of incorrect answers. The proposed specialized DPO finetuning can be applied in any scenario with a labeled dataset.

We compare our finetuning results with various deep learning-based vision models and show that our method performs substantially better in predicting student engagement markers. We also evaluate the generalizability of our fine-tuned model to out-of-distribution samples by applying it to a different dataset. We focus on analyzing still images as opposed to videos since recent image-based VLMs are able to run on end-user devices in real-time, e.g., Llama 3.2 (Meta) and Molmo (Allen AI).

The rest of the paper is organized as follows. Section 2 describes the essential finetuning details of our method on the student engagement dataset. Section 3 explains the experimental setup, compares our method to other vision and VLM models, and analyzes the results. Finally, Sect. 4 concludes the paper.

2   Methodology

2.1   Reinforcement Learning with Human Feedback

One challenge in training LLMs and VLMs is that it is difficult to measure the quality of output from these models automatically. Reinforcement Learning with Human Feedback (RLHF) provides a way for humans to be a source of the reward model without explicitly modeling the rewards. In LLMs and VLMs, RLHF is used to finetune the models, getting humans to choose between various outputs and learning a model to estimate the rewards.

Pretraining stage trains the language model to predict the next token given the prior text information using the large collection of datasets for natural language processing tasks. The loss used in the pretraining stage is normally the cross-entropy loss.

Supervised Fine-tuning (SFT) stage finetunes the pre-trained language model on datasets for specific downstream tasks. The datasets are normally high-quality datasets with appropriate instructions (prompts) and reasonable responses. The model obtained after the SFT finetuning is denoted as π_{ref}.

Preference sampling and Reward Learning: For a dataset D, for each input x, there is a pair of preferred/dispreferred answers (y_ω, y_ι), where y_ω denotes the answer that human labelers expressed preference for and y_ι denote the dispreferred answer. We can model this by a latent reward model r^* that generates the underlying preferences. A commonly used approach to model preferences is assuming the probability $p^*(y_\omega \succ y_\iota | x)$, which represents the probability of preferring answer y_ω over answer y_ι given the input x, as a sigmoid σ of the reward difference.

$$p^*(y_\omega \succ y_\iota | x) = \sigma(r^*(x, y_\omega) - r^*(x, y_\iota)) \tag{1}$$

We need to use a reward model r_ϕ to estimate the true reward of human preference via maximum likelihood since the true reward function is not accessible. For this, we minimize the negative log-likelihood loss.

$$\mathcal{L}_R(r_\phi, D) = -\mathbb{E}_{x, y_\omega, y_\iota \sim D}[\log \sigma(r_\phi(x, y_\omega) - r_\phi(x, y_\iota))] \tag{2}$$

Reinforcement Learning Optimization: This phase uses the learned reward function to further optimize the language model. To prevent model collapse and maintain the proximity to the distribution of the reference model π_{ref}, a KL divergence penalty is added.

$$\max_{\pi_\theta} \mathbb{E}_{x \sim D, y \sim \pi_\theta(y|x)}[r_\phi(x, y)] - \beta \mathbb{D}_{KL}[\pi_\theta(y|x) \parallel \pi_{ref}(y|x)]$$

where π_θ is the policy model that RLHF is optimizing, and β is the hyperparameter to restrict how far the model can deviate from the base reference model. Due to the non-differentiability, the objective is normally optimized with an RL algorithm such as Proximal Policy Optimization (PPO) [13]. PPO is a policy gradient algorithm that uses an objective function to find the best policy. The objective function minimizes the difference between the new and old policy. As a result, PPO avoids too large a policy update which may destabilize the learning.

2.2 Direct Preference Optimization

Direct Preference Optimization (DPO) [11] provides a simplification of traditional feedback-aligned LLMs. DPO avoids the Reinforcement Learning (RL) loop as RLHF and instead proposes a closed-form loss using a special choice of reward model parameterization in Eq. 2. Equation 3 shows the DPO loss, where y_ω is the preferred answer, y_ι the dispreferred answer, π_θ the policy model being optimized, and π_{ref} is the constraining reference model. The reference model provides stability in training and baseline for the policy model to improve against.

$$L_{DPO}(\pi_\theta, \pi_{ref}) = -\mathbb{E}\left[\log \sigma \left(\beta \log \frac{\pi_\theta(y_\omega|x)}{\pi_{ref}(y_\omega|x)} - \beta \log \frac{\pi_\theta(y_\iota|x)}{\pi_{ref}(y_\iota|x)}\right)\right] \quad (3)$$

One can rewrite the DPO loss to show that it maximizes the margin between preferred and dispreferred answer pairs. Setting $r_\omega = \log(\pi_\theta(y_\omega|x)) - \log(\pi_{ref}(y_\omega|x))$ and $r_\iota = \log(\pi_\theta(y_\iota|x)) - \log(\pi_{ref}(y_\iota|x))$ we can reformulate L_{DPO} as:

$$L_{DPO}(\pi_\theta, \pi_{ref}) = -\mathbb{E}[\log \sigma(\beta(r_\omega - r_\iota))] \quad (4)$$

For the student engagement detection problem, we work with multi-class labeled datasets where the labels indicate the different engagement behaviors of the student. Therefore, we propose to transform the labeled dataset to a preference dataset $D = \{x^i, y_\omega^i, y_\iota^i\}_{i=1}^N$, where N denotes the number of data samples. For each input image x, we assign an answer pair of preferred answer and dispreferred answer (y_ω, y_ι). A simple approach is to utilize the given ground truth to generate a preference dataset by using the given label of each image as the preferred answer y_ω and using the other labels in the dataset as the dispreferred answer y_ι. However, this approach will introduce bias to the vision-language model since the distribution of possible dispreferred answers has a much wider range.

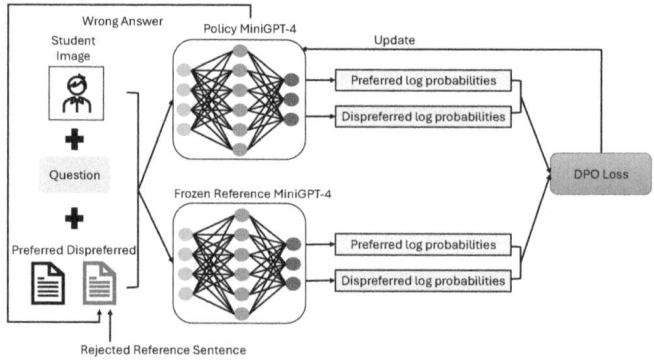

Fig. 1. DPO based Finetuning using a frozen reference & policy model.

In order to yield better answers, we propose to extend the reference data pairs D so that the dispreferred answers y_ι^i come from a more general distribution than ground truth. Since we do not have the distribution of generated answers of the vision-language model, **we use the wrong answers generated from the vision-language model itself to generate the dispreferred answers.** In practice, we form the dispreferred answers during training in the following manner. For a percentage of the time, if the generated answers from the Policy

MiniGPT-4 are incorrect, we use these as dispreferred answers. Otherwise, if the generated answers are correct, we randomly choose from other labels besides the correct label as the dispreferred answers. For our experiments, this was done 30% of the time. The architecture of the DPO based finetuning is shown in Fig. 1. We give example training pairs in the next section.

The probability $\pi_\theta(y|x)$ of any generated answer y is simply the product of the probabilities of its generated tokens. In contrast, when the answer y is provided, in order to compute $\pi_\theta(y|x)$, we only need to score how likely the LLM is to generate y if prompted with x. For this, we input x concatenated with y to the LLM and collect the log probabilities of all y tokens token by token.

2.3   Finetuning MiniGPT-4 on Engagement Datasets

We finetune MiniGPT-4 on the dataset using a set of prompts that are relevant to the student engagement problem. In order to achieve more general performance, we optimize the MiniGPT-4 model with DPO using automatically generated preference pairs.

In our model, we introduce a finetuning that aims to generate specific image descriptions to describe behaviors or affective states indicating engagement levels of students. To be more specific, our approach uses student engagement categorization from the Student Engagement Dataset (SED) [5], the DAiSEE dataset [6], and the EngageNet dataset [14]. SED captures three student engagement markers. DAiSEE and EngageNet has four levels of engagement labels.

The following is an example of how we finetune with the SED dataset. SED uses the labels: 'looking at the paper', 'looking at the screen', and 'wandering'. We set the corresponding correct reference sentence as follows:

The person is looking down at the paper
The person is looking straight at the screen
The person is looking away

Since DAiSEE and EngageNet are composed of videos only, we subsampled them with a frequency of 1 frame per second (fps), resulting in 10 frames per video clip. The 10 images are concatenated together and inserted into the finetuning and the VLM is asked to classify the student into one of "Highly-Engaged", "Engaged", "Barely-Engaged", and "Not-Engaged" classes. Now we give two examples of preferred and dispreferred answers used for DPO finetuning. The VLM generated a wrong answer: 'Yes, the person is looking away. The blue headband is tied around their hair ...' We use it as a dispreferred answer, while the correct answer, 'The person is looking straight at the screen.' is used as the preferred answer. On the other hand, if the VLM generated a correct answer for a different image, 'The person is looking away.', we use it as a preferred answer and randomly select one of the remaining two answers as a dispreferred answer, such as'The person is looking down at the paper.' We use the same prompt template as MiniGPT-4. The instruction of MiniGPT-4 is randomly sampled from a predefined instruction set which contains different instructions for the image caption task. It is well-known that prompting can influence the output of LLMs [12]. In our finetuning, we use the instruction set containing questions that are used specifically to determine the students' behavior. After extensive experimentation with Vicuna, the instruction set (when our goal is a natural language output) is chosen due to the consistently better performance. The following shows one example of our prompts:

Given the following image: <Img>ImageContent</Img>. You will be able to see the image once I provide it to you. Please answer my questions.

###Human: <Img><ImageFeature></Img> Is the person looking straight at the screen? Is the person looking down at the paper? Is the person looking away?
###Assistant:

During the finetuning, one of the questions is selected randomly from the instruction set. We found experimentally that using questions in this way produced a higher baseline.

Additionally, since the output in JSON format is easy to evaluate with keyword search, we use the following query to force Vicuna to respond in JSON format:

<Img><ImageHere></Img> Given label set:['looking at the screen','looking at the paper','wandering'] Question: What is the type of activity in the image and which category from the given label set would you use to describe this activity type? Answer me in the JSON format like {'label': 'activity_type'}

We evaluated two different versions of the MiniGPT-4 model in our experiments. One version is that we perform our finetuning using the MiniGPT-4 (Vicuna) checkpoint. Another version is that we perform our finetuning after using the MiniGPT-4 (Llama2) checkpoint. MiniGPT-4 (Vicuna) achieves an accuracy of 95.2% after our finetuning whereas MiniGPT-4 (Llama2) yields 88.6%. Therefore, we use MiniGPT-4 (Vicuna) as our base model.

3   Performance Evaluation

3.1   Datasets

Our evaluation utilizes three publicly available datasets, Student Engagement Dataset (SED) [5], DAiSEE dataset [6], and EngageNet [14]. SED contains both an unbalanced and balanced component. The unbalanced one contains 18,721 frames sampled at one fps from 400 videos collected from 19 students. It has samples divided into 3 categories 'looking at the paper', 'looking at the screen', or 'wandering'. The first two are considered as "engaged" since the completion of their tasks requires one of those two activities. Note that "wandering" means that the student is not engaged (it does not mean "wandering-around"). The balanced dataset is a smaller 1973 frame version that removes similar samples for each of the three classes, resulting in a more balanced number of frames across the classes.

DAiSEE dataset is a large labeled student engagement level dataset that is collected by a web camera during the period of a student watching educational and recreational videos. The dataset contains 9068 video snippets collected from

32 female and 80 male subjects aged 18 to 30. The dataset is labeled with four different student engagement levels: Very Low, Low, High, and Very High. In addition to these engagement labels, we relabeled a subset of DAiSEE with SED labels for out-of-distribution evaluation with the model trained using SED dataset. The goal is to evaluate whether the model can apply the knowledge learned from the student engagement dataset to an out-of-distribution dataset under the same premise that both the DAiSEE and SED datasets have students in front of a web camera. With this premise, we annotated the DAiSEE dataset according to the evaluation framework of SED with labels 'looking at their paper', 'looking at their screen', and 'wandering'.

Fig. 2. Example hard samples that contain images of students' faces not contained within the image.

We have also identified 85 samples (hard samples) from SED that were selected based on misidentification by MiniGPT-4, which we call the hard SED dataset. These hard samples tend to contain images of students that face one direction whilst their gazes face one another, or students' faces are not contained within the image. Examples of these hard Samples are shown in Fig. 2. We further evaluated GPT-4V [10] as opposed to our finetuning methodology on this handpicked dataset.

EngageNet is a large-scale, multimodal dataset designed for user engagement prediction in real-world, in-the-wild settings. It comprises over 11.3K ten-second video clips (approximately 31 h of data) from 127 participants recorded under diverse illumination conditions, and it captures both behavioral cues such as facial expressions, head pose, eye gaze, and cognitive responses from interactive questionnaires. Annotated into four engagement levels (Not Engaged, Barely Engaged, Engaged, Highly Engaged) using expert labels and self-reports, EngageNet provides a rich resource for developing and benchmarking deep learning models aimed at enhancing user experience in education, human-computer interaction, and related domains.

3.2   Data Preprocessing and Training

We used 80% of the data samples from the balanced SED dataset for finetuning and 20% of the samples for our evaluation with balanced data. We also evaluated our model with the raw SED dataset. We excluded the training balanced sample from the raw SED dataset and used the rest of the samples for evaluation. For both training and testing data samples, we combined the labels for the various categories and created an annotation JSON file that is a collection of data pairs where each image ID is associated with a reference sentence based on the category of that image. The detail of the reference sentence is discussed in Sect. 2.3. We also compared the results of our finetuning prompts with the original prompts (Orig-P) used by MiniGPT-4 [16]. Additionally, the evaluation results, using the original MiniGPT-4 checkpoint without finetuning (Orig-M) on the SED dataset, are shown as the baseline of the MiniGPT-4 model.

For the out-of-distribution evaluation on the DAiSEE dataset, we manually labeled 1046 frames with the three SED categories, assigned image IDs for each frame, and constructed an annotation JSON file. The annotation file for this dataset was constructed in a similar fashion to SED.

The model setup with reference and policy components is memory intensive. It was run on a system with two Nvidia RTX A6000 GPUs. Training ran with a global batch size of four. For the finetuning, we use the AdamW optimizer. The learning rate is controlled with a cosine learning rate scheduler. The initial learning rate is $3e^{-5}$, minimum learning rate is $1e-5$ and warmup learning rate is $1e^{-6}$. The warmup steps is set to 200. The β used in DPO loss calculation is set to 0.1. The training time per epoch is about 25 min.

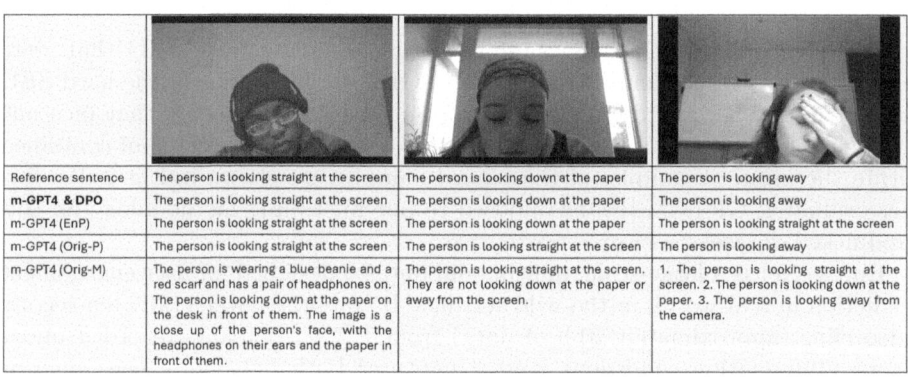

Reference sentence	The person is looking straight at the screen	The person is looking down at the paper	The person is looking away
m-GPT4 & DPO	The person is looking straight at the screen	The person is looking down at the paper	The person is looking away
m-GPT4 (EnP)	The person is looking straight at the screen	The person is looking down at the paper	The person is looking straight at the screen
m-GPT4 (Orig-P)	The person is looking straight at the screen	The person is looking straight at the screen	The person is looking away
m-GPT4 (Orig-M)	The person is wearing a blue beanie and a red scarf and has a pair of headphones on. The person is looking down at the paper on the desk in front of them. The image is a close up of the person's face, with the headphones on their ears and the paper in front of them.	The person is looking straight at the screen. They are not looking down at the paper or away from the screen.	1. The person is looking straight at the screen. 2. The person is looking down at the paper. 3. The person is looking away from the camera.

Fig. 3. Comparison of the generated sentences using four different MiniGPT-4 based finetuning models for the three student engagement behavior markers. The image samples are from SED dataset.

3.3 Assessing Correctness of Generated Answers

We use the output of the model to categorize the student engagements into one of the engagement labels. To evaluate MiniGPT4's output from the still-images, we use three different methods: keyword evaluation, sentence similarity (SS) [2] evaluation, and the Video-ChatGPT [9] evaluation benchmark of Correctness and Consistency. For the keyword evaluation, we consider the output as the correct answer if the generated sentence contains the desired keywords (i.e. paper, screen, away) representing the reference sentence. If no keywords match, the response is considered wrong. Our model, finetuned with DPO, did not return any results with multiple keywords matching. Though the other models we evaluated did. Therefore, we need a way to evaluate the meaning of the sentences. For example, an output generated by MiniGPT-4 (Orig-M) in Fig. 3 says, "The person is looking straight at the screen. They are not looking down at the paper or away from the screen." Based on its meaning, it should be counted as "The person is looking straight at the screen" instead of one of the other categories. A keyword approach would incorrectly categorize the response. To address this issue, we opted for two other methods that have a deeper understanding of sentences.

For the sentence similarity (SS) evaluation, we use a pretrained sentence transformer, BGE-M3 [2], to determine whether the generated sentence conveys the same meaning as the reference sentences. We do this by comparing the embedding of the generated sentences and the candidate reference sentences representing the ground truth classes. If the reference sentence has the highest similarity score among all possible reference sentences, it is marked as correct, otherwise, it is marked as incorrect. We use cosine similarity to capture the semantic similarity of sentence embeddings generated by BGE-M3 [2].

3.4   Results

First, we evaluate the performance of our model using both the accuracy and F1 score obtained by keyword and SS evaluation. We evaluate the performance of the finetuned m-GPT4 & DPO models (MiniGPT-4 model finetuned using DPO) against m-GPT4 models which were finetuned differently depending on the dataset.

For the SED dataset, we evaluate the performance of the finetuned m-GPT4 & DPO models with both the natural language prompts and JSON prompts (JSP). These are compared against m-GPT4 models, which include m-GPT4 model finetuned using the engagement specific prompt (EnP) we discussed in the previous section, m-GPT4 finetuned using the original prompts (Orig-P) of the MiniGPT-4 paper [16], and the original unfinetuned m-GPT4 model (Orig-M). We also compare our results against the deep learning vision model results. We finetuned MobileNet [7] and Xception [4], both pretrained on ImageNet (Pre-IN), on the balanced SED dataset and obtained similar accuracy results on the balanced SED dataset to those presented in [5].

Table 1. Results on balanced SED.

Method	Acc	F1	SS Acc	SS F1
m-GPT4 & DPO	96.7	96.7	96.7	96.7
m-GPT4 (EnP)	95.2	95.2	95.2	95.2
m-GPT4 (JSP/DPO)	96.0	95.9	✕	✕
m-GPT4 (JSP)	95.2	95.2	✕	✕
m-GPT4 (Orig-P)	87.6	87.4	87.6	87.4
m-GPT4 (Orig-M)	58.6	58.9	40.2	39.7
MobileNet (Pre-IN)	94	-	✕	✕
Xception (Pre-IN)	88	-	✕	✕
VGG16 (Pre-IN)	85	-	✕	✕

Table 2. Results on hard SED samples

Method	Acc	F1	SS Acc	SS F1
m-GPT4 & DPO	84.7	84.2	84.7	84.2
m-GPT4 (EnP)	81.2	80.8	81.2	80.8
m-GPT4 (JSP/DPO)	85.9	85.6	✕	✕
m-GPT4 (JSP)	82.4	82.8	✕	✕
m-GPT4 (Orig-P)	70.6	72.1	70.6	72.1
m-GPT4 (Orig-M)	56.5	61.3	31.8	34.7
GPT-4V	74.2	72.6	✕	✕
MobileNet (Pre-IN)	82.3	83.5	✕	✕
Xception (Pre-IN)	84.7	85	✕	✕

Table 3. Results on raw SED.

Method	Acc	F1	SS Acc	SS F1
m-GPT4 & DPO	94.6	95.8	94.6	95.8
m-GPT4 (EnP)	90.8	92.8	90.8	92.8
m-GPT4 (JSP/DPO)	94.3	95.6	✕	✕
m-GPT4 (JSP)	94.1	95.6	✕	✕
m-GPT4 (Orig-P)	89.5	90.0	89.2	89.8
m-GPT4 (Orig-M)	50.5	61.2	48.6	57.8
MobileNet (Pre-IN)	89.9	91.3	✕	✕
Xception (Pre-IN)	87	87.9	✕	✕

Table 4. Results on relabeled DAiSEE samples.

Method	Acc	F1	SS Acc	SS F1
m-GPT4 & DPO	88.4	87.9	88.5	88.0
m-GPT4 (EnP)	87.1	86.9	87.1	86.9
m-GPT4 (JSP/DPO)	87.2	87.1	✕	✕
m-GPT4 (JSP)	86.8	87.0	✕	✕
m-GPT4 (Orig-P)	88.0	87.7	87.2	87.2
m-GPT4 (Orig-M)	62.2	70.7	56.7	67.3
MobileNet (Pre-IN)	26.7	33.2	✕	✕
Xception (Pre-IN)	55.9	65.5	✕	✕

Table 5. DAiSEE engagement levels results

Method	Acc	F1	SS Acc	SS F1
m-GPT4 & DPO	77.1	75.7	71.8	73.0
m-GPT4 (Inst tune)	55.5	40.3	52.5	46.3
ViT Facial Exprn. recog.	54.5	49.3	✕	✕
ViT	53.8	47.9	✕	✕
EmotionNet	51.1	✕	✕	✕
DAiSEE	57.9	✕	✕	✕

Table 6. EngageNet engagement levels.

Method	Acc	F1	SS Acc	SS F1
m-GPT4 & DPO	65.1	64.2	62.6	64.4
m-GPT4 (Inst tune)	53.4	48.2	46.3	50.8
ViT Facial Exprn. Recog.	41.1	43.5	✕	✕
ViT	45.8	45.5	✕	✕
EngageNet	67.6	✕	✕	✕

Table 7. Evaluation with GPT-3.5 turbo

SED Balanced	Correctness ↑	Consistency ↑
m-GPT4 & DPO	**4.84**	**3.67**
m-GPT4 (EnP)	4.77	3.36
m-GPT4 (Orig-P)	4.42	3.58
SED Raw		
m-GPT4 & DPO	**4.77**	**3.70**
m-GPT4 (EnP)	4.69	3.27
m-GPT4 (Orig-P)	4.54	3.48
DAiSEE		
m-GPT4 & DPO	**4.47**	3.64
m-GPT4 (EnP)	4.41	3.09
m-GPT4 (Orig-P)	4.43	**3.68**
SED Hardsamples		
m-GPT4 & DPO	**4.24**	**3.54**
m-GPT4 (EnP)	4.16	3.45
m-GPT4 (Orig-P)	4.2	3.46

We show the evaluation results on the balanced SED dataset in Table 1. As we can see from this table, our method achieved 96.7% accuracy and F1 score, substantially better than the ones by the original deep learning vision models. Interestingly, the JSON prompts achieved almost the same result, though they did perform a bit worse. The results of VGG16 are taken from [5]. Since they only report the accuracy, the other results for these methods are left blank. In all the tables, an ✗ denotes that the sentence similarity is not relevant to the model.

On the evaluation of hard samples of SED (see Table 2), m-GPT4 & DPO model with JSON output performed the best, and this time it is better than m-GPT4 & DPO model with natural language output. Both m-GPT4 & DPO models significantly outperformed GPT-4V results. We also maintain superior performance compared to other m-GPT4 based models. At the same time, we find it interesting that the hard samples cause all the m-GPT4 models to perform worse. This performance hit does not manifest in the purely vision models.

For the out-of-distribution evaluation on the raw SED dataset, our model outperformed the other models across all four measures (see Table 3). On the unbalanced, raw dataset, our model performs a bit worse compared to itself on the balanced dataset. Interestingly, the results of m-GPT4 (Orig-P) outperform itself on the unbalanced raw dataset compared to the balanced dataset. We suspect that even though m-GPT4 (Orig-P) is trained on the balanced dataset, it is skewing toward the predominant class found in the raw unbalanced dataset.

Table 4 shows the out-of-distribution evaluation on the relabeled DAiSEE dataset. Again, both versions of m-GPT4 & DPO exhibit the best performance across all evaluation measures, and the JSON prompt performs slightly worse than the engagement specific prompt. This shows the generalization ability of the proposed approach since DAiSEE is an out-of-distribution dataset. In contrast, the results of MobileNet and Xception drop significantly, which clearly shows that they are not able to generalize well to out-of-distribution samples.

Tables 5 and 6 show our results on the DAiSEE and EngageNet datasets using their original labels. Again, m-GPT4 & DPO model exhibits the best performance. Interestingly, it outperforms EmotionNet [1], which aims to recognize emotions in video data, even though we used only 10 frames from each video. m-GPT4 & DPO model does a bit worse than EngageNet. We believe that is because EngageNet was specifically designed to capture Eye Gaze, Head Pose, and Facial Action Units, which results in better performance when determining the four levels of engagement, but that may not translate.

We also used GPT-3.5 turbo to evaluate the correctness of the natural language answers following the protocol described in [9]. The evaluation with GPT-3.5 turbo (see Table 7) outputs scores range from 0 to 5 for *Correctness* and *Consistency*, signifying the level of alignment between the model output and the ground truth. We find that these evaluation results are consistent with the keyword and Sentence Similarity evaluation results for both the SED and DAiSEE datasets. The proposed method (m-GPT4 & DPO) consistently scores higher across almost all evaluated datasets and performance measures.

4   Conclusions and Future Work

In this paper, we focused on the task of accurate recognition of engagement relevant visual behavior markers using VLMs. We exploited the direct preference optimization (DPO) approach and proposed a modification to its finetuning that uses the model's responses to strengthen its performance. Unlike other preference alignment models, the proposed DPO finetuning generates preference data pairs using the wrong answers generated by the policy model during finetuning. This approach, which is finetuned to the student engagement domain, can leverage pretrained VLMs. This makes the proposed approach easily extensible. We showed that our model's performance is superior to both pure vision models and other finetuning methods. We also demonstrated generalizability to out-of-distribution samples, which is important for real-life applications.

Appendix

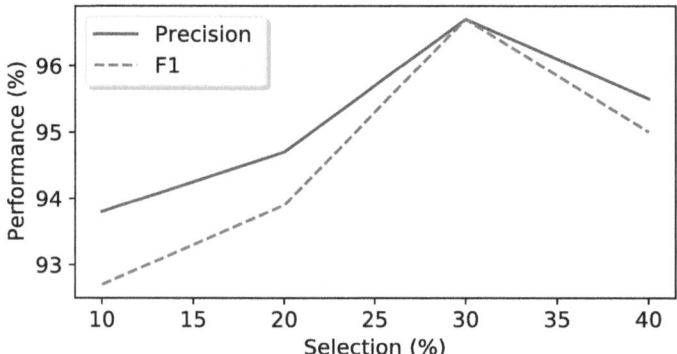

Fig. 4. The performance change when varying the percentage of generated text, used as rejected samples, in DPO tuning on the balanced SED dataset.

In our previous discussion in Sect. 2.2, we mentioned that we use the incorrectly generated answers as dispreferred answers in DPO tuning for 30% of the time. This percentage is selected based on the keyword matching evaluation results when tuning on the balanced SED dataset shown in Fig. 4.

References

1. Benitez-Quiroz, C.F., Srinivasan, R., Martinez, A.M.: Emotionet: an accurate, real-time algorithm for the automatic annotation of a million facial expressions in the wild. In: 2016 IEEE Conference on Computer Vision and Pattern Recognition (CVPR), pp. 5562–5570 (2016). https://doi.org/10.1109/CVPR.2016.600

2. Chen, J., Xiao, S., Zhang, P., et al.: BGE M3-Embedding: Multi-Lingual, Multi-Functionality, Multi-Granularity Text Embeddings Through Self-knowledge Distillation (2024). https://arxiv.org/abs/2402.03216

3. Chiang, W.L., Li, Z., Lin, Z., et al.: Vicuna: An Open-Source Chatbot Impressing GPT-4 with 90%* ChatGPT Quality (2023). https://lmsys.org/blog/2023-03-30-vicuna/

4. Chollet, F.: Xception: deep Learning with Depthwise Separable Convolutions. In: 2017 IEEE Conference on Computer Vision and Pattern Recognition (CVPR), pp. 1800–1807 (2017). https://doi.org/10.1109/CVPR.2017.195

5. Delgado, K., Origgi, J.M., Hasanpoor, T., et al.: Student engagement dataset. In: Proceedings of the IEEE/CVF International Conference on Computer Vision (ICCV) Workshops, pp. 3628–3636 (2021)

6. Gupta, A., D'Cunha, A., Awasthi, K., Balasubramanian, V.: DAiSEE: Towards User Engagement Recognition in the Wild (2022). https://arxiv.org/abs/1609.01885

7. Howard, A.G., Zhu, M., Chen, B., et al.: MobileNets: Efficient Convolutional Neural Networks for Mobile Vision Applications (2017). https://arxiv.org/abs/1704.04861

8. Lei, H., Cui, Y., Zhou, W.: Relationships between student engagement and academic achievement: a meta-analysis. Soc. Behav. Pers. Int. J. **46**, 517–528 (2018). https://doi.org/10.2224/sbp.7054

9. Maaz, M., Rasheed, H., Khan, S., Khan, F.S.: Video-ChatGPT: Towards Detailed Video Understanding via Large Vision and Language Models (2024). https://arxiv.org/abs/2306.05424

10. OpenAI: GPT-4V(ision) system card (2023). https://cdn.openai.com/papers/GPTV_System_Card.pdf

11. Rafailov, R., et al.: Direct preference optimization: your language model is secretly a reward model. In: Oh, A., et al. (eds.) Advances in Neural Information Processing Systems, vol. 36, pp. 53728–53741. Curran Associates, Inc. (2023). https://proceedings.neurips.cc/paper_files/paper/2023/file/a85b405ed65c6477a4fe8302b5e06ce7-Paper-Conference.pdf

12. Schulhoff, S., Ilie, M., Balepur, N., et al.: The Prompt Report: A Systematic Survey of Prompting Techniques (2025). https://arxiv.org/abs/2406.06608

13. Schulman, J., Wolski, F., Dhariwal, P., et al.: Proximal policy optimization algorithms (2017). https://arxiv.org/abs/1707.06347

14. Singh, M., Hoque, X., Zeng, D., et al.: Do i have your attention: a large scale engagement prediction dataset and baselines. In: Proceedings of the 25th International Conference on Multimodal Interaction, pp. 174–182 (2023)

15. Touvron, H., Lavril, T., Izacard, G., et al.: Llama: Open and efficient foundation language models (2023). https://arxiv.org/abs/2302.13971

16. Zhu, D., Chen, J., Shen, X., Li, X., Elhoseiny, M.: MiniGPT-4: enhancing vision-language understanding with advanced large language models (2023). https://arxiv.org/abs/2304.10592

LLM-Generated Semantic Co-occurrences for Multi-label Food Recognition

Daniel Ponte[1]([✉]) [ID], Eduardo Aguilar[1,2] [ID], Mireia Ribera[1] [ID],
and Petia Radeva[1,3] [ID]

[1] Deptartment de Matemàtiques i Informàtica, Universitat de Barcelona, Gran Via
de les Corts Catalanes 585, Barcelona 08007, Spain
dponteva163@alumnes.ub.edu, {eduardo.aguilar,ribera,petia.ivanova}@ub.edu
[2] Deptartment de Ingeniería de Sistemas y Computación, Universidad Católica del
Norte, Angamos 0610, Antofagasta 1270398, Chile
[3] Computer Vision Center Campus UAB, Edifici O, Cerdanyola,
08193 Barcelona, Spain

Abstract. Multi-label learning in food image recognition presents a
promising avenue for understanding the visual composition of meals
through joint ingredient prediction. In this article, we improve an existing
GCN-based framework by replacing its standard co-occurrence matrix
with a novel semantic variant, constructed using large language models
(LLMs). Unlike traditional approaches that derive co-occurrence statistics solely from the training data which often introducing biases and
limiting generalization, our method leverages prior knowledge extracted
from LLMs to build an adjacency matrix that captures broader and
more contextually grounded ingredient relationships. We evaluated our
approach on the MAFood-121 and VireoFood-172 datasets, significantly
outperforming the benchmark method that relies on dataset-conditioned
co-occurrence graphs. On MAFood-121, our model improved the mean
average precision (mAP) from 82.77% to 87.46%, while on VireoFood-
172, it increased from 60.88% to 65.28%. The results demonstrate the
effectiveness of integrating LLM-derived semantic structure into graph-
based multi-label models for structured food recognition.

Keywords: Multi-label · LLMs · Co-occurrence matrix · GCN

1   Introduction

Multi-label learning for food image recognition has gained increasing attention,
driven by the need to automatically understand the complex composition of
meals in real-world scenarios. Unlike traditional single-label classification, food
images often require the simultaneous prediction of multiple ingredients, many
of which co-occur naturally or share semantic relationships. Early efforts focused
on dish classification, but limitations in capturing internal ingredient structures
led to a shift toward fine-grained multi-label frameworks, accelerated by the
success of convolutional neural networks (CNNs) [6].

© The Author(s), under exclusive license to Springer Nature Switzerland AG 2026
M. Castrillón-Santana et al. (Eds.): CAIP 2025, LNCS 15622, pp. 220–230, 2026.
https://doi.org/10.1007/978-3-032-05060-1_19

The emergence of vision-language models (VLMs) (e.g. CLIP [12]) has expanded image recognition by aligning visual and textual information in a shared space, allowing flexible prompt-based recognition. Nevertheless, in food analysis, VLMs often treat labels independently, ignoring ingredients relationships (e.g. the presence of "rice" may imply "seafood" or "vegetable", a dependency not directly modeled in traditional approaches).

To capture label dependencies, GCNs [15] have been introduced into multi-label pipelines. DualCoOp [16] and SCPNet [9] combine VLMs with graph structures to refine predictions. Nevertheless, a major limitation persists: the co-occurrence matrices used are derived from training data, reflecting dataset-specific biases and limiting generalization across diverse cuisines and contexts.

To overcome this, we propose LLM-MLR (Large Language Model Multi-Label Recognition), a novel framework that incorporates external semantic knowledge extracted from LLMs. Instead of relying solely on training data, we generate a semantic co-occurrence matrix mined from over three million global recipes, guided by empirical co-occurrence frequency, culinary compatibility, shared cooking techniques, and multicultural diversity.

This matrix allows the GCN module to refine initial predictions from a CLIP-based backbone more effectively, resulting in improved performance and robustness. We validate LLM-MLR on two public datasets, MAFood-121 [2] and VireoFood-172 [5], which present challenges due to inter-class similarity, intra-class variability, and limited data. Our experiments show consistent improvements compared to baseline methods that rely solely on co-occurrence statistics extracted from the training data, highlighting the benefits of integrating external structured knowledge into food ingredient recognition.

Our main contributions in this paper are as follows:

1. We propose LLM-MLR, a novel framework that leverages large language models to build a semantic ingredient co-occurrence matrix, mitigating biases inherent in dataset-conditioned graphs.
2. We demonstrate the effectiveness of LLM-MLR through comprehensive experiments on two public food datasets, achieving superior generalization compared to the Multi-label Recognition (MLR) [13] baseline method.
3. We introduce a carefully designed prompt-based approach, explicitly defining culinary categories and their relationships, enabling LLMs to generate precise and semantically coherent ingredient co-occurrence probabilities.

2 Methodology

We present the LLM-MLR framework for multi-label ingredient recognition in food images. Our method integrates a frozen CLIP-based backbone for visual and textual representation extraction with a GCN for semantic refinement, an externally constructed semantic co-occurrence matrix derived using LLMs.

2.1   Overall Architecture

Our work builds upon the architecture proposed in the Multi-label Recognition (MLR) [13], which operates in two stages: feature extraction and semantic refinement. In the first stage, visual and textual features are extracted using a frozen CLIP backbone, which preserves the pretrained vision-language alignment. Instead of static prompts, MLR employs learnable positive and negative prompt learners, following recent advances in prompt tuning for vision-language models. These learnable prompts allow the model to adapt contextually to the specifics of the ingredient recognition task. The extracted features are then processed by a Graph Convolutional Network (GCN) that refines the initial logits, modeling dependencies between ingredient labels. The key contribution of our method lies in the construction of the semantic co-occurrence matrix (see Table 1) used by the GCN. While MLR relies on co-occurrence statistics derived directly from the training data, our approach replaces this with a semantically enriched matrix generated through prompt-based querying of a large language model (LLM), offering broader generalization and mitigating dataset-induced biases.

2.2   Semantic Co-occurrence Matrix Construction

Traditional approaches often derive co-occurrence matrices directly from training data, making them susceptible to dataset biases. To overcome this, we introduce a semantic co-occurrence matrix generated using a LLM. The matrix construction is based on information extracted from more than three million culinary recipes across international platforms like Allrecipes [3] and Epicurious [11], according to specific culinary and statistical criteria.

We used OpenAI's ChatGPT (GPT-4) [1] to generate the semantic co-occurrence matrix via prompt-based querying. The LLM was instructed to assign co-occurrence probabilities between ingredients based on culinary and statistical principles:

– Frequency of co-occurrence in recipe databases (e.g., meat + pasta with 0.82; meat + rice with 0.70).
– Culinary compatibility (e.g., soup + vegetable at 0.86; egg + bread at 0.60).
– Shared cooking techniques (e.g., friedfood + seafood at 0.76).
– Restrictions or dietary preferences, penalizing rare or redundant combinations (e.g., bread + dumpling at 0.24; rice + bread at 0.12).
– Multicultural and the global diversity (e.g. noodle + soup at 0.66).

The resulting matrices for MAFood-121 (10×10) and VireoFood-172 (18×18) (see Fig. 1) are normalized between 0 and 1, symmetric, and enforce 1 s on the diagonal. A further normalization is applied to stabilize graph propagation: row sums (excluding diagonals) are scaled to 0.2, while the diagonal is adjusted to 0.8. This ensures that the adjacency structure is numerically stable for GCN processing without overwhelming node self-representations.

Table 1. Prompt developed to instruct the LLM for matrix generation

Calculate the probability that two food groups appear together in the same dish or culinary preparation.

Food groups and what they include:
- *bread: wheat flour, bread, pizza dough, etc.*
- *dumpling: stuffed pasta, wheat semolina, etc.*
- *egg: whole eggs, egg whites, etc.*
- *friedfood: any fried food (tempura, croquettes, French fries, etc.).*
- *meat: red meats, chicken, turkey, bacon, sausage, etc.*
- *noodle/pasta: spaghetti, noodles, macaroni, ramen, etc.*
- *rice: rice in all its variations.*
- *seafood: fish, seafood, octopus, shrimp, etc.*
- *soup: light or dark poultry, beef or vegetable broths.*
- *vegetable: vegetables, legumes, mushrooms, avocado and fruits.*

Rules when generating the matrix:
- *a. The matrix is 10×10 and symmetric.*
- *b. The main diagonal is always 1.00.*
- *c. All values must be between 0.00 and 1.00 with two decimal places.*
- *d. Base the numbers on plausible culinary data and compatibility.*
- *e. Return only the matrix without explanations.*
- *f. Use tabs or spaces to separate columns.*

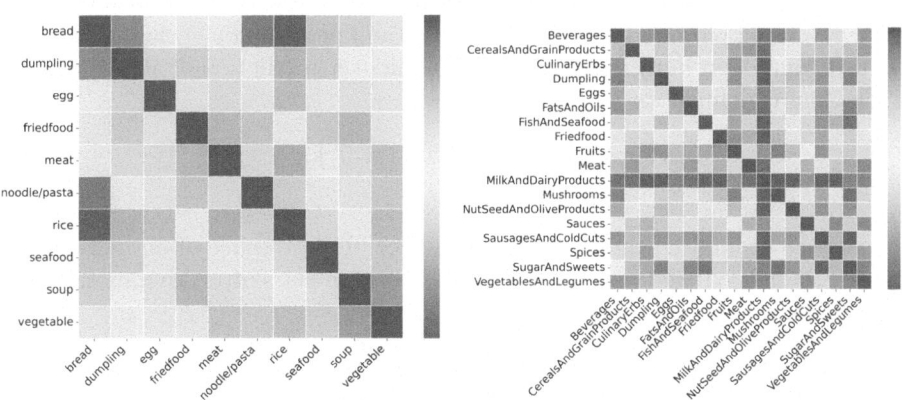

Fig. 1. Semantic Co-occurrence Matrices generated for MAFood-121 (left) and VireoFood-172 (right), based on structured knowledge extracted using a LLM. These matrices serve as the relational backbone of the graph-based ingredient prediction model.

2.3 GCN-Based Refinement Module

The semantic refinement stage employs a three-layer GCN architecture inspired by recent advances in graph-based multi-label learning [4]. The GCN processes the initial CLIP-based logits, integrating information from related ingredients through the adjacency matrix. Specifically, the architecture consists of:

- A first graph convolution layer (GCL) maps the input dimension (1) to an intermediate representation (4).
- A second GCL maintains the same intermediate dimension (4).
- A final GCL reduces the dimension back to 1.

Each of the first two layers applies a LeakyReLU activation with a negative slope of 0.2, introducing non-linearity and improving information propagation through the graph. No dropout is applied to preserve information across the small number of nodes. The propagation rule for each layer follows:

$$H^{(l+1)} = \sigma(\hat{A} H^{(l)} W^{(l)})$$

where $H^{(l)}$ is the feature matrix at layer l, $W^{(l)}$ is the learnable weight matrix, $\hat{A}$ is the normalized adjacency matrix, and $\sigma(\cdot)$ denotes the LeakyReLU activation function. The final refined logits are added residually to the original logits from CLIP to preserve strong visual signals.

2.4 Loss Function and Training Strategy

Given the class imbalance inherent in ingredient recognition datasets, we adopt the Asymmetric Loss function [14] to optimize the model. This loss introduces distinct focusing parameters for positive (γ^+) and negative (γ^-) classes to prioritize rare ingredient detection while controlling over-prediction of absent classes.
The loss is defined as:

$$\mathcal{L}(x, y) = -y \log p^+ - (1 - y) \log p^-$$

where:
- y is the ground truth label (1 for presence, 0 for absence of an ingredient),
- p^+ is the predicted probability for the positive class after asymmetric clipping and focusing,
- p^- is the predicted probability for the negative class,
- $\log p^+$ and $\log p^-$ are the log-likelihoods emphasizing correct predictions for positives and negatives, respectively.

The asymmetric clipping reduces the impact of correctly predicted negative examples, while the asymmetric focusing dynamically scales the contribution of hard-to-classify positive examples.

The model is trained using Stochastic Gradient Descent (SGD) with momentum. A cosine annealing learning rate schedule with a warmup phase is employed. Only the prompt learners, GCN parameters, and optionally the attention heads are updated during training. Data augmentation strategies such as resizing, Cutout [8], and RandAugment [7] are used to enhance robustness.

3 Experimental Setting

We first describe the datasets used, followed by the evaluation metrics, the experimental setup, and finally, the model settings employed to validate our approach.

3.1 Datasets

The two public datasets selected to validate our approach are MAFood-121 and VireoFood-172, each supporting multi-label food ingredient recognition with distinctive characteristics.

MAFood-121. consists of 21,175 images distributed across 121 traditional dishes from 11 global cuisines. The dishes are organized into 10 major food groups, such as "bread", "meat", and "seafood", enabling analysis across both common and culturally significant dishes.

VireoFood-172. contains 110,241 images covering 172 Chinese dishes. Unlike MAFood-121, this dataset provides ingredient-level annotations, allowing a detailed analysis of the components present in each dish. Ingredients (353 types) were grouped into 18 food categories based on the HELIS ontology [10] (Healthy Eating and Lifestyle ontology), which provides a structured categorization for food ingredients, facilitating consistent grouping of related items into broader food groups. For instance, in HELIS, ingredients such as Bread, Rice, and Spaghetti are grouped under *CerealsAndGrainProducts*; Pork chunks, Chicken legs, and Beef slices are grouped under *Meat*; and Asparagus, Zucchini slices, and Green soybeans are grouped under *VegetablesAndLegumes*. This structured organization enables the ingredient recognition task in VireoFood-172 to be aligned with the grouping structure used in MAFood-121.

For both datasets, the images were partitioned into 80% for training and 20% for testing. In all experiments, input images were resized to 224×224 pixels to match the requirements of the model backbone.

3.2 Metrics

We evaluate our models using standard multi-label classification metrics, computed at both macro and micro levels based on the per-class predictions. Specifically, we report **Mean Average Precision (mAP)**, **Precision**, **Recall**, **F1-Score**, and **Jaccard Index (J)**.

3.3 Experimental Setup

We trained end-to-end a CLIP-based architecture enhanced with a GCN (see Sect. 2). The visual encoder uses a frozen ResNet-101 backbone, with only the prompt learners and GCN parameters being optimized during training.

Two model variants were compared:

- MLR using a **train-conditioned co-occurrence matrix** (biased to training data statistics).

Table 2. Comparison between train-conditioned co-occurrence matrix and LLM-derived semantic co-occurrence matrix (LLM-MLR) on MAFood-121 and VireoFood-172 datasets.

		MACRO		MICRO		
Dataset	Method	F1	J	F1	J	mAP
MAFood-121	MLR	72.48	65.85	57.38	60.46	82.77
	LLM-MLR	**78.49**	**74.35**	**64.89**	**67.81**	**87.46**
VireoFood-172	MLR [13]	57.42	68.36	42.89	61.24	60.88
	LLM-MLR	**60.94**	**73.61**	**46.72**	**66.30**	**65.28**

- LLM-MLR using an **LLM-derived semantic co-occurrence matrix** (based on external culinary knowledge).

All models were initialized from pretrained CLIP weights. The training process used SGD with a momentum of 0.9 and a weight decay of 5×10^{-4}. The initial learning rate was set to 0.001, following a cosine annealing schedule with a warmup phase during the first epoch. Models were trained for 150 epochs, using a batch size of 32 for training and 100 for testing. No validation set was used; only training and testing splits were considered.

All experiments were conducted on NVIDIA GPUs with CUDA acceleration using PyTorch.

4   Results and Discussion

Table 2 presents the quantitative evaluation of our proposed LLM-MLR model compared to the MLR baseline, which relies solely on a co-occurrence matrix derived from the training set. The evaluation was conducted on the official test splits of MAFood-121 and VireoFood-172, two benchmark datasets for multi-label food ingredient recognition. Results show consistent improvements across all macro and micro metrics when replacing the biased co-occurrence graph with our LLM-derived semantic prior. In MAFood-121, macro-F1 improves from 72.48% to 78.49%, and mAP from 82.77% to 87.46%. For VireoFood-172, gains are also notable, with macro-F1 increasing from 57.42% to 60.94% and micro-mAP from 60.88% to 65.28%. The greater improvements in macro metrics demonstrate better handling of minority classes, which are typically harder to model under heavy class imbalance.

To understand the behavior of each model in more depth, Fig. 2 shows the per-class multi-label confusion matrices for MAFood-121 and VireoFood-172, respectively. Each matrix visualizes true positives (TP), false positives (FP), false negatives (FN), and true negatives (TN) for all food groups. The matrices on the top correspond to the baseline (MLR), and those on the bottom to our model (LLM-MLR). In MAFood-121, LLM-MLR improves both precision and recall across most food groups. For instance, *bread* increases from 975

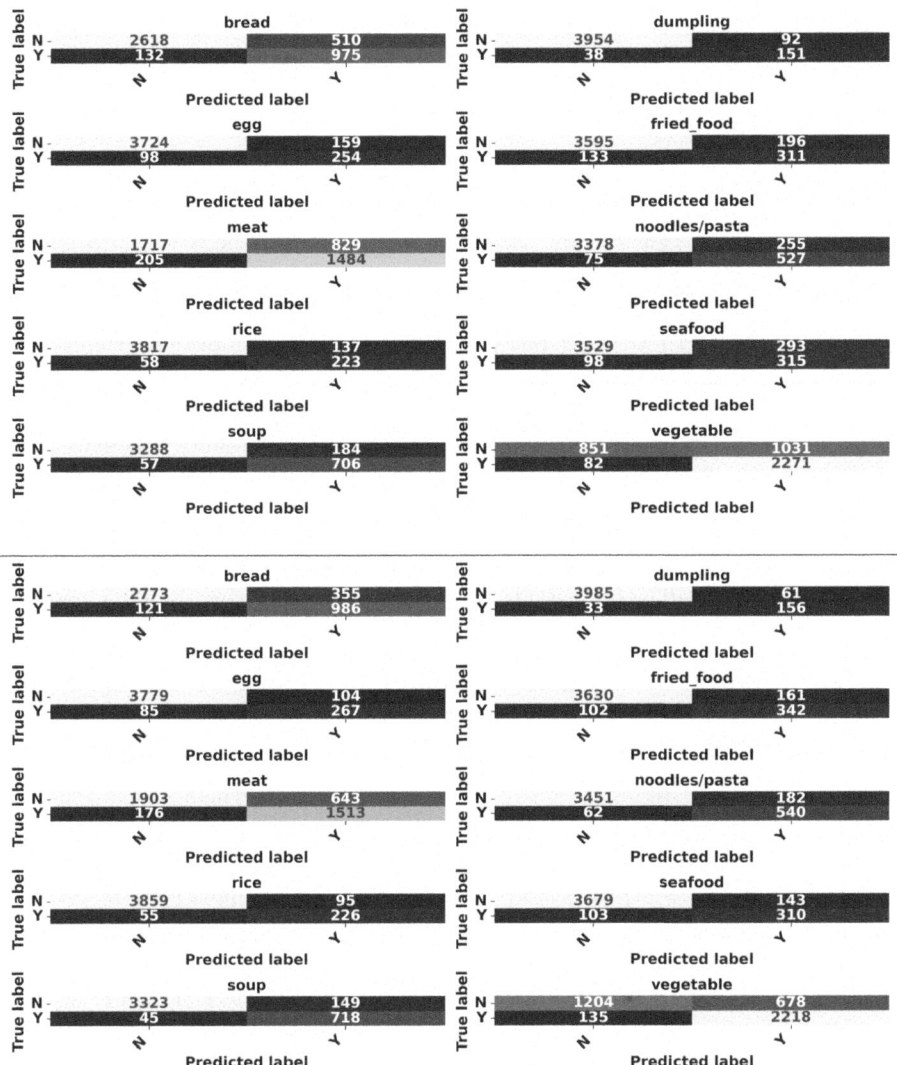

Fig. 2. Per-class multi-label confusion matrices for MAFood-121. Top: MLR baseline using a train-conditioned co-occurrence matrix. Bottom: LLM-MLR with an LLM-derived semantic matrix. The proposed model yields fewer false positives and improved true positive rates across most food groups. Notable gains are observed in *bread*, *meat* and *vegetable*, demonstrating stronger discriminative capacity and better context modeling.

to 986 TP while reducing FP from 510 to 355. Similar trends appear in *meat* and *vegetable*, the latter seeing a substantial FP reduction from 1,031 to 678 while maintaining high TP counts. These shifts suggest that the semantic prior improves discriminative capacity, especially for food groups that are visually or contextually confusable. On the other hand, Table 3 summarizes the qualitative differences between the baseline and the proposed method on the MAFood-121 and VireoFood-172 datasets.

Beyond per-class metrics, we performed a per-sample agreement analysis to determine how often each model succeeded or failed independently. Table 4 categorizes the predictions into 4 cases: both correct (YY), both incorrect (NN), only LLM-MLR correct (Y/N), and only MLR correct (N/Y). As expected, most test samples fall into the YY group, especially in high-frequency classes. However, in both datasets, the Y/N group is substantially larger than N/Y, demonstrating that LLM-MLR recovers harder examples more effectively than the baseline (e.g. in MAFood-121, there are 811 samples correctly predicted only by LLM-MLR, versus just 462 where the baseline was superior). This indicates that semantic priors help resolve visually ambiguous cases more reliably.

Furthermore, we observe that certain food groups exhibit stronger inter-class disentanglement under the semantic graph. In MAFood-121, ingredient pairs like *bread* and *fried_food*, which often co-occur in recipes but are visually distinct, are better separated by LLM-MLR. Likewise, in VireoFood-172, confusions between *meat* and *CerealsAndGrainProducts*, commonly caused by overlapping textures

Table 3. Comparison of model predictions across representative food groups. Classes highlighted where LLM-MLR improves over the baseline (MLR), as well as where performance remains stable or slightly degraded.

Dataset	MLR (Baseline)	LLM-MLR (Ours)
MAFood-121 [2]	*Bread*: 975 TP, 510 FP, 132 FN	*Bread*: 986 TP, 355 FP, 121 FN
	Vegetable: 2,271 TP, 1,031 FP	*Vegetable*: 2,218 TP, 678 FP
	Meat: 1,484 TP, 829 FP	*Meat*: 1,513 TP, 643 FP
VireoFood-172 [5]	*Meat*: 10,065 TP, 4,886 FP, 887 FN	*Meat*: 10,821 TP, 3,978 FP, 671 FN
	FishAndSeafood: 2,647 TP, 1,490 FP	*FishAndSeafood*: 2,688 TP, 1,186 FP
	Spices: 7,430 TP, 4,925 FP	*Spices*: 7,560 TP, 4,165 FP
Case Agreement	Correct: both TP in frequent classes (e.g., *Eggs*, *Dairy*) Incorrect: both FP in ambiguous pairs (e.g., *Sauces* vs *Spices*)	Improves rare/overlapping classes (e.g., *Vegetables*, *Fish*) Few degradation cases, often on borderline predictions

Table 4. Per-image agreement analysis between MLR (baseline) and LLM-MLR. YY: both correct, NN: both incorrect, Y/N: only LLM-MLR correct, N/Y: only baseline correct.

Dataset	YY	NN	Y/N	N/Y
	both correct	both incorrect	only ours	only baseline
MAFood-121	3,942	728	811	462
VireoFood-172	15,184	4,367	2,143	1,002

in composite dishes, are significantly reduced. These outcomes reinforce our core hypothesis: integrating a linguistically structured prior into a GCN enables better generalization and contextual reasoning, especially in scenarios affected by class imbalance and semantic overlap. These findings are consistent with broader trends in vision-language research, where pretrained language models provide strong inductive biases for downstream recognition tasks.

5   Conclusions

In this paper, we proposed LLM-MLR, a novel approach that integrates semantic ingredient relationships derived from LLMs into GCN for multi-label food ingredient recognition. Unlike traditional models that rely solely on training-data-specific co-occurrence matrices, our method leverages external structured knowledge, effectively mitigating dataset biases and significantly enhancing generalization capabilities.

Extensive experiments on the MAFood-121 and VireoFood-172 datasets demonstrate consistent and substantial improvements in multiple metrics such as macro-F1, Jaccard Index, and mean Average Precision. Specifically, our method shows notable advantages for minority classes, effectively reducing common misclassifications between visually similar food groups. These results clearly highlight the importance and efficacy of embedding linguistically informed semantic priors into recognition models, reinforcing their potential as robust inductive biases in complex multi-label recognition scenarios.

In future work, we aim to extend our framework beyond the prediction of food groups by simultaneously incorporating ingredient-level and dish-level tasks within a multi-task learning approach. This multi-task strategy could further exploit the structured semantic knowledge, benefiting from shared representations across closely related tasks and leveraging the inherent semantic relationships between dishes and ingredients. Additionally, exploring hierarchical food ontologies integrated directly into the GCN structure could yield even finer-grained and more accurate predictions, enhancing not only model accuracy, but also interpretability in food image recognition tasks.

Acknowledgments. This work has been partially supported by the Spanish project PID2022-136436NB-I00 (AEI-MICINN), Horizon EU project MUSAE (No. 01070421), 2021-SGR-01094 (AGAUR), Icrea Academia'2022 (Generalitat de Catalunya), Robo STEAM (2022-1-BG01-KA220-VET-000089434, Erasmus+ EU), DeepSense (ACE053/22/000029, ACCIÓ), DeepFoodVol (AEI-MICINN, PDC2022-133642-I00), PID2022-141566NB-I00 (AEI-MICINN), CERCA Programme / Generalitat de Catalunya and Beatriu de Pinós Programme (2022 BP 00257). D. Ponte acknowledges the support of Secretaría Nacional de Ciencia, Tecnología e Innovación Senacyt Panamá (Scholarship No. 270-2022-125).

Disclosure of Interests. The authors have no competing interests to declare that are relevant to the content of this article.

References

1. Achiam, J., et al.: Gpt-4 Technical Report (2023). arXiv preprint arXiv:2303.08774
2. Aguilar, E., Bolaños, M., Radeva, P.: Regularized uncertainty-based multi-task learning model for food analysis. JVCI **60**, 360–370 (2019)
3. Allrecipes, I.: Allrecipes (2023). https://www.allrecipes.com/
4. Bei, Y., et al.: Correlation-aware graph convolutional networks for multi-label node classification (2024). arXiv preprint arXiv:2411.17350
5. Chen, J.J., Ngo, C.W.: Deep-based ingredient recognition for cooking recipe retrieval. ACM Multimedia (2016)
6. Chen, J., et.al.: Zero-shot ingredient recognition by multi-relational graph convolutional network. In: AAAI CAI. vol. 34, pp. 10542–10550 (2020)
7. Cubuk, E.D., Zoph, B., Shlens, J., Le, Q.V.: Randaugment: practical automated data augmentation with a reduced search space. In: CVPRW, pp. 702–703 (2020)
8. DeVries, T., Taylor, G.W.: Improved regularization of convolutional neural networks with cutout (2017). arXiv preprint arXiv:1708.04552
9. Ding, Z., et.al.: Exploring structured semantic prior for multi label recognition with incomplete labels. In: CVPR, pp. 3398–3407 (2023)
10. Donadello, I., Dragoni, M.: Ontology-driven food category classification in images. In: ICIAP, Part II 20, pp. 607–617. Springer (2019)
11. Jimenez-Mavillard, A., Suarez, J.L.: Diffusion of elbulli's innovation: rate of adoption in allrecipes and epicurious. IJGFS **22**, 100243 (2020)
12. Radford, A., et.al.: Learning transferable visual models from natural language supervision. In: ICML, pp. 8748–8763. PmLR (2021)
13. Rawlekar, S., Bhatnagar, S., Srinivasulu, V.P., Ahuja, N.: Improving multi-label recognition using class co-occurrence probabilities. In: ICPR, pp. 424–439 (2025)
14. Ridnik, T., et.al.: Asymmetric loss for multi-label classification. In: ICCV, pp. 82–91 (2021)
15. Singh, I.P., Ghorbel, E., Oyedotun, O., Aouada, D.: Multi-label image classification using adaptive graph convolutional networks: from a single domain to multiple domains. Comput. Vis. Image Underst. (2024)
16. Sun, X., Hu, P., Saenko, K.: Dualcoop: Fast adaptation to multi-label recognition with limited annotations. NIPS **35**, 30569–30582 (2022)

Tracing Information Flow in LLaMA Vision: A Step Toward Multimodal Understanding

Alessia Saporita[1,2], Vittorio Pipoli[1,3], Federico Bolelli[1(✉)], Lorenzo Baraldi[1], Andrea Acquaviva[2], and Elisa Ficarra[1]

[1] University of Modena and Reggio Emilia, Modena, Italy
federico.bolelli@unimore.it
[2] University of Bologna, Bologna, Italy
[3] University of Pisa, Pisa, Italy

Abstract. Multimodal Large Language Models (MLLMs) have recently emerged as a powerful framework for extending the capabilities of Large Language Models (LLMs) to reason over non-textual modalities. However, despite their success, understanding how they integrate visual and textual information remains an open challenge. Among them, LLaMA 3.2-Vision represents a significant milestone in the development of open-source MLLMs, offering a reproducible and efficient architecture that competes with leading proprietary models, such as Claude 3 Haiku and GPT-4o mini. Motivated by these characteristics, we conduct the first systematic analysis of the information flow between vision and language in LLaMA 3.2-Vision. We analyze three visual question answering (VQA) benchmarks, covering the tasks of VQA on natural images—using both open-ended and multiple-choice question formats—as well as document VQA. These tasks require diverse reasoning capabilities, making them well-suited to reveal distinct patterns in multimodal reasoning. Our analysis unveils a four-stage reasoning strategy: an initial semantic interpretation of the question, an early-to-mid-layer multimodal fusion, a task-specific reasoning stage guided by the resulting multimodal embedding, and a final answer prediction stage. Furthermore, we reveal that multimodal fusion is task-dependent: in complex settings such as document VQA, the model postpones cross-modal integration until semantic reasoning over the question has been established. Overall, our findings offer new insights into the internal dynamics of MLLMs and contribute to advancing the interpretability of vision-language architectures. Our source code is available at https://github.com/AImageLab/MLLMs-FlowTracker.

Keywords: Multimodal LLMs · Cross-Modal Fusion · VQA

1 Introduction

Multimodal Large Language Models (MLLMs) [1,11,13,33] represent a significant evolution in artificial intelligence, extending traditional language models [2,6,26,27] to process and reason over non-textual modalities. They typically consist of a vision encoder [9,25,28] that extracts image features, a fusion mechanism that integrates visual and textual representations into a unified space, and a language model backbone. By fusing multimodal information, MLLMs enable more advanced capabilities essential for a broad range of real-world applications.

LLaMA 3.2-Vision [11] marks a significant milestone, offering an open-source, efficient, and scalable architecture that achieves competitive performance with proprietary models such as Claude 3 Haiku [3] and GPT-4o mini [21]. Given its architectural transparency and strong performance, we conduct a systematic analysis of the information flow between visual and textual modalities in LLaMA 3.2-Vision to shed light on the internal mechanisms of MLLMs, which have not been sufficiently explored in the literature.

Recent studies investigated the inner mechanisms of MLLMs by analyzing information storage in model parameters [5], knowledge encoding in model parameters [19], object-level visual grounding [20,22], visual signal decay across layers [29], and token-level redundancy [31]. Basu *et al.* [5] reveals that specific visual tokens are responsible for transferring information from the image to causal blocks within the architecture. Zhang *et al.* [31] identifies redundancy in early visual token representations and proposes a truncation strategy to improve model efficiency. Nevertheless, the internal mechanisms of MLLMs, including the information flow between modalities, have yet to be thoroughly investigated.

In this paper, we address this gap through a systematic analysis of the visual–textual interactions in LLaMA 3.2-Vision. We conduct experiments on three datasets—VQAv2 [10], Visual7W [34], and DocVQA [17]—which are well-suited for revealing distinct patterns in multimodal fusion due to their diverse reasoning requirements. VQAv2 and Visual7W are visual question answering datasets on natural images, differing in their question formats. VQAv2 employs open-ended questions, which require free-form generation and broad contextual understanding, whereas Visual7W adopts a multiple-choice format, which explicitly conditions the model's reasoning on the provided candidate answers. On the other hand, DocVQA is a benchmark for document visual question answering, which requires Optical Character Recognition (OCR) capabilities to extract and reason about text embedded in the image.

Our results reveal a four-stage reasoning strategy in LLaMA 3.2-Vision: an initial phase dedicated to the semantic interpretation of the input question, followed by an early-to-mid-layer multimodal integration phase, a task-specific reasoning stage, and a final stage for answer generation. Furthermore, we demonstrate that multimodal fusion is task-dependent. Hence, for simpler tasks such as VQA on natural images, cross-modal interactions emerge in early layers. In contrast, for more complex tasks like document VQA, the model postpones integration until semantic reasoning over the question has been established. Moreover, in the DocVQA dataset, we observe that visual information has a direct influence on response generation, as the model requires explicit access to fine-grained visual features during the answer decoding stage, due to the need to interpret structured text embedded within the image. To the best of our knowledge, this is the first study to explicitly trace the information flow between visual and textual modalities in LLaMA 3.2-Vision, offering novel insights into its cross-modal dynamics and contributing to the interpretability of MLLMs.

Contributions. The contributions of the paper can be summarized as follows:

- We present a systematic analysis of the information flow between visual and textual modalities in LLaMA 3.2-Vision;
- We investigate the information flow across three datasets that demand varying reasoning capabilities, encompassing tasks such as natural image understanding (in both open-ended and multiple-choice formats) and document visual question answering;
- We uncover a four-stage reasoning strategy comprising question interpretation, early multimodal fusion, task-specific reasoning, and answer generation;
- We demonstrate that multimodal fusion is task-dependent, meaning that in more complex tasks the model delays visual grounding until semantic reasoning over the question has been established.

2 Related Work

Multimodal Large Language Models (MLLMs). Multimodal Large Language Models (MLLMs) [1,11,13,33] have emerged as a significant advancement in artificial intelligence, combining the reasoning capabilities of Large Language Models (LLMs) [2,6,26,27] with the ability to interpret and generate content across multiple modalities [12,24]. The architecture of MLLMs consists of modality-specific encoders [8,9,25] and a large language model backbone. Notable examples include models like Flamingo [1], BLIP-2 [14], MiniGPT-4 [33], LLaVA [16], and OpenFlamingo [4]. LLaMA 3.2-Vision [11] represents a significant milestone in the development of open-access MLLMs, combining efficiency, scalability, and reproducibility while rivaling proprietary counterparts, such as Claude 3 Haiku [3] and GPT-4o mini [21].

Analyzing Internal Mechanisms of MLLMs. While significant progress has been made in interpreting large language models (LLMs) [7,18,23,30], the multimodal domain remains less explored. Recent works have started to investigate the internal states of MLLMs by analyzing information storage [5] and knowledge encoding [19] in model parameters, object-level visual grounding [20,22], visual signal decay across layers [29], token-level redundancy [31], and cross-modal interactions [32]. For instance, Basu *et al.* [5] analyzes how visual information is stored and transferred within the model's architecture. Yin *et al.* [29] studies shifting patterns of visual information flow across layers to enable more efficient inference in MLLMs. Zhang *et al.* [31] investigates redundancy in the early-stage visual representations of MLLMs and proposes a token truncation strategy that reduces computational overhead while preserving task performance. Despite recent progress, the internal mechanisms of multimodal integration in MLLMs remain poorly understood. This work addresses this gap with an in-depth analysis of the inner information flows in LLaMA 3.2-Vision, contributing to a deeper understanding of MLLMs.

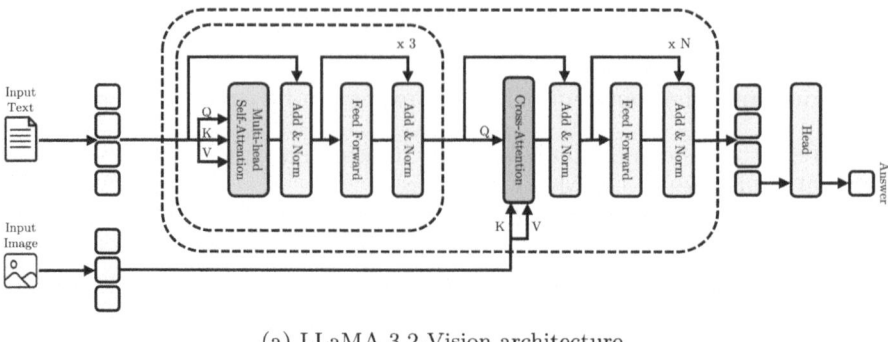

(a) LLaMA 3.2-Vision architecture.

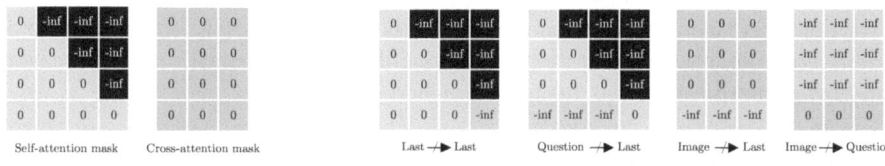

(b) Default masks.   (c) Masking to analyze different information flows.

Fig. 1. Overview of the proposed method. (a) LLaMA 3.2-Vision architecture; (b) default attention masking mechanism used in self- and cross-attention layers; (c) modified attention masks enabling analysis of distinct information flows, including last-to-last, question-to-last, image-to-last, and image-to-question pathways.

3   Method

In this work, we analyze the interaction between visual and textual modalities in LLaMA 3.2-11B-Vision [11]. To this end, we iteratively mask the attention between specific groups of tokens across sets of layers and measure the corresponding changes in output probabilities. A change in output probability indicates that the masking intervention disrupted a pathway of information flow that the model used to perform the task. In other words, this enables us to trace the information flow between modalities as a function of these probability changes, thereby identifying where and how multimodal fusion occurs and quantifying the direct contribution of each modality to the model's predictions.

3.1   Preliminaries

LLaMA 3.2-Vision Architecture. LLaMA 3.2-Vision is a multimodal decoder-only transformer that integrates visual and textual features via cross-attention into a LLaMA language model backbone [27], enabling vision-language reasoning. An overview of LLaMA 3.2-Vision architecture is reported in Fig. 1a. As shown, the model receives both visual and textual inputs. Specifically, an image x is partitioned into fixed-size patches and passed through an image encoder to produce a sequence of visual embeddings $V = [v_i]_{i=1}^{N_V}$, where N_V denotes the number of visual tokens and each $v_i \in \mathbb{R}^d$ represents a d-dimensional feature vector corresponding to a patch. The model leverages a ViT-H/14 [8], which processes images using a patch size of 14×14. Concurrently, a text sequence t, consisting of N_T tokens, is transformed into embeddings via a word embedding lookup table, resulting in textual representations $T = [t_i]_{i=1}^{N_T}, t_i \in \mathbb{R}^d$. The text tokens are provided as input to the MLLM and processed through L transformer layers. Each layer primarily consists of two components: a masked multi-head attention (MHA) [28] module and a feed-forward network (FFN). The hidden representation at layer l, denoted by $H^l = [h_i^l]_{i=1}^N \in \mathbb{R}^{N \times d}$, where N is the length of the input sequence, can be expressed as follows:

$$H^l = \text{FFN}(Z^l) + Z^l, \text{ where } Z^l = \text{MHA}(H^{l-1}) + H^{l-1}. \tag{1}$$

H^{l-1} represents the hidden representation of the previous layer and Z^l denotes the output of the MHA module. The masked MHA modules of each transformer layer l harness three projection matrices, $W_Q^l, W_K^l, W_V^l \in \mathbb{R}^{d \times d}$, to project the input H^{l-1} into query (Q^l), key (K^l) and value (V^l) matrices, respectively. These matrices are then split across N_H different attention heads, resulting in $\{Q^{l,j}\}_{j=1}^{N_H}, \{K^{l,j}\}_{j=1}^{N_H}$, and $\{V^{l,j}\}_{j=1}^{N_H}$, each of dimension $\mathbb{R}^{N \times \frac{d}{N_H}}$. The MHA operation at layer l is defined as follows:

$$\text{MHA}(Q^l, K^l, V^l) = \text{Concat}(head_1^l, \ldots, head_{N_H}^l)W_O^l, \tag{2}$$

$$head_j^l = \text{softmax}\left(\frac{Q^{l,j}(K^{l,j})^T}{\sqrt{d/N_H}} + M^{l,j}\right)V^{l,j}, \tag{3}$$

where $W_O^l \in \mathbb{R}^{d \times d}$ refers to the output projection matrix, and $M^{l,j}$ denotes the attention mask, indexed by layer l and head j, which governs token-to-token visibility during the MHA operation. The feed-forward network (FFN) computes the output representation through the following operation:

$$\text{FFN}(Z^l) = W_U^l \, \sigma\left(W_B^l(Z^l)\right), \tag{4}$$

where $W_U^l, W_B^l \in \mathbb{R}^{d \times d_{ff}}$ are projection matrices with inner-dimensionality d_{ff}, and σ represents a nonlinear activation function.

Visual-Textual Integration. In LLaMA 3.2-Vision model, the MHA module is used in both self-attention and cross-attention configurations. The architecture follows a structured design, inserting cross-attention layers at fixed intervals—specifically at layers 4, 9, 14, 19, 24, 29, 34, and 39—within a 40-layer decoder-only transformer. Within the multi-head self-attention modules, the query, key, and value matrices are computed via learned linear projections of the same textual hidden states. Conversely, cross-attention layers use textual hidden states as queries and the visual embedding as keys and values, enabling the integration of visual context into the evolving textual representations. Furthermore, in self-attention modules, a strictly upper triangular attention mask is applied, ensuring that each token can attend only to preceding positions in the sequence. In contrast, the cross-attention layers utilize a fully permissive attention mask, allowing unrestricted interactions between tokens across modalities. The default configurations of both attention masks are illustrated in Fig. 1b.

Autoregressive Decoding. The hidden representation h_N^L corresponding to the last position N in the input sequence at the final layer L is projected by an unembedding matrix $E \in \mathbb{R}^{|\mathcal{V}| \times d}$. The probability distribution over all words in the vocabulary $\mathcal{V}$ is then computed as:

$$P_N = \text{softmax}(E h_N^L), \tag{5}$$

where the token with the highest probability in P_N is selected as the final prediction. If the model is required to generate an answer composed of multiple tokens (*e.g.*, a phrase or sentence), the decoding proceeds autoregressively. Specifically, the selected token is appended to the input sequence, and the updated sequence is reprocessed to produce the next token.

3.2   Information Flow Analysis via Attention Masking

We investigate modality interactions in LLaMA 3.2-Vision by selectively masking attention connections between specific token groups. This intervention enables us to trace information flow by measuring the change in the output probability. A significant change indicates that the masked connections disrupt a critical pathway used by the model to perform the task.

To investigate various types of information flow, we modify the self- and cross-attention masks. The multi-head self-attention is the only module that enables communication across different positions within the textual input sequence. In contrast, the cross-attention module is responsible for integrating visual and textual information. To assess the direct influence of each modality on the final prediction, we intervene at each layer l by selectively preventing the final textual token—responsible for generating the answer—from attending to (i) preceding textual tokens (question-to-last flow), (ii) visual tokens (image-to-last flow), or (iii) itself (last-to-last flow). Additionally, to examine the integration of visual and textual modalities, at each layer l, we block the attention from the question tokens (excluding the final token) to the visual tokens (image-to-question flow). Each intervention is applied within a local window of $k = 9$ layers centered around layer l. Formally, we block the information flow from a source token group (*e.g.*, question, image, and last token) to a target group (*e.g.*, question and last token) by setting the corresponding entries in the self- or cross-attention mask to $-\infty$, as illustrated in Fig. 1c. Moreover, we quantify the information flow between the source and target groups as the relative change in output probability. Specifically, given an image–question pair, the model produces an answer with probability p_1. After masking the attention between the source and target token groups according to the aforementioned strategies, the model generates the same answer with a probability p_2. The relative change in probability is computed as $p_c\% = ((p_2 - p_1)/p_1) \times 100$. A significant change in output probability indicates that the pathway was critical to the model's reasoning process.

4   Experimental Results

4.1   Datasets

This work examines the internal dynamics of LLaMA 3.2-Vision across three benchmark datasets that involve diverse reasoning capabilities, encompassing tasks such as visual question answering (VQA) and document visual question answering. We use VQAv2 [10], a VQA benchmark comprising open-ended questions paired with real-world images from the COCO dataset [15]. The dataset includes three question types: binary (*yes/no*), numerical, and open-ended (*other*). Visual7W [34] is a multiple-choice VQA dataset based on COCO images, with questions grouped into six categories. The DocVQA dataset [17] is a benchmark dataset for visual question answering on scanned documents, comprising nine distinct question types that require fine-grained reasoning over text extracted via OCR, as well as an understanding of the document's visual and spatial layout. We report results for three selected categories in Visual7W—*what, when* and *why*—and in DocVQA—*free text, handwritten*, and *layout*—as the remaining categories exhibited similar information flow patterns. We use only correctly answered samples to avoid noise from disrupted or irrelevant pathways in incorrect predictions. For VQAv2 and Visual7W, we randomly sample 1,000 correctly answered image-question pairs for each question category; for DocVQA, we use all the 3,133 correctly answered samples. Data is sourced from the validation sets of VQAv2 and DocVQA, and the test set of Visual7W.

4.2   Main Results

In our experiments, we examine whether each modality directly contributes to the prediction and how the model integrates visual and textual representations. To this end, we mask attention from a target token group to a source group at each layer l, within a sliding window of $k = 9$ layers.

Evaluation of Information Flows via Attention Masking. Our results are presented in Fig. 2, which illustrates the patterns of information flow—last-to-last, image-to-last, question-to-last, and image-to-question—across three categories for each datasets. The x-axis denotes the layer index, while the y-axis represents the relative change in output probability resulting from blocking the attention between specific token groups. Values near zero suggest minimal information flow, whereas larger negative values indicate a stronger dependence on the masked interaction. We identify a four-stage reasoning process in LLaMA 3.2-Vision: (1) an initial semantic interpretation of the prompt, (2) an early-to-mid-layer visual integration, (3) a task-specific reasoning stage, and (4) a final answer prediction. As shown in the aforementioned figure, during the early self-attention layers, there is a strong information flow from the question to the last token, suggesting that the model initially focuses on interpreting the semantics of the prompt. Visual features are incorporated predominantly in the early-to-mid cross-attention layers, where the image-to-question information flow is most evident. Notably, as multimodal integration progresses, the influence of the question

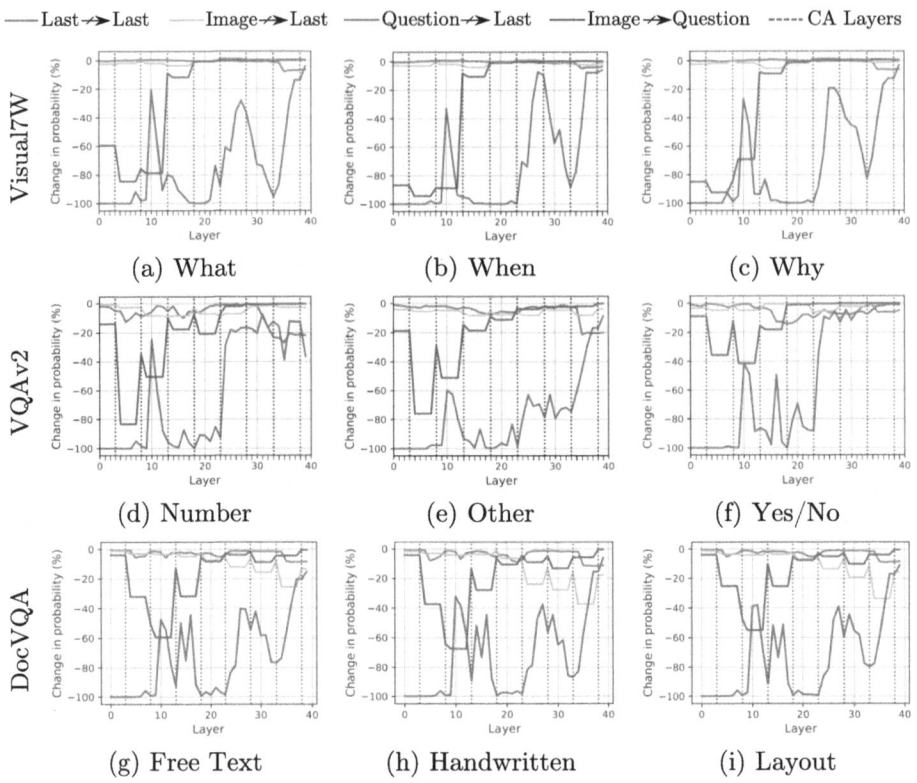

Fig. 2. Relative changes in prediction probability across three datasets—Visual7W, VQAv2, and DocVQA—each analyzed by question category. Dotted blue lines indicate cross-attention layers. (Color figure online)

on the last token decreases, indicating that the model is processing visual and textual information into a joint representation. After this integration, the flow from the question to the last token increases again, marking a task-specific reasoning phase in which the grounded multimodal representation is used to derive the answer. Finally, in the last few layers, this flow declines, signaling the final stage of answer generation. This reasoning process remains consistent across all three datasets, with similar trends observed across all question categories within each dataset. Additionally, in DocVQA, we observe an image-to-last information flow in the last layers, suggesting that the model directly extracts visual information to answer the question. This behavior aligns with the task's inherent reliance on OCR capabilities, where answers depend on accurately interpreting and localizing textual content within complex documents.

Comparative Visualization of Multimodal Information Flows. Moreover, we compare the three most relevant information flows—question-to-last, image-to-last, and image-to-question—across the datasets, since the last-to-last

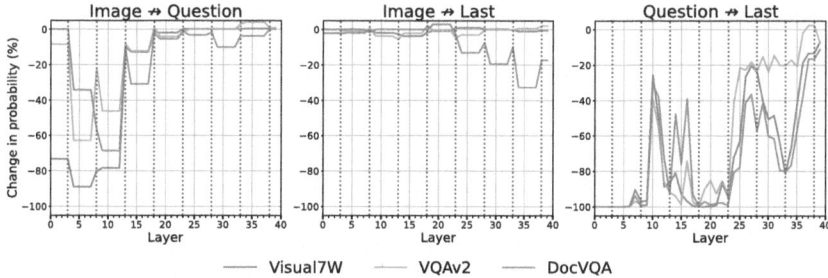

Fig. 3. Visualizations of different information flows—image-to-question, image-to-last, question-to-last flows—across Visual7W, VQAv2, and DocVQA datasets. Each curve represents the average change in output probability, computed over the entire dataset.

information flow is negligible in all the tasks, *i.e.*, the last token does not contribute to the answer prediction. Results are presented in Fig. 3, which shows the average information flow across all question categories for each dataset. An analysis of the image-to-question information flow reveals that multimodal integration occurs at different depths depending on the task. For relatively simple VQA datasets involving natural images, such as VQAv2 and Visual7W, the model predominantly integrates visual and textual information around layer 5. In contrast, for the more complex DocVQA dataset, this integration is noticeably delayed, emerging closer to layer 10. This shift likely reflects the increased demand for higher-level textual reasoning. Furthermore, the magnitude of the global minimum in the change in output probability, reflecting the absolute strength of information flow, is found to be task-dependent. The Visual7W dataset, which adopts a multiple-choice question format, exhibits a stronger image-to-question information flow compared to open-ended tasks such as VQAv2 and DocVQA. In multiple-choice question formats, the need to assess a fixed set of answer candidates likely induces the model to focus directly on relevant visual features at earlier layers of the architecture. Regarding the image-to-last information flow, a notable effect is observed exclusively in the final cross-attention layers for the DocVQA dataset. This behavior reflects DocVQA's reliance on direct visual input for answer generation, likely due to the task's dependence on OCR-specific content. In contrast, the question-to-last flow exhibits a consistent pattern across all three datasets, supporting the presence of a shared four-stage reasoning process, in line with the progression described in the preceding paragraph.

4.3   Ablation on Window Size K

In the previous experiments, we adopted a window size of $k = 9$. To assess the impact of window size on model behavior, we report the relative change in answer probability for the *Other* question category of VQAv2 across various window sizes: $k = 1, 5, 7, 9, 11, 15$. Results are shown in Fig. 4. Overall, the observed patterns of information flow remain consistent across these different values of

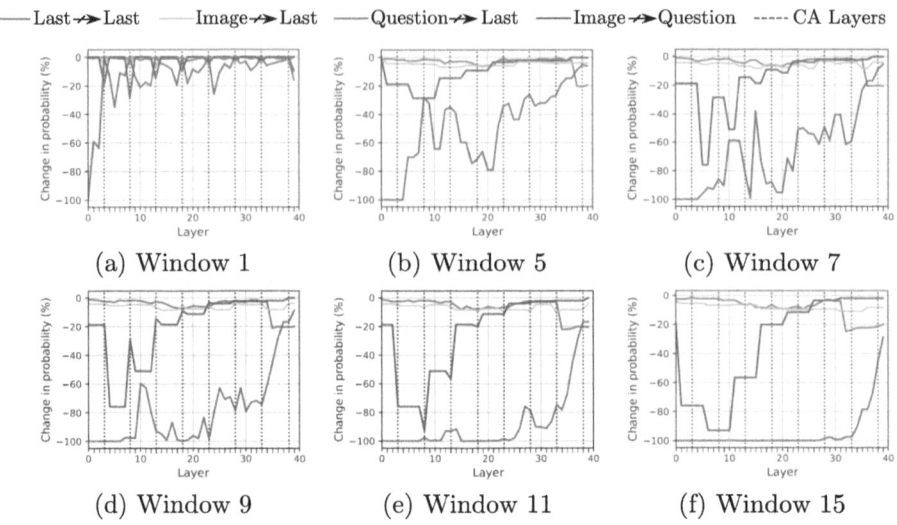

Fig. 4. Relative changes in prediction probability on LLaMA 3.2-Vision for *other* questions from the VQAv2 dataset across different attention windows.

k. The multimodal integration still occurs in early-to-mid cross-attention layers, and interestingly, with $k = 1$, we can observe the importance of each cross-attention layer, confirming that the earliest ones are where most of the multimodal integration occurs. However, the change in output probability becomes increasingly pronounced as k increases, and the four-stage reasoning process becomes less visible. This trend is expected, as restricting more attention edges during computation impairs the model's ability to effectively contextualize the input.

5   Conclusion

In this paper, we present a comprehensive analysis of the information flow between visual and textual tokens in LLaMA 3.2-Vision. We conduct experiments on three datasets—VQA v2, Visual7W, and DocVQA—covering the tasks of VQA on natural images with both open-ended and multiple-choice questions and document VQA. We reveal a four-stage reasoning process that includes question interpretation, multimodal integration, task-specific reasoning, and answer generation. We demonstrate that the model's multimodal fusion strategy is task-dependent. More complex tasks involve delayed integration after establishing semantic reasoning over the prompt. Additionally, we uncover the direct influence of the image on the final prediction in OCR-based tasks, where fine-grained visual information significantly influences the model's ability to generate accurate answers. This work advances the understanding of cross-modal dynamics within MLLMs and contributes to their interpretability.

Acknowledgments. This work was supported by Key Digital Technologies Joint Undertaking (KDT JU) in EdgeAI "Edge AI Technologies for Optimised Performance Embedded Processing" project, GA No. 101097300 and by the University of Modena and Reggio Emilia and Fondazione di Modena through the "Fondo di Ateneo per la Ricerca - FAR 2024" (CUP E93C24002080007). This work also received funding from DECIDER, the European Union's Horizon 2020 research and innovation programme under GA No. 965193 and "AIDA: explAinable multImodal Deep learning for person-Alized oncology" (Project Code 20228MZFAA).

Disclosure of Interests. The authors have no conflicts of interest to declare.

References

1. Alayrac, J.B., et al.: Flamingo: a Visual Language Model for Few-Shot Learning. In: Advances in Neural Information Processing Systems (2022)
2. Almazrouei, E., et al.: The falcon series of open language models (2023). arXiv preprint arXiv:2311.16867
3. Anthropic: The Claude 3 Model Family: Opus, Sonnet, Haiku. In: OpenAI (2024)
4. Awadalla, A., et al.: OpenFlamingo: An open-source framework for training large autoregressive vision-language models (2023). arXiv preprint arXiv:2308.01390
5. Basu, S., et al.: Understanding information storage and transfer in multi-modal large language models (2024). arXiv preprint arXiv:2406.04236
6. Chowdhery, A., et al.: PaLM: Scaling Language Modeling with Pathways. J. Mach. Learn. Res. **24**(240) (2023)
7. Clark, K., et al.: What Does BERT Look At? An Analysis of BERT's Attention (2019). arXiv preprint arXiv:1906.04341
8. Dosovitskiy, A., et al.: An image is worth 16x16 words: transformers for image recognition at scale (2020). arXiv preprint arXiv:2010.11929
9. Fang, Y., et al.: EVA: Exploring the limits of masked visual representation learning at scale. In: Computer Vision and Pattern Recognition (2023)
10. Goyal, Y., et al.: Making the V in VQA matter: elevating the role of image understanding in visual question answering. In: Computer Vision and Pattern Recognition (2017)
11. Grattafiori, A., et al.: The Llama 3 Herd of Models (2024). arXiv preprint arXiv:2407.21783
12. Han, J., et al.: OneLLM: One framework to align all modalities with language. In: Proceedings of the IEEE/CVF Conference on Computer Vision and Pattern Recognition (CVPR) (2024)
13. Hui, B., et al.: Qwen2. 5 - Coder Technical Report (2024). arXiv preprint arXiv:2409.12186
14. Li, J., et al.: BLIP-2: Bootstrapping language-image pre-training with frozen image encoders and large language models. In: International Conference on Machine Learning (2023)
15. Lin, T.Y., et al.: Microsoft COCO: Common Objects in Context. In: European Conference on Computer Vision (2014)
16. Liu, H., et al.: Visual instruction tuning (2023)
17. Mathew, M., et al.: DocVQA: A dataset for VQA on document images. In: Winter Conference on Applications of Computer Vision (2021)

18. Meng, K., et al.: Locating and editing factual associations in GPT. In: Advances in Neural Information Processing Systems (2022)
19. Meng, K., et al.: Mass-Editing Memory in a Transformer (2022). arXiv preprint arXiv:2210.07229
20. Neo, C., et al.: Towards interpreting visual information processing in vision-language models (2024). arXiv preprint arXiv:2410.07149
21. OpenAI: GPT-4o mini: advancing cost-efficient intelligence (2024). https://openai.com/index/gpt-4o-mini-advancing-cost-efficient-intelligence/
22. Palit, V., et al.: Towards vision-language mechanistic interpretability: a causal tracing tool for BLIP. In: International Conference on Computer Vision (2023)
23. Park, K., et al.: The linear representation hypothesis and the geometry of large language models (2023). arXiv preprint arXiv:2311.03658
24. Pipoli, V., et al.: MissRAG: Addressing the Missing Modality Challenge in Multimodal Large Language Models. In: International Conference on Computer Vision (2025)
25. Radford, A., et al.: Learning transferable visual models from natural language supervision. In: International Conference on Machine Learning (2021)
26. Team, G., et al.: Gemma: open models based on gemini research and technology (2024). arXiv preprint arXiv:2403.08295
27. Touvron, H., et al.: LLaMA: Open and efficient foundation language models (2023). arXiv preprint arXiv:2302.13971
28. Vaswani, A., et al.: Attention Is All You Need. Adv. Neural Inf. Process. Syst. **30** (2017)
29. Yin, H., et al.: Lifting the veil on visual information flow in MLLMs: unlocking pathways to faster inference. In: Computer Vision and Pattern Recognition Conference (2025)
30. Zhang, L., et al.: Mechanistic unveiling of transformer circuits: self-influence as a key to model reasoning (2025). arXiv preprint arXiv:2502.09022
31. Zhang, X., et al.: From redundancy to relevance: information flow in LVLMs across reasoning tasks (2024). arXiv preprint arXiv:2406.06579
32. Zhang, Z., et al.: Cross-modal information flow in multimodal large language models (2024). arXiv preprint arXiv:2411.18620
33. Zhu, D., et al.: MiniGPT-4: Enhancing vision-language understanding with advanced large language models (2023). arXiv preprint arXiv:2304.10592
34. Zhu, Y., et al.: Visual7w: Grounded question answering in images. In: Computer Vision and Pattern Recognition (2016)

Robotics, Interaction and Intelligent Systems

Multimodal Audio-Visual Emotion Recognition for Social Robotics

Giuseppe De Simone(✉)⬛, Luca Greco⬛, Alessia Saggese⬛, and Mario Vento⬛

University of Salerno, Salerno, Italy
{gidesimone,lgreco,asaggese,mvento}@unisa.it

Abstract. Thanks to recent advances in deep learning based algorithms, humanoid social robots are increasingly exhibiting human-like behaviors. In this context, the analysis of soft biometrics, particularly emotion recognition, is crucial for enhancing communication between social robots and humans, facilitating emotion-aware dialogues. In light of these considerations, we propose a multimodal emotion recognition system tailored for social robotics applications. The system processes both video and audio data to classify the emotion in one among six different classes, employing 3D convolutional operations that eliminate the need for transformer-based architecture, effectively reducing the model's size and making the network able to run over low power embedded devices mounted directly on board of the robot. The proposed approach was trained on the CREMA-D dataset and demonstrates impressive performance when compared to video-only and audio-only counterparts, outperforming state-of-the-art methods both unimodal and multimodal.

Keywords: Social Robotics · 3D Multimodal Architecture

1 Introduction

Humanoid social robots have been defined as "human-made technologies that can take physical or digital form, resemble people in form or behavior to some degree, and are designed to interact with people" [12]. Within this context, an important and not negligible capability that the robot needs to have in order to effectively mimic human behavior is to automatically recognize soft biometrics [6,17], such as gender [15], age [3] and emotion [11]. Emotion is particularly important for social robots, as it helps in improving interactions between humans and robots [18], thus boosting user acceptance [28], and allowing robots to create emotion-aware conversations [5,7,26]. In Zhang et al. [35], for instance, the emotion is recognized by analyzing the voice of the speaker; the authors propose a Conformer-based architecture [16], pre-trained on a large dataset collected from YouTube (and composed by about 500k hours of audio samples) to solve the Automatic Speech Recognition (ASR) task. The pre-trained model was subsequently fine-tuned to address the specific emotion recognition tasks.

M. Castrillón-Santana et al. (Eds.): CAIP 2025, LNCS 15622, pp. 245–255, 2026.
https://doi.org/10.1007/978-3-032-05060-1_21

Even if achieving state of the art performance on the well known CREMA-D dataset [2], it is important to highlight that the model contains about 0.6 billion parameters, leading to high inference times. This limitation significantly affects system responsiveness, making the model unsuitable for social robotics applications, where real-time responses are essential to ensure effective human-robot interaction. Furthermore, this kind of approach requires that the person speaks in order to identify the emotion, while it is important to also understand people emotion when the robot (and thus not the interlocutor) is speaking. In order to face this issue, another possible approach is to instead analyze the face. Kim et al. [21] proposed a teacher-student paradigm using the Vision Transformer (ViT) [8] as an encoder. Differently from the method in [35], their model contains a relatively limited number of parameters (about 1.5 million). Anyway, in order to obtain state-of-the-art accuracy on the CREMA-D dataset the authors required that the person is already known to the system, which is not a feasible constraint when dealing with social robotic applications. Furthermore, it is important to highlight that in applications such as surveillance video may be the exclusive source of information; vice-versa, social robotics may benefit from both audio and video as complementary sources of data. As a result, it is increasingly important to transition from unimodal approaches, where either video or audio is analyzed, to multimodal methods that jointly consider both modalities. This shift enables more comprehensive analysis and better-informed decision-making. It is important to highlight that in the recent literature, even if in different contexts, multimodal approaches have demonstrated impressive performance in uncontrolled and complex scenarios, since enhancing robustness to noise and disturbances [31]. For instance, in the field of autonomous vehicle driving, multimodal systems have been employed to improve motion planning and ensure safety in self-driving applications [25]. In the specific context of emotion recognition, a pioneering multimodal approach has been proposed by Goncalves et al. [14], with a versatile module capable of predicting emotions. The method employs the Wav2Vec2.0 model as the audio encoder, pre-trained on multiple audio datasets such as Libri-Light and CommonVoice. Additionally, the authors use an EfficientNet [29] pre-trained on the AffectNet corpus [24] as the video encoder. To analyze the temporal sequences of audio and video data extracted by the encoders, two separate Conformer modules are employed for audio and visual processing. Additionally, a third Conformer module is used as a fusion layer to integrate information from both modalities. Specifically, video data are processed using 2D convolutional networks designed for spatial feature extraction, resulting in a sequence of embeddings for individual frames. These temporal features are further refined by the Visual Conformer module. Although this method achieves impressive performance, a promising direction would involve simultaneously analyzing spatial and temporal information through more effective and efficient operators, such as 3D convolutions [10]. Furthermore, it is important that the model is lightweight, so as to be able to run on board of the robot. A possible solution for maintaining a limited size of the model in multimodal approaches can be found in the literature for solving a different task, namely the Active Speaker Detection [27]. Specifically, Liao et al. [22] designed a lightweight model that takes as input both audio and video data, integrating spatial and temporal analysis. This model is capable of achieving state-of-the-art performance for the task while using a significantly reduced number of parameters, namely less than 1 million.

Inspired by the approaches outlined in [14] and [22], in this paper we proposed a system able to process two input modalities, namely video and audio, to estimate the interlocutor's emotional state. We address both the challenges related to explicitly analysis of the temporal evolution and model complexity, by proposing an efficient and effective multimodal emotion recognition system tailored for social robots, employing 3D convolutional layers to capture both temporal and spatial features simultaneously during convolution operations, effectively bypassing the computational overhead associated with multi-head attention and transformer-based architectures. The proposed multimodal architecture is trained and tested on the widely-used CREMA-D dataset [2], ensuring that speakers are exclusive to each data split. The proposed model achieves state-of-the-art performance, outperforming both prior literature and its unimodal (video-only and audio-only) variants.

2   Proposed Method

An overview of the proposed method is illustrated in Fig. 1. The system takes video and audio data as inputs; both of them are preprocessed before being represented through two independent encoders. The outputs of these encoders, representing the corresponding embeddings, are then fused using a widely adopted concatenation approach. Finally, the fused embedding is passed to a classifier module, where the classification task is performed using a Multi-Layer Perceptron (MLP). The overall model is trained using the Cross-Entropy loss function, jointly optimizing the two independent encoders, as proposed in [9,30].

It is important to highlight that using two independent encoders allows the proposed system to learn complementary features from audio and video separately, ensuring that each modality is processed in its optimal feature space before fusion. This separation enables thus the model to capture diverse emotional cues, such as facial expressions from video and vocal tone variations from audio, leading to better generalization across different environments and users. Furthermore, a joint training optimizes both encoders in a way that enhances their cooperation, making the model more adaptable to real-world situations.

2.1   Video Encoder

The video encoder takes as input the preprocessed data from the video preprocessing modules and extracts spatiotemporal features using 3D convolutional layers. Specifically, the video encoder processes a video tensor with dimensions $B \times T_v \times C \times H \times W$, where B denotes the batch size, T_v represents the temporal dimension, C is the number of channels, and H and W are the height and

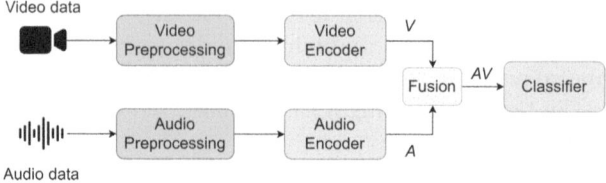

Fig. 1. High-level architecture of the proposed method. Video and audio data are first passed through their respective preprocessing modules, before being processed by the video and audio encoders, and subsequently fused using the proposed fusion method. Finally, the resulting fused embedding is given as input to the classifier for prediction.

width of the frames, respectively. The 3D convolutional layers extract features across both spatial and temporal axes [32], resulting in an embedding vector V with dimensions $B \times E_v$, where E_v represents the size of the embedding space corresponding to the video information. Specifically, based on the performance achieved in other video analytics tasks [19], we adopt the ResNet3D architecture [32], as detailed in the SlowFast framework [10]. The SlowFast model is a dual-path neural network architecture designed for video understanding tasks. It captures both slow-changing spatial semantics and fast-changing motion dynamics by processing input at different frame rates, leveraging a slow branch for detailed spatial analysis and a fast branch for motion-sensitive temporal analysis. In our implementation, we use only the slow branch, referred to as `SlowR50` to design a lightweight video encoder, pretrained on Kinetics dataset [19]. This approach reduces the number of parameters compared to its dual-branch counterparts, resulting in a more efficient model. The input tensor, with the previously described dimensions, is processed through a series of ResNet blocks, enabling the extraction of both spatial and temporal features.

2.2   Audio Encoder

The audio encoder processes a time-frequency representation of the audio stream so as to extract spatio-temporal features. Specifically, the audio stream is represented using the widely adopted Mel-Frequency Cepstral Coefficient (MFCC), a representation inspired to the human auditory system [1]. Consequently, the input data processed by the audio model results in an audio tensor with dimensions $B \times F \times T_a$, where B denotes the batch size, and F and T_a represent the frequency and time axes, respectively. The encoder uses $(1+1)$D convolutional layers to extract meaningful audio features, producing an output tensor A with dimensions $B \times \tilde{T}_a \times E_a$, where $\tilde{T}_a$ is the new temporal axis after the convolution operations and E_a represents the embedding space corresponding to the audio information. Inspired by the impressive results achieved on the Active Speaker Detection task, we use the lightweight audio model from LightASD [22]. LightASD features a compact architecture with two encoders, one for audio and one for video. In this context, given the above mentioned requirement to employ 3D

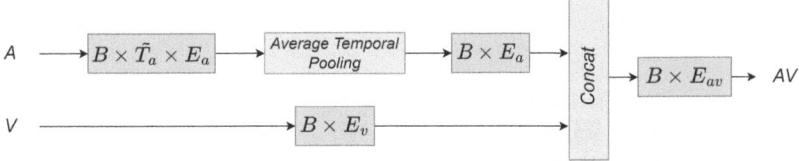

Fig. 2. Architecture of the fusion module. A temporal analysis is applied to the audio embedding A using an *average temporal pooling* operation. The resulting tensor is then concatenated with the video tensor V, producing the final multimodal embedding AV.

convolutions for explicitly characterized the temporal information in the video encoder, we employ only the audio encoder from LightASD. It is applied to the time-frequency representation employing convolutional layers to extract spatial and temporal features.

2.3   Fusion

The fusion operation illustrated in Fig. 2 aims to combine audio and video embeddings into a unified representation, denoted as AV, with dimensions $B \times E_{av}$, where E_{av} represents the fused embedding space for audio and video. Specifically, the layer takes as input the video and audio embeddings, denoted as V and A, obtained from the video and audio encoders, respectively. An average temporal pooling operation is performed on audio embedding to aggregate temporal information, resulting in a tensor of dimensions $B \times E_a$. Note that the video tensor lacks a temporal axis due to the use of 3D convolution operations in the video encoder, that does not require additional processing. The two embeddings are fused using concatenation operator to produce a single representation of shape $B \times E_{av}$. The resulting fused embedding, named AV, is then fed into a classifier for recognition tasks.

3   Experimental Setup

3.1   Dataset

The proposed architecture has been trained using the widely utilized CREMA-D dataset [2], a publicly available multimodal dataset designed for emotion recognition tasks. The dataset comprises over 7.000 audio-visual clips recorded by 91 actors performing scripted sentences in various emotional states, including happiness, anger, sadness, and neutrality. Each clip includes both audio and video modalities. In this paper, we partitioned the dataset into training, validation, and test sets with a ratio of 80%, 10%, and 10%, respectively. This resulted in a training set comprising 72 actors (34 female, 38 male), a validation set of 9 actors (4 female, 5 male), and a test set of 10 actors (5 female, 5 male). The dataset includes the following emotion classes: neutral (NEU), happy (HAP), sad (SAD), angry (ANG), fear (FEA), and disgust (DIS). The distribution of the dataset among the different emotions is illustrated in Fig. 3.

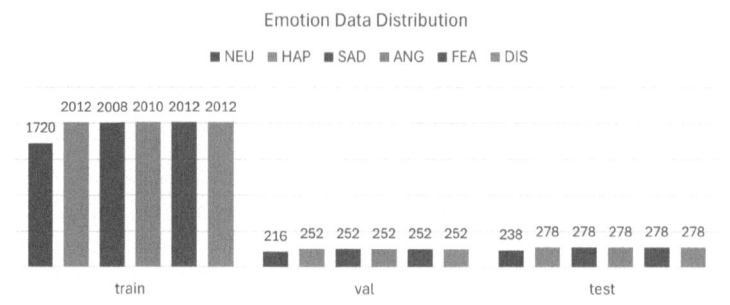

Fig. 3. Distribution of emotion classes in the dataset across training, validation, and test sets.

3.2   Implementation Details

Preprocessing Procedure.

Video Preprocessing As a first step, the S3DF face detector [34] is applied to each frame to identify and extract the face region within the image. Subsequently, a central crop and a side scale crop with a size of 256 are performed to preserve the aspect ratio of the input data obtaining a video tensor with dimensions $T_v \times C \times H \times W$, where H and W correspond to the height and width required by the input of the model used for training. Finally, a uniform temporal sub-sampling is conducted to ensure that T_v is equal to 8.

Audio Preprocessing Initially, the audio sample is loaded in the time domain with a sampling rate of 16 kHz. Subsequently, the audio sample is transformed into an MFCC representation, with the number of MFC Coefficients set to 128. The temporal dimension in the time-frequency domain, denoted as T_a (see Sect. 2.2), is computed as the number of STFT windows: $T_a = \frac{T'_a}{\text{hop\_length}}$ where T'_a represents the total number of audio samples (including padding) in the time domain, and **hop_length** refers to the hop size between windows, set to 200.

Training Procedure. The models were trained using the PyTorch framework with a learning rate of 5×10^{-5}, optimized using the AdamW optimizer. A *Reduce on Plateau* learning rate scheduler was employed, monitoring the validation loss. If the validation loss did not improve for 4 epochs, the learning rate was reduced by a factor of 0.1. Additionally, *Early Stopping* was applied to halt the training process if no improvement in validation loss was observed for 10 consecutive epochs. To mitigate overfitting, the *RandAugment* [4] algorithm was used on the videos as data augmentation protocol with 3 layers at magnitude 3. RandAugment applies a randomized combination of augmentation operations (e.g., rotation, flipping, color adjustments) enhancing the model's robustness and generalization capabilities. Moreover, considering that the background in the video is a green screen, it was replaced with a random background sampled from the MIT Indoor Scenes database [23] to further reduce overfitting and increase generalization capabilities. The best model weights were saved at the end of each epoch based on the best validation loss.

3.3   Results

In order to evaluate the performance of the proposed method, the following metrics widely adopted in the literature [13] have been used: Accuracy - (Eq. 1), Micro/Macro F1-Score (Eq. 2), Weighted Average Recall - WAR, Unweighted Average Recall - UAR - (Eq. 3):

$$\text{Accuracy} = \frac{\sum_{i=1}^{C} TP_i}{\sum_{i=1}^{C} N_i} \tag{1}$$

$$\text{Micro-F1} = \frac{2 \cdot \sum_{i=1}^{C} TP_i}{2 \cdot \sum_{i=1}^{C} TP_i + \sum_{i=1}^{C}(FP_i + FN_i)} \qquad \text{Macro-F1} = \frac{\sum_{i=1}^{C} F1_i}{C} \tag{2}$$

Here, TP_i, FP_i, and FN_i represent the true positive, false positive, and false negative instances for class i, respectively. C denotes the total number of classes in the dataset, while $F1_i$ corresponds to the F1-Score of the i-th class.

$$\text{recall}_i = \frac{TP_i}{TP_i + FN_i} \qquad \text{WAR} = \frac{\sum_{i=1}^{C}(\text{recall}_i \cdot N_i)}{\sum_{i=1}^{C} N_i} \qquad \text{UAR} = \frac{\sum_{i=1}^{C} \text{recall}_i}{C} \tag{3}$$

In these equations, recall_i and N_i represent the recall of i-th class and the number of instances in that class, respectively.

Note that, as we can see, there is a high correlation among the metrics, thus we could have chosen just one. Anyway, there is not agreement in the literature on which specific metric using, thus to foster future comparisons we have included all of them.

A preliminary evaluation was conducted to compare the proposed multimodal method with its unimodal counterparts: the video-only approach employing the SlowR50 model and the audio-only approach using the LightASD model. The results are summarized in Table 1. We can observe that the proposed method outperforms both unimodal counterparts (about 80% for the video modality and 59% for the audio modality), achieving a score of 88% across all metrics, thereby confirming the advantages of the multimodal approach over the counterpart unimodal (+8% with respect video, +29% with respect to audio).

Table 1. Performance of the proposed method compared across modalities (Mod.) to unimodal counterparts: video only (V) and audio only (A), in terms of Accuracy (Acc.), Micro/Macro F1-Score, Unweighted Average Recall (UAR), and Weighted Average Recall (WAR).

Method	Mod.	Acc.	Micro-F1	Macro-F1	UAR	WAR
SlowR50 [10]	V	0.805	0.805	0.804	0.802	0.805
LightASD [22]	A	0.595	0.595	0.598	0.591	0.595
Proposed Method	A+V	**0.882**	**0.882**	**0.880**	**0.879**	**0.882**

Table 2. Performance of the proposed method compared across modalities (Mod.) to state of the art methods, in terms of Accuracy (Acc.), Micro/Macro F1-Score, Unweighted Average Recall (UAR), and Weighted Average Recall (WAR).

Method	Mod	Acc.	Micro-F1	Macro-F1	UAR	WAR
VAVL [14]	A+V	-	0.779	0.829	-	-
MTCAE-DFER [33]	V	-	-	-	0.847	0.850
CoordViT [20]	A	0.830	-	-	-	-
Proposed Method	A+V	**0.882**	**0.882**	**0.880**	**0.879**	**0.882**

In order to also verify the effectiveness of the proposed approach with respect to the state of the art, we also analyze the results on the same dataset of the recent literature. The detailed comparison is presented in Table 2. It is important to note that each study employs distinct metrics for evaluation; therefore, only the metrics available for each method are included in the table. Analyzing the results, we can conclude that the proposed method demonstrates impressive performance also compared to the existing literature, encompassing both unimodal (`CoordViT` and `MTCAE-DFER`) and multimodal (`VAVL`) approaches. Specifically, the proposed method achieves an improvement of 10.3% and 5.1% in terms of Micro-F1 and Macro-F1, respectively, compared to the multimodal method `VAVL`. Additionally, it demonstrates superior performance with a 3.2% increase in UAR and WAR compared to the video-based method `MTCAE-DFER`. Furthermore, the proposed approach outperforms the audio-based method `CoordViT` with a 5.2% improvement in accuracy.

4   Conclusions

This paper presents a novel multimodal emotion recognition method that integrates a 3D convolutional video encoder and a lightweight audio encode. By leveraging both spatial and temporal features from video and audio modalities, the proposed model achieves state-of-the-art performance on the CREMA-D dataset. The results demonstrate significant improvements over unimodal methods, with the proposed system achieving an accuracy and Micro-F1 score of 0.882, Macro-F1 score of 0.880, Unweighted Average Recall (UAR) of 0.879, and Weighted

Average Recall (WAR) of 0.882. These results surpass those achieved by both video-only and audio-only approaches, specifically `MTCAE-DFER` and `CoordViT`, respectively, as well as the existing multimodal method `VAVL` from the literature. The efficiency of the model demonstrates its suitability for practical applications, particularly in resource-constrained environments like social robotics.

Acknowledgements. This work has been partially supported by MIUR PRIN – PNRR,

No. E63C22002150007 - MUMBLE: MUltimodal, Multitask, unBalanced Learning Environments

References

1. Abdul, Z.K., Al-Talabani, A.K.: Mel Frequency cepstral coefficient and its applications: a review. IEEE Access **10**, 122136–122158 (2022)
2. Cao, H., Cooper, D.G., Keutmann, M.K., Gur, R.C., Nenkova, A., Verma, R.: CREMA-D: crowd-sourced emotional multimodal actors dataset. IEEE Trans. Affect. Comput. **5**(4), 377–390 (2014)
3. Carletti, V., Greco, A., Percannella, G., Vento, M.: Age from faces in the deep learning revolution. IEEE Trans. Pattern Anal. Mach. Intell. **42**(9), 2113–2132 (2020)
4. Cubuk, E.D., Zoph, B., Shlens, J., Le, Q.V.: Randaugment: practical automated data augmentation with a reduced search space. In: Proceedings of the IEEE/CVF Conference on Computer Vision and Pattern Recognition Workshops, pp. 702–703 (2020)
5. Dantcheva, A., Elia, P., Ross, A.: What else does your biometric data reveal? A survey on soft biometrics. IEEE Trans. Inf. Forensics Secur. **11**(3), 441–467 (2016)
6. De Carolis, B., Macchiarulo, N., Palestra, G.: Soft biometrics for social adaptive robots. In: Wotawa, F., Friedrich, G., Pill, I., Koitz-Hristov, R., Ali, M. (eds.) Advances and Trends in Artificial Intelligence. From Theory to Practice, pp. 687–699. Springer International Publishing, Cham (2019). https://doi.org/10.1007/978-3-030-22999-3_59
7. De Simone, G., Saggese, A., Vento, M.: Empowering human interaction: a socially assistive robot for support in trade shows. In: 2024 33rd IEEE International Conference on Robot and Human Interactive Communication (ROMAN), pp. 901–908. IEEE (2024). https://doi.org/10.1109/RO-MAN60168.2024.10731165
8. Dosovitskiy, A.: An image is worth 16x16 words: transformers for image recognition at scale (2020). arXiv preprint arXiv:2010.11929
9. Ephrat, A., et al.: Looking to listen at the cocktail party: a speaker-independent audio-visual model for speech separation. ACM Trans. Graph. **37**(4), 112:1–112:11 (2018).https://doi.org/10.1145/3197517.3201357
10. Feichtenhofer, C., Fan, H., Malik, J., He, K.: SlowFast Networks for Video Recognition. In: 2019 IEEE/CVF International Conference on Computer Vision (ICCV), pp. 6201–6210 (2019).https://doi.org/10.1109/ICCV.2019.00630. https://ieeexplore.ieee.org/document/9008780?denied=, iSSN: 2380-7504
11. Foggia, P., Greco, A., Saggese, A., Vento, M.: Multi-task learning on the edge for effective gender, age, ethnicity and emotion recognition. Eng. Appl. Artif. Intell. **118**, 105651 (2023)

12. Fox, J., Gambino, A.: Relationship Development with Humanoid Social Robots: Applying Interpersonal Theories to Human-Robot Interaction. Cyberpsychology, behavior and social networking (2021).https://doi.org/10.1089/cyber.2020.0181

13. George, S.M., Muhamed Ilyas, P.: A review on speech emotion recognition: A survey, recent advances, challenges, and the influence of noise. Neurocomputing **568**, 127015 (2024)

14. Goncalves, L., Leem, S.G., Lin, W.C., Sisman, B., Busso, C.: Versatile audio-visual learning for handling single and multi modalities in emotion regression and classification tasks (2023). arXiv preprint arXiv:2305.07216 **4**

15. Greco, A., Saggese, A., Vento, M., Vigilante, V.: Gender recognition in the wild: a robustness evaluation over corrupted images. J. Ambient. Intell. Humaniz. Comput. **12**(12), 10461–10472 (2021)

16. Gulati, A., et al.: Conformer: convolution-augmented transformer for speech recognition. Interspeech 2020 (2020)

17. Hegel, F., Muhl, C., Wrede, B., Hielscher-Fastabend, M., Sagerer, G.: Understanding Social Robots. In: 2009 Second International Conferences on Advances in Computer-Human Interactions,. pp. 169–174 (2009). https://doi.org/10.1109/ACHI.2009.51. https://ieeexplore.ieee.org/document/4782510

18. Heredia, J., et al.: Adaptive multimodal emotion detection architecture for social robots. IEEE Access **10**, 20727–20744 (2022). https://doi.org/10.1109/ACCESS.2022.3149214

19. Kay, W., et al.: The kinetics human action video dataset (2017). arXiv preprint arXiv:1705.06950

20. Kim, J.Y., Lee, S.H.: CoordViT: A novel method of improve vision transformer-based speech emotion recognition using coordinate information concatenate. In: 2023 International Conference on Electronics, Information, and Communication (ICEIC), pp. 1–4 (2023). https://doi.org/10.1109/ICEIC57457.2023.10049941. https://ieeexplore.ieee.org/document/10049941, iSSN: 2767-7699

21. Kim, J.Y., Lee, S.H.: Accuracy enhancement method for speech emotion recognition from spectrogram using temporal frequency correlation and positional information learning through knowledge transfer. IEEE access **12**, 128039–128048 (2024)

22. Liao, J., Duan, H., Feng, K., Zhao, W., Yang, Y., Chen, L.: A light weight model for active speaker detection. In: Proceedings of the IEEE/CVF Conference on Computer Vision and Pattern Recognition (CVPR), pp. 22932–22941 (2023)

23. MIT: MIT Indoor Scene Recognition. https://web.mit.edu/torralba/www/indoor.html

24. Mollahosseini, A., Hasani, B., Mahoor, M.H.: AffectNet: A database for facial expression, valence, and arousal computing in the wild. IEEE Trans. Affect. Comput. **10**(1), 18–31 (2019)

25. Pan, C., et al.: VLP: Vision language planning for autonomous driving. In: 2024 IEEE/CVF Conference on Computer Vision and Pattern Recognition (CVPR) (2024)

26. Poria, S., Majumder, N., Mihalcea, R., Hovy, E.: Emotion recognition in conversation: research challenges, datasets, and recent advances. IEEE Access **7**, 100943–100953 (2019). https://doi.org/10.1109/ACCESS.2019.2929050

27. Roth, J., Chaudhuri, S., et al.: Ava active speaker: an audio-visual dataset for active speaker detection. In: ICASSP 2020 - 2020 IEEE International Conference on Acoustics, Speech and Signal Processing (ICASSP), pp. 4492–4496 (May 2020), iSSN: 2379-190X

28. Ruiz-Garcia, A., Elshaw, M., Altahhan, A., Palade, V.: A hybrid deep learning neural approach for emotion recognition from facial expressions for socially assistive robots. Neural Comput. Appl. **29**(7), 359–373 (2018). https://doi.org/10.1007/s00521-018-3358-8

29. Tan, M., Le, Q.: EfficientNet: rethinking model scaling for convolutional neural networks, pp. 6105–6114. PMLR (2019)

30. Tao, R., Pan, Z., Das, R.K., Qian, X., Shou, M.Z., Li, H.: Is someone speaking? Exploring long-term temporal features for audio-visual active speaker detection. In: Proceedings of the 29th ACM International Conference on Multimedia, pp. 3927–3935 (2021)

31. Tian, H., Tao, Y., Pouyanfar, S., Chen, S.C., Shyu, M.L.: Multimodal deep representation learning for video classification. World Wide Web **22**(3), 1325–1341 (2019). https://doi.org/10.1007/s11280-018-0548-3

32. Tran, D., Wang, H., Torresani, L., Ray, J., LeCun, Y., Paluri, M.: A closer look at spatiotemporal convolutions for action recognition. In: Proceedings of the IEEE Conference on Computer Vision and Pattern Recognition, pp. 6450–6459 (2018)

33. Xiang, P., Wu, K., Lin, C., Bai, O.: Mtcae-dfer: Multi-task cascaded autoencoder for dynamic facial expression recognition (2024). arXiv preprint arXiv:2412.18988

34. Zhang, S., Zhu, X., Lei, Z., Shi, H., Wang, X., Li, S.Z.: S3fd: Single shot scale-invariant face detector. In: Proceedings of the IEEE International Conference on Computer Vision, pp. 192–201 (2017)

35. Zhang, Y., et al.: BigSSL: exploring the frontier of large-scale semi-supervised learning for automatic speech recognition. IEEE J. Sel. Topics Signal Process. **16**(6), 1519–1532 (2022)

Enhancing Collaborative Image Classification via Spatio-Temporal Graph Neural Networks: A Proof-of-concept Study on Human Group Decisions

I. Mateos-Aparicio-Ruiz[✉] [iD], P. Montealegre-Macias[iD], O. Deniz[iD], A. Pedraza[iD], and G. Bueno[iD]

VISILAB, E.T.S. Ingeniería Industrial, University of Castilla-La Mancha, Ciudad Real, Spain
{Israel.MateosAparici,Pedro.Montealegre,Oscar.Deniz,Anibal.Pedraza, Gloria.Bueno}@uclm.es

Abstract. Collaborative decision-making is essential in expert-driven image classification tasks, where individual assessments may be inconsistent or limited. We propose a task- and label-independent spatio-temporal graph neural network (STGNN) framework to model real-time interactions among human participants during group classification. The architecture combines graph neural networks (GNNs) and recurrent units to capture relational and temporal dependencies across dynamic graph sequences, with an auxiliary contrastive loss encouraging alignment among agreeing participants, coherence with chosen options and separation from alternatives. Experiments on a collaborative web platform covered five expert classification tasks of varying complexity, including cyanobacteria and diatom identification, Ki67 scoring, HER2 grading and glomerulonephritis diagnosis. From 1,369 group classification instances by 34 participants, multiple STGNN configurations were tested, varying GNN architecture, feature initialization and temporal granularity. Stratified 5-fold cross-validation showed several configurations outperforming the majority voting (MV) baseline in global top-1 accuracy, with the best (GIN+GRU, $T = 20$) achieving 0.7757 vs. 0.7633 for MV. Improvements were also observed in complex tasks such as glomerulonephritis (0.4778 vs. 0.4167), HER2 (0.6100 vs. 0.5633), and Ki67 (0.8261 vs. 0.7993), demonstrating the potential of STGNNs for enhancing collaborative image classification.

Keywords: Collaborative Decision-Making · Graph Neural Networks · Spatio-Temporal Graph Neural Networks · Image Classification · Dynamic Interaction Modeling

Supplementary Information The online version contains supplementary material available at https://doi.org/10.1007/978-3-032-05060-1_22.

1   Introduction

Collaborative decision-making (CDM) is critical in expert domains such as healthcare, aviation and finance, where complex decisions benefit from multiple expert perspectives. In medicine, interdisciplinary collaboration has been shown to enhance diagnostic accuracy and treatment outcomes [3, 19, 21]. Similar trends are observed in aviation, where CDM improves flight coordination [6] and in finance, where it mitigates risk-taking behaviors [16]. Extending these principles to image classification tasks with high inter-observer variability may likewise improve accuracy. However, existing aggregation methods, such as majority voting (MV), remain static and fail to capture the dynamic, real-time nature of human interactions in CDM tasks.

This work addresses this gap by proposing a structured, task- and label-independent framework that models the dynamic interactions among human participants during collaborative image classification. The primary goal is to assess whether spatio-temporal graph neural networks (STGNNs) can capture latent patterns in group decisions and leverage them to improve predictive accuracy beyond MV. The proposed framework combines graph neural networks (GNNs) and recurrent neural networks (RNNs) to jointly model relational dependencies (e.g., clusters of agreement among participants or consistent individual-option preferences) and temporal dependencies (e.g., gradual convergence to consensus or shifts in selection over time) within graph sequences. Unlike prior methods, our approach treats decision-makers and options as nodes in a sequence of graph snapshots to model a collaborative classification instance. Notably, the framework is designed to be independent of the classification task or number of labels, allowing broad applicability.

To evaluate the framework, we collected data from five expert-driven classification tasks using a web platform that supports real-time interaction. We assessed the ability of the framework to outperform MV by modeling the temporal structure of human collaboration.

The main contributions are: i) a general, task- and label-independent STGNN-based framework for real-time collaborative image classification; ii) a graph construction method representing participants and label options as dynamic interaction graphs, capturing both agreement relations and temporal decision trajectories; iii) an auxiliary loss promoting consistency among agreeing participants, alignment with selected options and separation from alternatives; and iv) a comprehensive evaluation across five expert-driven tasks, demonstrating improvements over MV in complex settings.

This paper is organized as follows: Sect. 2 reviews related work. Section 3 describes the experimental setup. Section 4 details graph construction, architecture and training. Section 5 presents results, and Sect. 6 discusses implications and limitations. Section 7 concludes.

2   Related Work

Collaborative modeling research spans static aggregation, hybrid intelligence and dynamic interaction learning. Early studies in collective intelligence show that interaction among decision-makers can outperform static aggregation methods such as MV [17,18]. However, these approaches are largely heuristic and lack structured modeling of real-time interaction dynamics.

Building on this, several works emphasize aggregating diverse human judgments to improve group outcomes. [5] adopt principles from portfolio theory to select human agents based on behavioral correlation, aiming to increase heterogeneity and reduce group error. While they model dependence statistically, decisions are made independently and without real-time influence modeling. Similarly, in human–AI collaboration, [10] demonstrate improvements when combining AI and therapist input in clinical assessments. However, collaboration is implemented through feedback-based refinement rather than real-time decision integration.

Trajectory-based modeling has also been explored. Lin *et al.* [11] use Long Short-Term Memory (LSTM) networks to predict human decisions in social dilemmas, leveraging behavioral sequences from interacting participants. Yet interpersonal effects are embedded as input features, not explicitly structured in the model. Semeraro *et al.* [20] review learning approaches for human–robot collaboration, noting that even in CDM-focused studies, the robot typically assists a human rather than modeling human-human interaction explicitly.

Other frameworks address collaboration from a statistical or learning perspective. The HAM framework [4] decomposes joint performance into individual accuracy and diversity but remains static and retrospective, inferring diversity from outcome correlations without modeling relational or temporal dynamics. Similarly, in multi-agent reinforcement learning, Xu *et al.* [26] propose COLA, which enables decentralized agents to infer latent states via contrastive learning. While this fosters emergent coordination, it targets policy learning and does not explicitly represent inter-agent influence as graph dynamics.

Our problem can also be framed as a label aggregation task. Mazzetto *et al.* [14] propose PGMV, a semi-supervised method that removes the common independence assumption among annotators and offers theoretical guarantees. However, it remains classification-task dependent and assumes offline aggregation. Majdi *et al.* [13] introduce Crowd-Certain, which adjusts labeler weights based on disagreement with a reference classifier. Like other aggregation methods, it models annotators in isolation, without temporal dynamics or relational modeling.

In summary, prior work addresses aspects of collaborative modeling (diversity, aggregation, prediction) but rarely unifies them or considers evolving interdependent behaviors. To our knowledge, our framework is the first to model CDM as a spatio-temporal process with structured agreement and option-choice dynamics.

3   Materials

We collected data from five expert-driven image classification tasks selected to capture diverse real-world challenges in biology and pathology: (1) classification of cyanobacteria and (2) diatom genera from microscopic images, (3) scoring of Ki67 expression into three predefined ranges, (4) HER2 grading with five labels, including the non-standard 1.5+ class proposed in [15], and (5) glomerulonephritis diagnosis on PAS-stained glomerular crops. Annotation consistency was ensured through standardized task instructions and training materials, with ground-truth labels defined by domain experts following established protocols. All participants provided informed consent, and their data were anonymized throughout.

Participants first completed individual assessments using standardized forms, then engaged in group sessions via a web-based platform after sufficient intervals to minimize recall bias. Expertise levels ranged from low to high (Supplementary Table S2.1), with the "Glomeruli" session including only high-expertise participants. Some individuals contributed to multiple sessions. AI bots (deep learning models with simulated cursor trajectories) also participated but were excluded from analysis, which focuses exclusively on human behavior. Table 1 summarizes key statistics per task.

Table 1. Summary of experimental data per classification task.

Session	# Questions	# Humans	# Options	# Labels
Cyanobacteria	299	14	6	11
Diatoms	291	11	6	47
Ki67	299	12	3	3
HER2	300	10	5	5
Glomeruli	180	4	6	12
All	1369	34	–	–

3.1   Collaborative Decision-Making Platform

Data collection used a web-based platform supporting real-time collaborative classification. Participants viewed the same image and dragged their cursors toward polygon vertices representing labels, while seeing others' cursor positions during a fixed interval. Final choices were recorded at the interval's end (Fig. 1). No identifying information, expertise levels or prior performance was shown; direct communication (audio, video, chat) was disabled. Interaction was limited to implicit visual feedback of group tendencies within the interface.

4   Methods

4.1   Graph Construction

Each classification instance is represented as a sequence of T graphs $\{G_1, G_2, \ldots, G_T\}$, where nodes correspond to human participants and classification options.

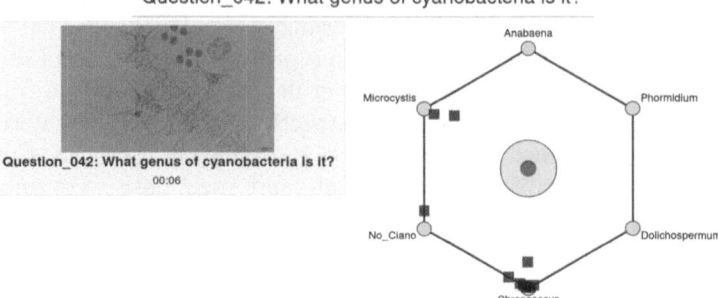

Fig. 1. Interface of the collaborative platform.

At time t, undirected edges connect participants selecting the same label (agreement), while directed edges link participants to their current label choice (Fig. 2). Graphs were sampled at $T = 20$ uniform time intervals; however, because early frames show limited participant movement (see Supplementary Fig. S3.1), we also tested a reduced variant discarding the first two frames, resampling into $T = 10$ intervals and collapsing repeated graphs.

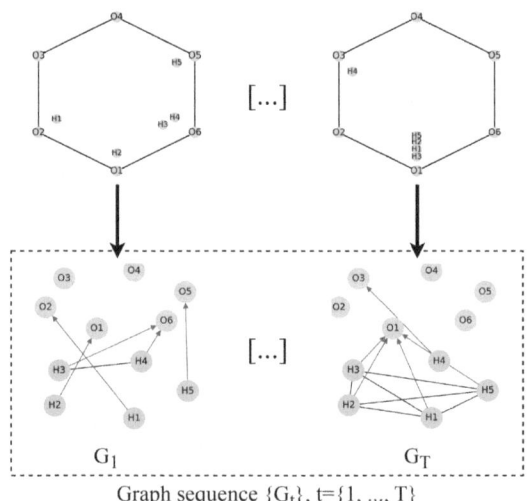

Fig. 2. Graph sequence construction for an image classification question. Human nodes $(H1, \ldots, HN)$ connect via undirected edges when in agreement; directed edges point to selected option nodes $(O1, \ldots, OM)$.

Human node features included pre-session accuracy and expertise level (0: low, 1: medium, 2: high), providing a coarse inductive bias to guide influence estimation during collaboration. Although static, these features offer the only consistent session-wide attributes available across all tasks. Option node features, in contrast, were initialized without intrinsic metadata to avoid bias: each feature vector $\mathbf{o}_m \in \mathbb{R}^d$ was sampled from $\mathbf{o}_m \sim \mathcal{N}(\mathbf{0}, 1.01\mathbf{I})$.

4.2   STGNN Framework

The proposed approach integrates a spatial module (GNN) with a temporal module, implemented here as a Gated Recurrent Unit (GRU) [1], to model evolving interactions over time. At each time step t, the graph snapshot is processed by a GNN to capture spatial relationships among human and option nodes. Human node features, defined in $\mathbb{R}^2$ based on pre-session accuracy and expertise level, are projected into a shared embedding space $\mathbb{R}^d$ via a multilayer perceptron (MLP) to align with option feature dimensionality. The GNN outputs are (i) passed to the GRU to model temporal dynamics across time steps and (ii) used to compute an auxiliary loss reinforcing agreement and decision structure at each t. GRU outputs are propagated as node features for G_{t+1}, although we also tested reinitializing each G_t with the original features (X_1) instead of the previous hidden state (h_t) to mitigate oversmoothing. At the final time step, option node embeddings are passed through an MLP and softmax layer to generate classification probabilities (Fig. 3).

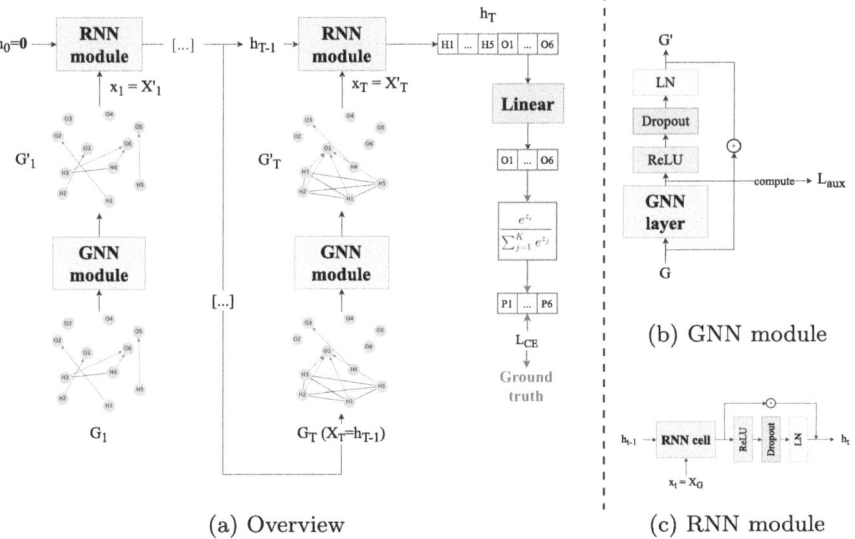

(a) Overview

(b) GNN module

(c) RNN module

Fig. 3. Overview of the proposed STGNN framework. At each time step, node features are processed by a GNN and passed to a GRU to capture temporal dynamics. Human features are projected into a shared space via an MLP and GRU outputs are reused or reinitialized in the next graph. An auxiliary loss is computed after each GNN pass. Final predictions are made from option node embeddings using an MLP and softmax.

Model Variants. The modular design of the framework enables flexible experimentation with different architectural components. For the spatial module, we evaluated four GNN layers, each providing distinct aggregation strategies and inductive biases:

– *Graph Convolutional Network (GCN)* [9]: performs convolutions by aggregating normalized neighbor information, extending classical convolutions to graphs via spectral graph theory. Node features are updated using a learnable weight matrix applied to the degree-normalized adjacency.
– *Graph Attention Network (GAT)* [23]: uses self-attention mechanisms on graphs, allowing nodes to compute weighted combinations of neighbor features and prioritize the most relevant ones.
– *Graph Isomorphism Network (GIN)* [25]: achieves high discriminative power, matching the Weisfeiler-Lehman test. Its additive aggregator and learnable residual parameter effectively distinguish graph structures.
– *GraphSAGE* [7]: supports inductive learning on large or evolving graphs by sampling neighbors and applying an aggregator (*e.g.*, mean, max, LSTM-based).

For temporal modeling, we adopted Gated Recurrent Units (GRUs) [1] due to their reduced parameter count and lower risk of overfitting compared to LSTMs. GRUs use gating mechanisms to balance past information retention and incorporation of new input, providing sufficient temporal expressiveness while mitigating oversmoothing and vanishing gradients. Preliminary experiments with LSTMs showed worse performance, supporting the suitability of GRUs for our setting.

Loss Function. The training objective combines cross-entropy loss with an auxiliary term. The auxiliary loss $\mathcal{L}_{\text{aux}}$ is defined over two edge sets: $\mathcal{E}_A$ (participant pairs in agreement) and $\mathcal{E}_D$ (participant–option selections). Let $\mathbf{p}_i$ and $\mathbf{o}_j$ denote participant and option embeddings, respectively, and $\mathcal{O}$ the set of option indices.

$$\mathcal{L}_{\text{aux}} = \mathcal{L}_{\text{agree}} + \mathcal{L}_{\text{decide}} + \mathcal{L}_{\text{cross}}, \tag{1}$$

where

$$\mathcal{L}_{\text{agree}} = \frac{1}{|\mathcal{E}_A|} \sum_{(p,q)\in\mathcal{E}_A} [1 - \cos(\mathbf{p}_p, \mathbf{p}_q)], \tag{2}$$

$$\mathcal{L}_{\text{decide}} = \frac{1}{|\mathcal{E}_D|} \sum_{(p,o)\in\mathcal{E}_D} [1 - \cos(\mathbf{p}_p, \mathbf{o}_o)], \tag{3}$$

$$\mathcal{L}_{\text{cross}} = \frac{1}{|\mathcal{E}_D|} \sum_{(p,o)\in\mathcal{E}_D} \sum_{\substack{k\in\mathcal{O} \\ k\neq o}} \max\{0, \mu - \cos(\mathbf{p}_p, \mathbf{o}_k)\}. \tag{4}$$

Here, $\mu = 0.5$ defines the contrastive margin. The cross-entropy loss with respect to the ground truth is defined as:

$$\mathcal{L}_{\text{CE}} = -\frac{1}{N} \sum_{i=1}^{N} \sum_{c=1}^{C} y_{i,c} \log(\hat{y}_{i,c}), \tag{5}$$

where N is the number of instances, C the number of classes, yi, c the ground truth indicator, and $\hat{y}_{i,c}$ the predicted probability. The final loss is:

$$\mathcal{L}_{\text{total}} = (1 - \lambda)\mathcal{L}_{\text{CE}} + \lambda\mathcal{L}_{\text{aux}}, \tag{6}$$

with $\lambda = 0.5$ empirically chosen. This formulation encourages embedding consistency among agreeing participants, alignment with selected options and separation from negatives.

4.3   Model Training and Evaluation

Models were trained using AdamW (learning rate 1.5×10^{-5}, weight decay 0.05, $\epsilon = 10^{-8}$) and a ReduceLROnPlateau scheduler (factor 0.1, patience 5, min delta 10^{-4}, cooldown 1, min LR 10^{-8}). Early stopping was applied if validation accuracy failed to improve by 0.01 within 10 epochs (maximum: 100). Due to sample heterogeneity, batch size was set to 1 with a scaled learning rate.

Stratified 5-fold cross-validation was used, reporting top-k accuracy for $k \in 1, 2, 3$ by aggregating predictions across test folds before metric computation. This approach enables direct comparability across tasks and avoids per-fold averaging, which would require uncertainty estimates. Top-2 and top-3 metrics were excluded for Ki67 (three classes) and for the global metric (cross-session label variability).

5   Results

Table 2 reports MV accuracy among human participants as a baseline for evaluating the proposed models. Global top-1 accuracy is 0.7633, with session accuracies ranging from 0.4167 ("Glomeruli") to 0.9622 ("Diatoms"). Ties were resolved randomly. Table 3 then summarizes top-1 accuracy across all STGNN variants for $T = 20$ and a reduced version ($T \leq 10$), using either hidden states (h_t) or initial features (X_1). Extended results (top-2, top-3) appear in Supplementary Table S2.2.

Global top-1 accuracy across configurations ranges from 0.7207 to 0.7757. Two variants exceed MV (0.7633), with GIN+GRU ($T = 20$) achieving the best result (0.7757) with either feature strategy. A one-sided bootstrap test (10^4 resamples) comparing this best model to MV yielded $\Delta = 0.0124$ with $p = 0.0467$, indicating significance at the 5% level (Fig. 4).

Session-level results show varied trends. In "Glomeruli", all models outperform MV (0.4167), with GCN+GRU and SAGE+GRU ($T = 20$) reaching 0.4778. In "HER2", GIN+GRU ($T = 20$, h_t) achieves 0.6100 vs. 0.5633 for MV. For

Table 2. Top-1, top-2 and top-3 accuracy achieved by majority vote among human participants.

Session	Top-1 accuracy	Top-2 accuracy	Top-3 accuracy
Cyanobacteria	0.9431	0.9933	0.9967
Diatoms	0.9622	0.9931	1.0000
Glomeruli	0.4167	0.6000	0.7333
HER2	0.5633	0.8267	0.9133
Ki67	0.7993	-	-
Global	0.7633	-	-

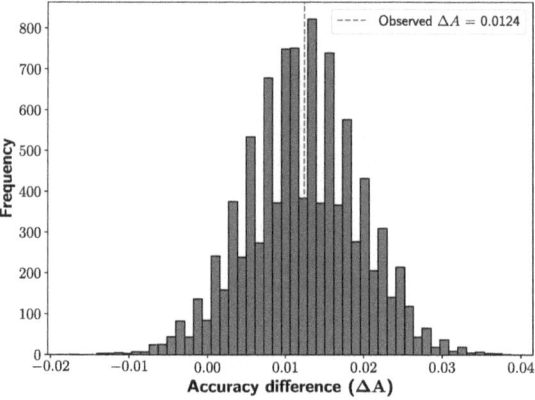

Fig. 4. Bootstrap distribution of accuracy differences (best model – MV baseline). $\Delta = 0.0124$ (red dashed line). (Color figure online)

"Ki67" (MV: 0.7993), GIN+GRU ($T = 20$, X_1) achieves 0.8261. Conversely, MV dominates "Cyanobacteria" and "Diatoms" in top-1 accuracy, though STGNNs match or exceed MV in top-2/top-3 metrics (see Supplementary Table S2.2). Overall, while MV remains strong in tasks with high inter-participant agreement, STGNNs can outperform it when collaboration involves ambiguity, disagreement or evolving group dynamics. Moreover, the high top-k accuracy achieved even in MV-dominated tasks suggests that STGNNs reliably assign high confidence to the correct label, even when it is not ranked first.

These task-specific results reveal contrasting behaviors across aggregation methods. In "Diatoms" and "Cyanobacteria", the combination of clear visual categories and high participant agreement means static aggregation is already near ceiling, leaving little room for improvement. In more difficult tasks, such as "Glomeruli", a challenging task due to overlapping morphological patterns, STGNNs outperform MV by more than 6 points. These patterns suggest that

Table 3. Top-1 accuracy across STGNN variants, with different GNN layers and configurations. Bold highlights the highest per-session and global results.

Time steps (T)	Next features (X_{t+1})	Session	GAT+GRU	GCN+GRU	GIN+GRU	SAGE+GRU
20	$\mathbf{h_t}$	Cyanobacteria	0.9097	0.9298	0.9264	0.9164
		Diatoms	0.9381	**0.9519**	0.9381	0.9175
		Glomeruli	0.4278	**0.4778**	0.4611	**0.4778**
		HER2	0.5567	0.5433	**0.6100**	0.5533
		Ki67	0.7793	0.7893	0.8227	0.7692
		Global	0.7465	0.7597	**0.7757**	0.7473
	$\mathbf{X_1}$	Cyanobacteria	0.9197	**0.9365**	0.9264	0.9197
		Diatoms	0.9278	**0.9519**	0.9347	0.9175
		Glomeruli	0.4389	**0.4778**	0.4722	**0.4778**
		HER2	0.5700	0.5533	0.6033	0.5533
		Ki67	0.7759	0.7793	**0.8261**	0.7692
		Global	0.7502	0.7611	**0.7757**	0.7480
Reduced ≤ 10	h_t	Cyanobacteria	0.9331	0.9331	0.9164	0.9130
		Diatoms	0.9244	0.9244	0.9244	0.9107
		Glomeruli	0.4333	0.4556	0.4333	0.4722
		HER2	0.5100	0.5000	0.5167	0.5167
		Ki67	0.7592	0.7793	0.7692	0.7391
		Global	0.7348	0.7400	0.7348	0.7297
	X_1	Cyanobacteria	0.9331	0.9331	0.9164	0.9130
		Diatoms	0.9244	0.9244	0.9244	0.9107
		Glomeruli	0.4333	0.4611	0.4333	0.4722
		HER2	0.5100	0.5033	0.5167	0.5167
		Ki67	0.7592	0.7793	0.7692	0.7391
		Global	0.7348	0.7414	0.7348	0.7297

STGNN-based modeling is most beneficial when individual decisions are noisy, inconsistent or sparse.

To contextualize these gains, we compared the best-performing STGNN (with GIN+GRU, $T = 20$) against a broader set of aggregation and inference baselines, including Dawid–Skene [2], Zero-Based Skill [22], MACE [8], GLAD [24], MMSR [12], Wawa [22], weighted MV (WMV) by accuracy or expertise and an MLP trained on final vote histograms (Suplementary S1 for details on each method). As shown in Table 4, our method achieves the highest global accuracy (0.7757) and outperforms all baselines in challenging sessions such as "HER2" and "Glomeruli" by a margin of at least 3–6 points.

Table 4. Top-1 accuracy of STGNN vs. baseline aggregation methods. Bold highlights best per-session and global results.

Model	Cyanobacteria	Diatoms	Glomeruli	HER2	Ki67	Global
GIN+GRU(T = 20, X$_{t+1}$ = X$_t$)	0.9264	0.9381	**0.4611**	**0.6100**	**0.8227**	**0.7757**
MV	**0.9431**	0.9622	0.4167	0.5633	0.7993	0.7633
WMV (accuracy)	0.9298	**0.9656**	0.4333	0.5833	0.7692	0.7611
WMV (expertise)	0.9365	0.9588	0.4167	0.5400	**0.8227**	0.7611
Dawid–Skene	0.9398	0.9588	0.4056	0.5833	0.7860	0.7619
Zero-Based Skill	0.9365	0.9622	0.4111	0.5633	0.7993	0.7611
MMSR	0.9365	0.9622	0.4111	0.5667	0.7993	0.7619
MACE	0.9365	**0.9656**	0.4167	0.5733	0.7993	0.7648
GLAD	0.9331	**0.9656**	0.4111	0.5667	0.7993	0.7619
Wawa	0.9365	0.9622	0.4111	0.5633	0.7993	0.7611
MLP (vote histogram)	0.8997	0.9072	0.3667	0.5733	0.8127	0.7407

6   Discussion

This study explored the use of STGNNs to model collaborative dynamics in expert-driven image classification. Results demonstrate that these models can outperform static aggregation strategies like MV by leveraging temporal and relational patterns. The best configuration (GIN+GRU, $T = 20$) achieved statistically significant improvements in global accuracy and excelled in tasks involving ambiguity or interpretive variability.

These gains likely stem from relational and temporal inductive biases. For example, in "HER2", the presence of borderline cases (subclass 1.5+) led to frequent disagreement, while in "Glomeruli", a small group of highly skilled participants worked under complex visual conditions. In both cases, STGNNs propagated local agreement and temporal consistency across the graph, stabilizing uncertain predictions. Even in "Ki67", a visually simpler task involving subjective thresholds, GIN+GRU achieved a 2.34% margin over MV, indicating that participant-specific dynamics are beneficial even when ambiguity is moderate.

In contrast, tasks like "Cyanobacteria" and "Diatoms" showed high participant agreement and strong individual accuracy, leaving little room for improvement. Here, MV performed near ceiling, but STGNNs still matched or exceeded MV in top-2 and top-3 metrics, suggesting their label rankings remain informative even when the top prediction offers no gain.

Among all models, GIN+GRU with a full temporal window ($T = 20$) achieved the highest global accuracy. Its MLP-based aggregation likely helps capture expressive neighborhood patterns, particularly when behavioral signals are subtle or inconsistent. Importantly, this advantage is not attributable to model size: all STGNN variants have comparable parameter counts (Supplementary Table S2.3), remaining lightweight. This highlights the role of task-specific inductive bias in driving performance.

Finally, comparisons to alternative aggregation methods (including probabilistic models, weighted MV and shallow classifiers) showed that most failed to surpass the MV baseline. In contrast, STGNNs consistently outperformed these baselines, suggesting that temporal and relational modeling captures collaborative signals such as evolving consensus or persistent dissent that static vote-based approaches cannot exploit.

6.1 Limitations and Future Work

This study has several limitations, each of which motivates specific directions for future research. First, node features were limited to pre-session accuracy and encoded expertise levels. These coarse features may fail to capture the complexity of human decision-making and contribute to oversmoothing, particularly in recurrent architectures. Future work could incorporate richer behavioral signals, such as interaction timing, uncertainty estimates or learned embeddings from participant histories.

Second, the model was designed for broad applicability without task-specific adjustments. While this increases flexibility, it may limit the ability to learn distinctive patterns and restricts evaluation to overall accuracy, which may not fully reflect performance on complex problems. Future extensions could explore adaptive components, such as task-aware attention mechanisms or content-conditioned message passing, that retain generality while enabling specialization.

Third, the dataset included few participants for some tasks, potentially affecting result reliability and generalizability. Expanding the number of tasks, sessions and participant profiles would support more robust analyses and richer modeling of human heterogeneity.

Finally, while the STGNN framework improves predictive performance, it adds modeling complexity and reduces interpretability compared to simpler baselines like MV. This trade-off is critical in practical deployments requiring human oversight. Future research should investigate interpretable STGNN variants to address this concern.

7 Conclusion

This proof-of-concept study explored spatio-temporal graph neural networks (STGNNs) for modeling collaborative decision-making (CDM) in expert-driven image classification. By integrating graph neural networks (GNNs) with recurrent units, the framework captures participant interactions over time while remaining task- and label-independent. Across five classification tasks, several STGNN configurations outperformed MV, with the best result (GIN+GRU, $T = 20$) achieving a statistically significant global top-1 accuracy of 0.7757 versus 0.7633 for MV ($\Delta = 0.0124$).

Session-level gains were most notable in "Glomeruli" and "HER2", where ambiguity and complex visual cues challenged static aggregation, and even in

"Ki67", where STGNNs reduced inter-observer variability despite its relative simplicity. These findings suggest temporal and relational modeling can enhance performance under conditions of disagreement or evolving group dynamics, though improvements remain modest and task-dependent. Future work should address these constraints by incorporating richer dynamic node features (*e.g.*, uncertainty, response timing, behavioral embeddings), exploring task-adaptive components to balance generality and specialization, and assessing integration into decision-support systems to support expert collaboration in practice.

Acknowledgments. This work is funded by the HANS (Ref. PID2021-127567NB-I00) and the DIAMOND (Ref. TED2021-132147B-I00) projects, both supported by the Spanish Ministry of Science, Innovation and Universities, and the European FEDER fund.

Disclosure of Interests. The authors have no competing interests to declare that are relevant to the content of this article.

References

1. Cho, K., et al.: Learning phrase representations using RNN encoder-decoder for statistical machine translation. arXiv preprint arXiv:1406.1078 (2014)
2. Dawid, A.P., Skene, A.M.: Maximum likelihood estimation of observer error-rates using the EM algorithm. J. Roy. Stat. Soc.: Ser. C (Appl. Stat.) **28**(1), 20–28 (1979)
3. Dogba, M.J., et al.: Enhancing interprofessionalism in shared decision-making training within homecare settings: a short report. J. Interprofessional Care (2020)
4. Gao, W., Feng, J., Wei, M., Zou, R., Sun, J.: Towards a multi-granulated statistical framework for human-machine collaboration in image classification. IEEE Transa. Multimedia (2025)
5. Geng, B., Cheng, X., Brahma, S., Kellen, D., Varshney, P.K.: Collaborative human decision making with heterogeneous agents. IEEE Transa. Comput. Soc. Syst. **9**(2), 469–479 (2021)
6. Geske, A.M., Herold, D.M., Kummer, S.: Integrating AI support into a framework for collaborative decision-making (CDM) for airline disruption management. J. Air Transp. Res. Soc. **3**, 100026 (2024)
7. Hamilton, W., Ying, Z., Leskovec, J.: Inductive representation learning on large graphs. Adv. Neural Inf. Process. Syst. **30** (2017)
8. Hovy, D., Berg-Kirkpatrick, T., Vaswani, A., Hovy, E.: Learning whom to trust with MACE. In: Proceedings of the 2013 Conference of the North American Chapter of the Association for Computational Linguistics: Human Language Technologies, pp. 1120–1130 (2013)
9. Kipf, T.N., Welling, M.: Semi-supervised classification with graph convolutional networks. arXiv preprint arXiv:1609.02907 (2016)
10. Lee, M.H., Siewiorek, D.P., Smailagic, A., Bernardino, A., Bermúdez i Badia, S.B.: A Human-AI collaborative approach for clinical decision making on rehabilitation assessment. In: Proceedings of the 2021 CHI Conference on Human Factors in Computing Systems, pp. 1–14 (2021)

11. Lin, B., Bouneffouf, D., Cecchi, G.: Predicting human decision making with LSTM. In: 2022 International Joint Conference on Neural Networks (IJCNN), pp. 1–8. IEEE (2022)
12. Ma, Q., Olshevsky, A.: Adversarial crowdsourcing through robust rank-one matrix completion. Adv. Neural. Inf. Process. Syst. **33**, 21841–21852 (2020)
13. Majdi, M.S., Rodriguez, J.J.: Crowd-Certain: label aggregation in crowdsourced and ensemble learning classification. arXiv preprint arXiv:2310.16293 (2023)
14. Mazzetto, A., Sam, D., Park, A., Upfal, E., Bach, S.: Semi-supervised aggregation of dependent weak supervision sources with performance guarantees. In: International Conference on Artificial Intelligence and Statistics, pp. 3196–3204. PMLR (2021)
15. Pedraza, A., Gonzalez, L., Deniz, O., Bueno, G.: Deep neural networks for HER2 grading of whole slide images with subclasses levels. Algorithms **17**(3), 97 (2024)
16. Piehlmaier, D.M.: The one-man show: the effect of joint decision-making on investor overconfidence. J. Consumer Res. **50**(2), 426–446 (2023)
17. Rosenberg, L., Lungren, M., Halabi, S., Willcox, G., Baltaxe, D., Lyons, M.: Artificial swarm intelligence employed to amplify diagnostic accuracy in radiology. In: 2018 IEEE 9th Annual Information Technology, Electronics and Mobile Communication Conference (IEMCON), pp. 1186–1191. IEEE (2018)
18. Rosenberg, L.B.: Human Swarms, a real-time method for collective intelligence. In: Artificial Life Conference Proceedings, pp. 658–659. MIT Press One Rogers Street, Cambridge, MA 02142-1209, USA journals-info ...(2015)
19. Saint-Pierre, C., Herskovic, V., Sepúlveda, M.: Multidisciplinary collaboration in primary care: a systematic review. Fam. Pract. **35**(2), 132–141 (2018)
20. Semeraro, F., Griffiths, A., Cangelosi, A.: Human-robot collaboration and machine learning: a systematic review of recent research. Robot. Comput. Integrated Manufacturing **79**, 102432 (2023)
21. Shams-Vahdati, N., Shams Vahdati, S., Samad-Soltani, T.: Design and evaluation of collaborative decision-making application for patient care in the emergency department. Heal. Sci. Reports **7**(2), e1931 (2024)
22. Toloka-AI: Crowd-Kit Documentation. https://toloka.ai/docs/crowd-kit (2023), toloka AI
23. Veličković, P., Cucurull, G., Casanova, A., Romero, A., Lio, P., Bengio, Y.: Graph Attention Networks. arXiv preprint arXiv:1710.10903 (2017)
24. Whitehill, J., Wu, T.f., Bergsma, J., Movellan, J., Ruvolo, P.: Whose vote should count more: optimal integration of labels from labelers of unknown expertise. Adv. Neural Inf. Process. Syst. **22** (2009)
25. Xu, K., Hu, W., Leskovec, J., Jegelka, S.: How Powerful are Graph Neural Networks? arXiv preprint arXiv:1810.00826 (2018)
26. Xu, Z., et al.: Consensus learning for cooperative multi-agent reinforcement learning. In: Proceedings of the AAAI Conference on Artificial Intelligence. vol. 37, pp. 11726–11734 (2023)

Synthesizing Images with Different Exposure Settings for Low-Light Image Enhancement

Ahmed Alhawwary$^{(\boxtimes)}$ ⓘ, Janne Mustaniemi ⓘ, and Janne Heikkilä ⓘ

University of Oulu, Oulu, Finland
Ahmed.Alhawwary@oulu.fi

Abstract. Capturing high-quality images under low-light conditions remains a significant challenge, particularly for mobile cameras, which have relatively small sensor sizes compared to DSLR cameras. A promising approach to enhance low-light imaging involves fusing information from two consecutive captures with different exposure settings: a long-exposure image, which is less noisy but often blurred, and a short-exposure image, which is sharp but noisy. Recent advances in deep learning (DL) have demonstrated significant improvements over classical methods. However, these approaches require large-scale datasets for training. Collecting real-world data is technically demanding and laborious, leading prior work to rely on synthetic data. Existing methods either generate blur using video frames, thus limiting the diversity of blur to the camera motion, or simulate blur-noise pairs from still images, assuming planar scenes and neglecting depth variations. In this paper, we propose a pipeline that leverages 3D Gaussian Splatting (3DGS) to simulate camera motion within the 3D scene, allowing for the rendering of realistic blurry and noisy image pairs. We compare the synthesized data to real-world images and demonstrate the fidelity of our approach. This work lays the foundation for generating large-scale datasets suitable for training DL models aimed at low-light image enhancement.

Keywords: Deblurring · Denoising · Noise Calibration · Synthetic Data

1 Introduction

Capturing images under low-light conditions is challenging, especially for mobile cameras, which have relatively smaller sensor sizes compared to DSLR cameras, limiting their ability to collect light photons. The captured images are typically noisy and grainy, and the colors are often distorted. Increasing the exposure time can reduce noise and enhance color fidelity, but may introduce blur due to camera shake or scene motion. One effective approach for enhancing the images is to fuse information from short- and long-exposure images [2,7,10,14,16–19]. In addition to improving image quality, fusing multiple exposure images also offers

© The Author(s), under exclusive license to Springer Nature Switzerland AG 2026
M. Castrillón-Santana et al. (Eds.): CAIP 2025, LNCS 15622, pp. 270–280, 2026.
https://doi.org/10.1007/978-3-032-05060-1_23

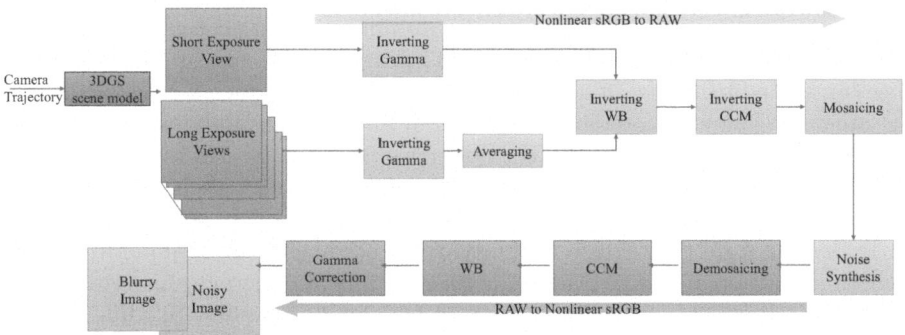

Fig. 1. Overview of our data synthesizing pipeline. It can be used to generate realistic short and long exposure images for training deep neural networks.

advantages in scenarios where a single exposure cannot adequately capture the high dynamic range of the scene.

Deep learning (DL)-based methods [7,10,14,19] have introduced significant improvements over conventional (non-DL) approaches [16]. However, training DL models typically requires large-scale datasets. Previous works have resorted to synthetic datasets, as acquiring real blurry-noisy image pairs is cumbersome and demands complex hardware setups [8,9,20]. In [7], synthetic blurry-noisy image pairs are generated from still images. Blur is synthesized by averaging multiple planar homography-warped versions of the original image, assuming the scene is a single plane and ignoring depth variations. Furthermore, the still images are assumed to be noise-free, which is generally not true, as real images are often contaminated by shot noise, read noise, and other noise sources [13]. Another line of work uses video sequences [19], mainly captured in outdoor daylight, to synthesize blurry images by averaging interpolated frames to avoid motion discontinuities. Static frames are averaged to get a pseudo-clean image, which is then used to synthesize the corresponding noisy image. While video-based synthesizing methods implicitly respect the 3D scene geometry and depth variations, the diversity of the resulting synthetic blur is limited to the motion patterns present in the captured video sequences.

In this paper, we present a pipeline for synthesizing images with different exposure settings exhibiting realistic motion blur and sensor noise using 3D Gaussian Splatting (3DGS) [5]. More specifically, we focus on generating long-short exposure pairs to be used in training models for restoring sharp and noise-free images. After reconstructing the scene with 3D Gaussians, we generate a random 3D camera trajectory to render both short- and long-exposure images. The long-exposure image is simulated by averaging an appropriate number of rendered views sampled along the camera trajectory. The availability of large-scale multi-view image datasets, such as [15], makes it feasible to leverage 3DGS for reconstruction and to build extensive datasets of long- and short-exposure image pairs.

The camera imaging pipeline includes an image signal processing unit (ISP) that transforms raw sensor data into a compressed JPEG image. Since blur and noise originate in the raw image space, we simulate this process by inverting the ISP pipeline, following previous works [1,8], and then reapplying it to convert the result into the nonlinear sRGB color space as illustrated in Fig. 1. Additionally, we carefully model the noise level differences between the blurry and noisy images, a factor that has been largely overlooked in prior works [7,19].

To demonstrate the realism of the synthesized image pairs, we modify Open-Camera [3], an open-source Android camera application, to capture two sets of images. The first set is used to reconstruct the scene using 3DGS, while the second set consists of long-short (blurry-noisy) exposure image pairs. During the capture of the real long-short exposure pairs, we also record the device's inertial measurements, including acceleration and angular velocity, using the onboard IMU sensor.

In our experiments, we compare the synthesized blur and noise against real captured blurry and noisy images. We further demonstrate that blur synthesized through 3DGS-based 3D trajectory rendering better approximates real-world blur than traditional planar homography warping. Our work establishes a new pathway for generating large-scale datasets tailored for training deep learning models targeting low-light image restoration.

2   Method

We leverage 3D Gaussian Splatting (3DGS) [5] to synthesize realistic long- and short-exposure image pairs. First, we briefly introduce the 3DGS framework, followed by a detailed description of our data generation pipeline. An overview of the pipeline is shown in Fig. 1.

2.1   3D Gaussian Splatting

3DGS [5] models the scene using a set of Gaussian primitives G, each characterized by a position μ, covariance Σ, opacity α, and spherical harmonics (SH) coefficients for view-dependent color encoding. Images are rendered by rasterizing these Gaussians onto the image plane. The color of each pixel $C_i(x, y, P_i, G)$ in the ith image, captured at camera pose P_i, is computed by α-blending the Gaussians overlapping the pixel position (x, y). The parameters of the Gaussians G are optimized by minimizing a loss function that compares the rendered images to the corresponding ground truth images.

The camera poses $P_i \in SE(3)$ are initialized using structure-from-motion (SfM) [11]. Following the original 3DGS framework, we use COLMAP [11] to estimate the camera poses and to initialize the model with a sparse set of 3D points.

By employing an explicit representation of the scene's radiance field, 3DGS enables faster training and real-time rendering, while maintaining image quality comparable to methods based on implicit representations, such as NeRF, where each image pixel must be decoded individually by a multilayer perceptron.

2.2   Camera ISP and Unprocessing

Cameras include an image signal processing (ISP) pipeline that transforms the captured raw sensor data into a final JPEG image. A typical ISP pipeline consists of several stages, including demosaicing, white balancing, camera color correction, noise reduction, linear or nonlinear photo finishing (color manipulation), tone mapping, gamma correction, and JPEG compression. Modern cameras also incorporate advanced post-processing steps, many of which are proprietary, making them difficult to reverse-engineer and accurately model.

Since noise and blur occur during the image capture process, it is important to simulate them in the raw image space. Therefore, we invert the ISP pipeline to unprocess the nonlinear sRGB-rendered images back to the linear raw space. This unprocessing involves the following steps: inverting gamma correction, reversing white balancing, undoing camera color correction, and applying mosaicing. In the next section, we briefly introduce the typical operations performed in a camera ISP pipeline.

Demosiacing. Camera sensors are equipped with a color filter array (CFA), commonly the Bayer filter, where each 2×2 block contains two green filters, one red filter, and one blue filter. Each photosensor is positioned beneath one of these color filters. The process of reconstructing the full RGB color for each pixel from the CFA data is known as demosaicing.

White Balance and Color Correction. The human eye can perceive the true color of a white paper under different lighting conditions. Similarly, cameras attempt to replicate this ability by applying white balance gains to correct the image's color. Modern cameras typically include auto white balance (AWB) algorithms, which estimate these gains by analyzing the scene's illumination.

White balance is both camera- and scene-dependent, making it challenging to estimate directly from the input image. However, it can be sampled from a typical range of real-world data. While white balance corrects for neutral white tones, it does not guarantee accurate correction for non-neutral colors. Thus, cameras also apply a color correction matrix (CCM) to transform pixel intensities into a common color space, typically the XYZ color space, which has a broad color gamut [12]. This correction matrix generally consists of two components: a Camera Color Transform (CST) that converts from the camera's color space to XYZ, and a fixed matrix that converts from XYZ to linear sRGB. The DNG (Digital Negative) file format for RAW images includes two matrices, 'colorMatrix1' and 'colorMatrix2', which are factory-calibrated for two specific illuminations: D65 (daylight) and Standard Light A (indoor incandescent).

Gamma Correction. Gamma correction is a nonlinear transformation applied to the image. It is often mistakenly explained as necessary for displaying content on screens properly. However, the primary purpose of gamma correction is to account for the human eye's nonlinear perception and sensitivity to light. The human eye is more sensitive to variations in dark areas than in bright areas. Gamma encoding takes advantage of this by allocating more bits of dynamic range to low-intensity pixels at the expense of brighter regions [1].

The camera ISP also includes tone mapping, a process used to preserve contrast and retain detail in both dark and bright areas when mapping high dynamic range (HDR) values to a low dynamic range (LDR). This tone mapping process is typically proprietary and irreversible. In [1], a generic tone mapping model is assumed, but its impact is relatively minor. In our work, we only consider gamma correction encoding to convert between linear and nonlinear sRGB spaces. Our pipeline generates images at both RAW and sRGB levels, providing flexibility for data generation.

2.3   Noise Modeling

3DGS employs spherical harmonics to model view-dependent RGB colors, enabling smooth renderings even when the input images contain slight noise [5,6]. We consider two primary sources of noise: shot noise and read noise. Read noise is signal-independent and originates from the internal camera circuitry, whereas shot noise is signal-dependent and arises from the stochastic nature of light photon arrival at the camera sensor.

A common misconception is that the noise level is solely determined by the ISO value. However, it is also influenced by the exposure time that controls the amount of incident photons. As illustrated in Fig. 2, two images captured with the same ISO value of 50, but with different exposure times, show that the image with the shorter exposure contains visible grainy noise.

The arrival of photons at the sensor is well modeled by a Poisson distribution $P(z \mid \lambda)$, where z and λ represent the observed and expected number of photons at a pixel, respectively. Read noise is typically modeled by a Gaussian distribution with zero mean. The noisy image formation can be expressed as:

$$I = kz + n \tag{1}$$

where k is the signal gain and n represents the read noise. Here, k corresponds to the ISO amplification of the signal. Camera manufacturers generally map ISO values linearly to the signal intensity gain, meaning that doubling the ISO doubles the signal intensity. Therefore, K is linearly proportional to ISO, so:

$$k = s \cdot \text{ISO} \tag{2}$$

where s is a constant that relates k to the ISO value. From Eq. 2, taking the expectation of I yields the clean signal, as:

$$\mu = E(I) = E(kz) + E(n) = k\lambda = s \cdot ISO \cdot \lambda \tag{3}$$

Thus, we can see that, given a clean signal, to synthesize shot noise, we need to know the value of s.

Parameter Calibration. Two main parameters in the noise model need to be calibrated to synthesize noise on a clean image: the s constant and the variance of the read noise σ. To estimate σ, we capture dark frames (also known as bias frames) at various ISO values, where the camera sensor is completely covered [13].

Taking the variance of (1), we obtain:

$$\text{Var}(I) = \text{Var}(kz) + \text{Var}(n) = k^2\lambda + \sigma \tag{4}$$

We observe that $\text{Var}(I)$ is linearly related to $k\lambda$, where k is the slope. Since we do not have direct access to $k\lambda$ during image capture, we approximate it by the mean intensity of a uniformly illuminated frame.

To mitigate the influence of background texture and other variations–thereby isolating the noise component–we capture two identical images I_1 and I_2 under the same illumination conditions, exposure time, and ISO settings, and compute their difference. From equations (1), (2), and (3), we have:

$$\begin{aligned}
\text{Var}(I_1 - I_2) &= \text{Var}(I_1) + \text{Var}(I_2) = 2\,\text{Var}(kz + n) \\
&= 2\,s^2 \cdot \text{ISO}^2 \cdot \lambda + 2\sigma^2 = 2\,s \cdot \text{ISO} \cdot \mu + 2\sigma^2
\end{aligned} \tag{5}$$

Hence, to calibrate s, we capture images of a white paper under fixed illumination at various ISO settings and exposure times, and then fit a linear model, as illustrated in Fig. 2.

After obtaining the value of s, we can simulate shot noise for a target image I_n with ISO_n and exposure time t_n as follows:

$$I_n = \mathcal{P}\left(\frac{I_r}{s \cdot \text{ISO}_r} \cdot \frac{t_n}{t_r}\right) \cdot s \cdot \text{ISO}_n \tag{6}$$

where I_r is the rendered 3DGS image, ISO_r and t_r are the ISO and exposure time corresponding to the images used for training the 3DGS model, and $\mathcal{P}(\cdot)$ denotes the per-pixel Poisson distribution sampler. Since $\lambda \propto t$, we have $\lambda_n/\lambda_r = t_n/t_r$, which justifies the exposure adjustment factor. An example is shown in Fig. 2, where we synthesize noise on a clean image (obtained by averaging multiple RAW captures at ISO 50 and exposure time 1/15 seconds) to simulate an image of ISO 1600 and exposure time 1/2000 seconds.

2.4   Blur Synthesis

The camera trajectory during the capture of short and long exposure images can be simulated using various types of polynomials or spline curves. Following [6], we use a random-order Bézier curve to model the trajectory. This random trajectory defines the camera poses for both short and long exposure captures.

The Bézier curve can be formulated as:

$$L(m) = \sum_{i=0}^{l} \binom{n}{i}(1 - m)^{l-i}m^i\,Cp_i, \quad \text{for } m \in [0, 1], \tag{7}$$

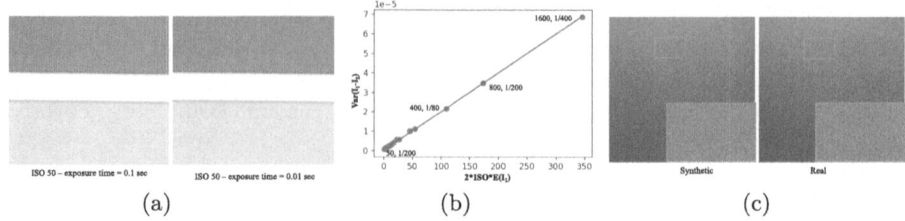

Fig. 2. (a) Two images captured with the same ISO value but different exposure times exhibit varying noise levels. (b) Estimation of s in Eq. 2 for the Samsung S23 Ultra phone. The tuples in the graph are ISO and exposure time in seconds, respectively. (c) An example of noise synthesis on a clean image captured by the S23 camera.

where l is the curve order, Cp_i is the i-th control point, and m is a parameter used to sample camera poses along the curve, with $m = 0$ corresponding to the initial pose and $m = 1$ to the final pose. For instance, a first-degree curve (i.e., a linear segment) is characterized by two control points (start and end).

The 3D spatial displacement between the short exposure and long exposure camera poses is determined by βm, where $\beta \in [0, 1]$. The segment corresponding to the long exposure is then uniformly sampled to obtain subposes for view rendering.

We render views from the 3DGS model at each sampled camera pose. For the short exposure image, we follow the pipeline presented in Sect. 2.2 to unprocess the rendered image and add noise. For the long exposure image, we first transform the rendered images to linear sRGB by inverting the gamma correction, then average the rendered images along the trajectory. While it would be ideal to perform averaging in the RAW space, it is not strictly necessary, since we can exploit the linearity of the unprocessing transformations (white balance, color correction) to simplify the process and avoid unnecessary computations across all rendered images. Noise is then added as described in the previous section. The overall process for synthesizing the short and long exposure images is illustrated in Fig. 1. To avoid floater artifacts, which are common when rendering novel views far from the training set, we constrain the simulated camera trajectories to remain close to the original training camera poses.

2.5   Real Data Capturing

To evaluate the quality of the synthetic dataset, we captured real long-short exposure image pairs. We modified an Android application, OpenCamera [3], to enable the capture of consecutive long and short exposure images.

We first captured still, sharp, and well-exposed images of the scene, which are used to train the 3DGS model. Camera poses for both the short exposure and the training images are jointly estimated using COLMAP. During the image capture process, we also recorded gyroscope and accelerometer readings from the device's IMU to synthesize blurred images and compare them against real blurry captures.

The gyroscope and accelerometer measurements are integrated–single integration for the gyroscope (rotation) and double integration for the accelerometer (translation)–to estimate the camera trajectory. Modern Android phones provide a linear acceleration sensor, which internally processes the raw accelerometer data to remove the gravity component through low-pass filtering and sensor fusion [3].

Since IMU readings are in metric units (meters), while COLMAP pose estimations are only determined up to an unknown scale, we recover the scale by measuring the physical length of an object in the scene and extracting its corresponding 3D length from COLMAP's sparse reconstruction.

It is important to note that the estimated translation and rotation suffer from drift due to noise in the IMU signals. This drift is particularly severe for translations because double integration amplifies accumulated errors over time [4].

3   Experiments

In the experiments, we first show a comparison of blur synthesis methods, followed by a comparison of noise synthesis to real noisy images.

3.1   Blur Synthesis Comparison

We compare the realism of blur synthesis with the method proposed in [7], which uses planar homography warping and random rotations to generate blurred images. Planar homography assumes a flat scene and therefore does not account for scene depth or surface normal variations. The homography for a planar scene with depth d and normal vector $\mathbf{n}$ is computed as:

$$\mathbf{H}(t) = \mathbf{K} \left(\mathbf{R}(t) + \frac{1}{d}\mathbf{T}(t)\mathbf{n}^\top \right) \mathbf{K}^{-1} \tag{8}$$

where $\mathbf{T}(t)$ and $\mathbf{R}(t)$ are the translation and rotation matrices at timestamp t, respectively, and $\mathbf{K}$ is the camera intrinsic matrix.

The homography is computed for each timestamp relative to the initial pose, and the resulting warped images along the trajectory are averaged to synthesize the blur. In Fig. 3, we utilize the recorded translations and rotations during the capture of real blurry-noisy image pairs to compare different blur synthesis approaches. We show a comparison between blur synthesized by moving the camera within the reconstructed 3DGS scene and blur synthesized via homography warping (HW), using two settings: (1) HW with rotation only (i.e., assuming zero translation) and (2) HW with both rotation and translation, using an approximate single depth value for the entire scene and assuming a surface normal aligned with the camera's Z-axis (i.e., perpendicular to the image plane).

The examples demonstrate that blur synthesized through 3DGS more closely resembles real-world blur compared to homography-based methods. It is important to note that the synthesized images are not perfectly aligned with the real

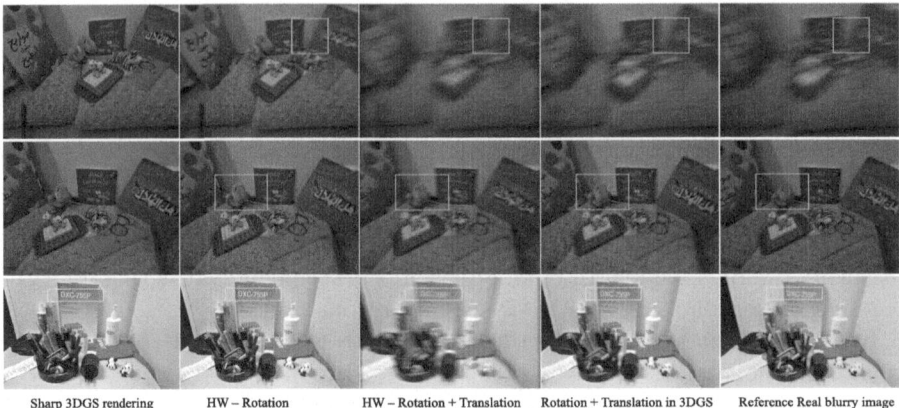

| Sharp 3DGS rendering | HW – Rotation | HW – Rotation + Translation | Rotation + Translation in 3DGS | Reference Real blurry image |

Fig. 3. Blur Synthsis comparison.

blurry images due to inaccuracies in the estimated rotations and translations from the noisy IMU readings. Nevertheless, the real blurry images serve as references for qualitative comparison.

3.2   Noise Synthesis

| Clean render | Noise addition | Real noisy image |

Fig. 4. Noise synthesis. The image brightness has been scaled up for visualization.

In Fig. 4, we present examples of noise synthesis applied to clean renderings from the 3DGS reconstruction alongside the corresponding real noisy images. Overall, the synthesized noise exhibits similar characteristics to those observed in the real images. Note that image brightness has been scaled for better visualization.

For synthesizing noise over the images, we trained 3DGS using raw images processed through our own ISP pipeline, shown in Fig. 1, instead of using the camera's RGB (JPEG) outputs. This choice is motivated by the fact that the camera's internal ISP alters both the noise and color characteristics, as discussed in Sect. 2.2. Specifically, we apply white balancing and camera color correction based on the metadata information from the DNG files, followed by gamma correction and bit-depth reduction from 10 bits to 8 bits. After that, we proceed with the standard 3DGS training procedure.

For completeness, we report the PSNR (Peak Signal-to-Noise Ratio) and SSIM (Structural Similarity Index) for testing images (i.e., the short exposure images) in Table 1. As expected, adding noise leads to a decrease in both PSNR and SSIM. However, the Jensen-Shannon Divergence (JSD) between the real and synthetic images indicates an increase in the similarity of the probability distribution of the synthetic images to their real noisy counterparts after noise addition.

Table 1. Quantitative comparison before and after noise synthesis. Although PSNR and SSIM decrease after adding noise–as expected–the drop in the JSD metric indicates that the synthesized images become more similar to the corresponding GT noisy images.

	Cow Scene			Pen Holder Scene		
	PSNR ↑	SSIM ↑	JSD ↓	PSNR ↑	SSIM ↑	JSD ↓
Clean & GT	32.18	0.77	0.31	24.61	0.63	0.13
Noisy & GT	31.27	0.67	0.27	24.27	0.55	0.10

4   Conclusion and Future Work

We presented a procedure for synthesizing images with different exposure settings, enabling the generation of large-scale datasets for training deep neural networks for low-light image enhancement. We demonstrated that the blurry and noisy images synthesized by simulating camera motion in the 3D Gaussian space closely resemble real-world images. In the Future, we can explore the benefit of training DL models with a dataset generated by this pipeline.

Acknowledgments. This research was supported by the European Union through the EU Interreg Aurora project IMMERSE (20366448).

References

1. Brooks, T., Mildenhall, B., Xue, T., Chen, J., Sharlet, D., Barron, J.T.: Unprocessing images for learned raw denoising. In: Proceedings of the IEEE/CVF conference on computer vision and pattern recognition, pp. 11036–11045 (2019)
2. Chang, M., Feng, H., Xu, Z., Li, Q.: Low-light image restoration with short-and long-exposure raw pairs. IEEE Trans. Multimedia **24**, 702–714 (2021)
3. Harman, M.: Open Camera. https://opencamera.sourceforge.io/ (2013), version available at Google Play and F-Droid
4. Joshi, N., Kang, S.B., Zitnick, C.L., Szeliski, R.: Image deblurring using inertial measurement sensors. ACM Trans. Graphics (TOG) **29**(4), 1–9 (2010)

5. Kerbl, B., Kopanas, G., Leimkühler, T., Drettakis, G.: 3d gaussian splatting for real-time radiance field rendering. ACM Trans. Graph. **42**(4), 139–1 (2023)

6. Lee, D., Park, J., Lee, K.M.: GS-blur: A 3d scene-based dataset for realistic image deblurring. In: The Thirty-eight Conference on Neural Information Processing Systems Datasets and Benchmarks Track (2024). https://openreview.net/forum?id=Awu8YlEofZ

7. Mustaniemi, J., Kannala, J., Matas, J., Särkkä, S., Heikkilä, J.: Lsd _2–joint denoising and deblurring of short and long exposure images with cnns. arXiv preprint arXiv:1811.09485 (2018)

8. Rim, J., Kim, G., Kim, J., Lee, J., Lee, S., Cho, S.: Realistic blur synthesis for learning image deblurring. In: European Conference on Computer Vision, pp. 487–503. Springer (2022)

9. Rim, J., Lee, H., Won, J., Cho, S.: Real-World Blur Dataset for Learning and Benchmarking Deblurring Algorithms. In: Vedaldi, A., Bischof, H., Brox, T., Frahm, J.-M. (eds.) ECCV 2020. LNCS, vol. 12370, pp. 184–201. Springer, Cham (2020). https://doi.org/10.1007/978-3-030-58595-2_12

10. Rim, J., Lee, J., Yang, H., Cho, S.: Deep hybrid camera deblurring for smartphone cameras. In: ACM SIGGRAPH 2024 Conference Papers, pp. 1–11 (2024)

11. Schonberger, J.L., Frahm, J.M.: Structure-from-motion revisited. In: Proceedings of the IEEE Conference on Computer Vision and Pattern Recognition, pp. 4104–4113 (2016)

12. Seo, D., et al.: Graphics2raw: mapping computer graphics images to sensor raw images. In: Proceedings of the IEEE/CVF International Conference on Computer Vision, pp. 12622–12631 (2023)

13. Wei, K., Fu, Y., Zheng, Y., Yang, J.: Physics-based noise modeling for extreme low-light photography. IEEE Trans. Pattern Anal. Mach. Intell. **44**(11), 8520–8537 (2021)

14. Xing, X., et al.: High quality reference feature for two stage bracketing image restoration and enhancement. In: Proceedings of the IEEE/CVF Conference on Computer Vision and Pattern Recognition (CVPR) Workshops, pp. 6267–6276 (2024)

15. Yu, X., et al.: Mvimgnet: A large-scale dataset of multi-view images. In: Proceedings of the IEEE/CVF Conference on Computer Vision and Pattern Recognition, pp. 9150–9161 (2023)

16. Yuan, L., Sun, J., Quan, L., Shum, H.Y.: Image deblurring with blurred/noisy image pairs. In: ACM SIGGRAPH 2007 Papers, pp. 1–es (2007)

17. Zhang, Z., Xu, R., Liu, M., Yan, Z., Zuo, W.: Self-supervised image restoration with blurry and noisy pairs. Adv. Neural. Inf. Process. Syst. **35**, 29179–29191 (2022)

18. Zhang, Z., et al.: Ntire 2024 challenge on bracketing image restoration and enhancement: Datasets methods and results. In: Proceedings of the IEEE/CVF Conference on Computer Vision and Pattern Recognition, pp. 6153–6166 (2024)

19. Zhao, Y., Xu, Y., Yan, Q., Yang, D., Wang, X., Po, L.M.: D2hnet: Joint denoising and deblurring with hierarchical network for robust night image restoration. In: European Conference on Computer Vision, pp. 91–110. Springer (2022)

20. Zhong, Z., Gao, Y., Zheng, Y., Zheng, B.: Efficient Spatio-Temporal Recurrent Neural Network for Video Deblurring. In: Vedaldi, A., Bischof, H., Brox, T., Frahm, J.-M. (eds.) ECCV 2020. LNCS, vol. 12351, pp. 191–207. Springer, Cham (2020). https://doi.org/10.1007/978-3-030-58539-6_12

DADO: A Depth-Attention Framework for Object Discovery

Federico Gonzalez[1,2](✉)●, Estefania Talavera[4]●, and Petia Radeva[1,3]●

[1] Universitat de Barcelona, Gran Via de les Corts Catalanes, 585, Barcelona 08007, Spain
fgonzalez@untdf.edu.ar, petia.ivanova@ub.edu
[2] Universidad Nacional de Tierra del Fuego, Fuegia Basket 251, Ushuaia 9410, Argentina
[3] Institut de Neurosciències, University of Barcelona, Passeig de la Vall dHebron, 171, Barcelona 08035, Spain
[4] University of Twente, Drienerlolaan 5, 7522 NB Enschede, The Netherlands
e.talaveramartinez@utwente.nl

Abstract. Unsupervised object discovery, the task of identifying and localizing objects in images without human-annotated labels, remains a significant challenge and a growing focus in computer vision. In this work, we introduce a novel model, DADO (Depth-Attention self-supervised technique for Discovering unseen Objects), which combines an attention mechanism and a depth model to identify potential objects in images. To address challenges such as noisy attention maps or complex scenes with varying depth planes, DADO employs dynamic weighting to adaptively emphasize attention or depth features based on the global characteristics of each image. We evaluated DADO on standard benchmarks, where it outperforms state-of-the-art methods in object discovery accuracy and robustness without the need for fine-tuning.

Keywords: object discovery · unseen object detection · depth-attention

1 Introduction

Object discovery is the unsupervised task of identifying and localizing objects in images or videos without prior knowledge of their categories or reliance on labeled data, distinguishing it from supervised object detection [18,30]. The field's terminology is evolving, often segmenting the process into stages: unsupervised saliency to separate foreground, single-object discovery to localize a primary object, multi-object discovery (or zero-shot unsupervised object detection) to identify multiple significant objects, and class-agnostic instance segmentation for individual object masks [29,33].

While large Visual Language Models (VLMs) like CLIP [23] or SAM [17] exhibit impressive capabilities in vision-language tasks, such as zero-shot classification and retrieval through associating visual content with semantic descriptions, they are not inherently equipped for unsupervised object discovery of

M. Castrillón-Santana et al. (Eds.): CAIP 2025, LNCS 15622, pp. 281–291, 2026.
https://doi.org/10.1007/978-3-032-05060-1_24

novel, unseen objects. Their abilities are tied to concepts seen during training, and lack autonomous spatial localization or object proposal generation mechanisms necessary for true unsupervised discovery without further adaptation.

Object discovery remains crucial for understanding visual data in a category-agnostic, cost-efficient manner, particularly for novel object detection or in data-scarce environments where labeled data is unavailable. It complements large pre-trained models by enabling broader generalization and adaptation in unconstrained, real-world settings.

Self-supervised learning (SSL) is fundamental to modern object discovery, allowing models to learn robust visual representations directly from unlabeled data, providing flexibility for dynamic environments. Foundational SSL methods like SwAV [3], SimCLR [5], BYOL [12], MAE [13], and MoCo [14] were instrumental, with DINO [4,22] becoming a key method for extracting high-quality unsupervised semantic features.

A central challenge in the unsupervised paradigm is inferring what constitutes an "object" and its relative importance without labels, avoiding potentially limiting assumptions about its visual characteristics, location, or size. Although classical definitions often describe objects as salient structures with distinct, separable boundaries [1,2,19], rigid adherence to such assumptions can hinder the design of versatile discovery systems.

To address these inherent challenges, we propose DADO, a novel unsupervised object discovery approach. DADO integrates attention and depth cues with a dynamic entropy-balancing mechanism for image-feature-based weighting, thereby improving accuracy and robustness. Our comprehensive evaluation on standard benchmarks shows that DADO consistently outperforms state-of-the-art object discovery methods.

2   Related Work

Unsupervised object discovery, identifying objects without appearance-based assumptions, remains an open challenge. Before the SSL era, methods such as LOD [35] and rOSD [34] achieved notable results in multi-object discovery. The widespread availability of SSL, Vision Transformers (ViT), and large pre-trained models, particularly DINO, has significantly advanced this field, with DINO features forming the basis for many recent approaches.

Prominent SSL-based methods include LOST [28], which leverages attention maps from pre-trained ViTs to identify a 'seed' patch and derive object masks and bounding boxes. TokenCut [36] constructs a graph from self-attention tokens and employs normalized graph cuts by spectral clustering to delineate foreground objects. However, both LOST and TokenCut often focus on the single most salient region, grappling with images containing multiple or overlapping objects.

Addressing this limitation, MOST [25] also uses DINO features, but applies fractal analysis through box-counting on patch similarities to identify and cluster multiple foreground object 'pools'. In contrast, FOUND takes an inverse approach by using DINO features to discover the background, considering everything else as the foreground.

In addition to appearance-based methods, self-supervised depth estimation has proven valuable for object localization [6,16,27]. This utility comes from the observation that depth discontinuities often align with object boundaries [15,32], facilitating the separation of foreground and background elements and helping in handling occlusion.

3   The DADO Method

DADO leverages SSL to extract attention insights and depth estimation techniques to capture structural features of the scene (see Fig. 1). Our pipeline takes a single RGB input image and computes both depth estimation and attention features. It then segments the scene into discrete depth layers and, in parallel, constructs a global attention map. By combining each depth layer with the attention map, DADO isolates candidate objects at varying depth ranges. The code is publicly available at link[1].

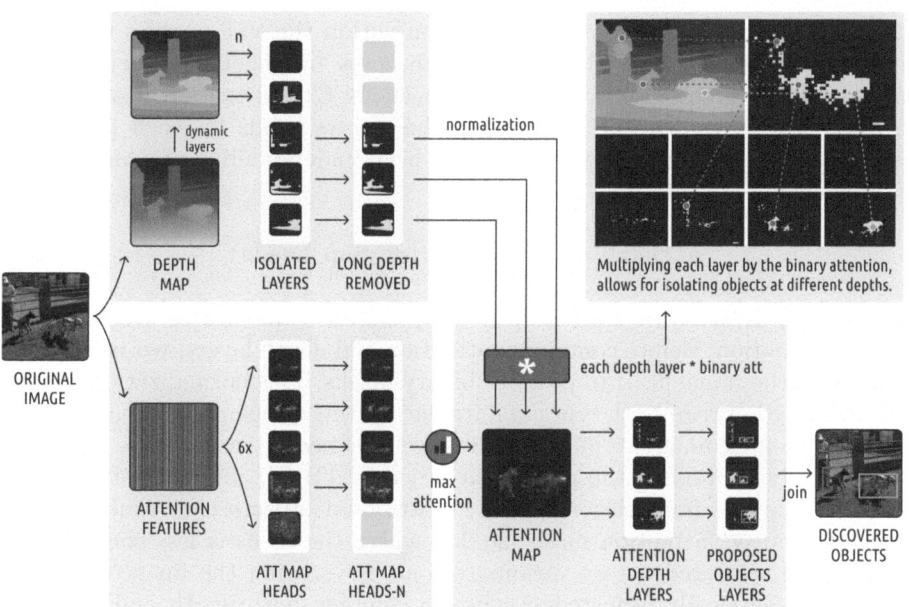

Fig. 1. Our proposed DADO framework.

Depth Map Estimation. We extract the depth maps using a Dense Prediction Transformer (DPT) model [26]. This model produces a dense depth representation of the scene, capturing its structural characteristics and spatial configuration. The depth map distinguishes between foreground and background elements, providing essential cues for object separation based on depth segmentation.

---

[1] https://github.com/fedegonzal/dado.

After generating the depth map, we process it to isolate distinct depth layers. This step is crucial for separating objects located at different depths, particularly when one object is positioned in front of another. To extract these isolated layers, we generate adaptive depth intervals from the depth map by analyzing the distribution of depth values using a histogram. Significant peaks are identified on the basis of their prominence, under the assumption that they correspond to dominant depth layers in the scene. Around each peak, a depth bin is defined with a configurable overlap (20% by default), resulting in a set of meaningful, data-driven depth ranges that reflect the underlying structure of the environment. This process is represented in Fig. 1 as *dynamic layers*.

We then generate n binary images, each corresponding to one of these depth ranges. In each binary image, pixels within the corresponding depth bin are set to ones, while all others are set to zero. Since the farthest layers typically correspond to background regions, we discard those layers. The result is a stack of binary images representing the foreground information at different depth levels.

Attention Map Generation. Concurrently, we extract attention maps using DINO ViT Small. The attention mechanism within the ViT enables the model to focus on distinct and relevant image features by generating weight maps that highlight these regions. DINO produces six CLS attention heads, which we aggregate by taking the maximum value across heads for each pixel. The resulting attention map provides a coarse visual understanding of which regions in the image are of interest.

$$\texttt{attMask} = \max(\texttt{attention\_map\_heads}).$$

Integrating Attention and Depth Maps. DADO integrates attention and depth information. Before combining attention and depth layers, **we normalize** each one to the range [0, 1] to produce binary masks. This normalization ensures compatibility between both types of data, facilitating a meaningful combination.

Object-centric images typically exhibit well-defined attention maps where the model clearly focuses on the prominent object(s) [21,31]. In contrast, complex images containing multiple or sparsely distributed objects can result in more diffuse or "noisy" attention maps [20,24], where the focus is less concentrated and may be scattered across various regions. Leveraging the image's entropy, we apply dynamically adjusted weights to combine both depth and attention maps, ensuring that the contribution of each model is balanced according to the structural complexity of the scene.

Depth-Attention-Driven Object Isolation. Given the pipeline described above, we define DADO as the combination of depth and attention information, weighted dynamically according to the complexity of the input image.

Let $\texttt{attMask}$ be the global attention map, and let $\texttt{depthLayer\_i}$ represent the i-th foreground depth layer, obtained by histogram-based segmentation of the depth map. For each layer, we compute a weighted combination with the attention map as follows:

$$\text{DADO}_i = w_a \cdot \texttt{attMask} \times w_d \cdot \texttt{depthLayer}_i, \tag{1}$$

where w_a and w_d are dynamically determined based on the entropy or structural complexity of the input image.

The cross-correlation CC between the attention map and the depth map is calculated as the mean of their element-wise product:

$$CC = \frac{1}{N} \sum_{i=1}^{N} \left(\texttt{atts\_normalized}[i] \times \texttt{depth\_normalized}[i] \right),$$

where N is the number of elements in the maps. If CC exceeds a threshold (e.g., 0.5), both weights are set to 0.5, indicating that both maps are reliable:

$$\text{if } \texttt{CC} > 0.5, \quad w_a = w_d = 0.5.$$

Otherwise, the weight w_a is calculated based on the attention sparsity:

$$w_a = \frac{1}{1 + \texttt{attention\_sparsity}},$$

where the lower attention sparsity (more concentrated attention) results in a higher weight. The weight w_d is determined by the depth gradient consistency:

$$w_d = \texttt{depth\_gradient\_consistency}.$$

Following Eq. (1), **DADO** is the set of all combined maps for each depth layer:

$$\mathbf{DADO} = \{ \text{DADO}_i \mid i = 1, \dots, n \}. \tag{2}$$

This formulation allows the model to emphasize either attention or depth features, adapting to the characteristics of each scene, and improving object isolation across different depth levels.

Adaptive Thresholding. To threshold the attention-depth map (*att_depth*), we compute the average of the mean and standard deviation of the attention-depth values:

$$\tau = \frac{\text{mean}(\texttt{att\_depth}) + \text{std}(\texttt{att\_depth})}{2}.$$

This threshold τ effectively sets a central value between the typical range and the variability of the map. Next, the attention-depth map is thresholded based on this value. Any element in the map with a value less than or equal to the threshold is set to 0, effectively removing or "masking" that part of the map. On the other hand, values greater than the threshold are set to 255, marking these parts of the map as the regions of interest or relevant features. The attention depth map is then binarized as follows:

$$\texttt{att\_depth\_masked}(i,j) = \begin{cases} 0, & \text{if } \texttt{att\_depth\_masked}(i,j) \leq \tau \\ 255, & \text{if } \texttt{att\_depth\_masked}(i,j) > \tau \end{cases}$$

This process results in a binary mask, where values above τ are highlighted as significant, while the rest of the map is suppressed. This thresholding technique helps to emphasize the most significant parts of the attention-depth map, based on its statistical properties (mean and standard deviation). τ is set at a midpoint between the mean and standard deviation, and the values above τ are considered important for further processing, while the values below it are masked out.

Bounding Box Estimation. From the thresholded composite map, potential object regions are delineated using morphological operations and contour detection methodologies. This dual-step approach facilitates the generation of robust bounding boxes, which are refined using Soft Non-Maximum Suppression to minimize overlap and optimize object localization.

4   Validation

Datasets. To facilitate a comparable evaluation, our work uses Pascal VOC 2007 (VOC07) [10], and VOC 2012 (VOC12) [11].

Evaluation Metrics. We evaluate our unsupervised object discovery method using two complementary metrics: Correct Localization (CorLoc) [9] and object discovery Average Precision (odAP) [7]. CorLoc measures the percentage of images where the dominant object is correctly localized, based on an Intersection over Union (IoU) threshold of 0.5. It only assesses whether at least one object is detected per image, without considering multiple object retrieval. To address this limitation, we also report results using odAP, which extends standard Average Precision to the unsupervised setting. odAP evaluates both the accuracy of object localization and the ability to retrieve all relevant objects, summarizing performance via precision-recall curves. We compare our method against state-of-the-art approaches. We benchmark against MOST [25], LOST [28], and TokenCut [36] using CorLoc. For odAP, we compare with MOST [25], rOSD [34], and LOD [35].

Results. DADO achieves competitive results in both single-object discovery and multi-object discovery. Table 1 shows the performance of our method with

Table 1. Object Discovery Average Precision (odAP) evaluation on Multiple Object Localization.

Method	Year	VOC07	VOC12
rOSD [34]	2020	4.3	5.27
LOD [35]	2021	4.5	5.34
MOST [25]	2023	**6.4**	–
DADO (ours)	2025	6.2	**5.9**

Table 2. CorLOC performance comparison of DADO with Related Work.

Method	Year	VOC07	VOC12
rOSD [34]	2020	54.5	55.3
LOD [35]	2021	53.6	55.1
LOST [28]	2021	61.9	64.0
TokenCut [36]	2023	68.8	72.1
MOST [25]	2023	74.8	**77.4**
DADO (ours)	2025	**78.3**	74.2

an odAP of 6.2 for VOC07 and 5.9 for VOC12. For CorLOC, DADO achieves 78.3% on VOC07 and 74.2% on VOC12, as shown in Table 2.

Large pre-trained models like DINO have significantly improved the quality of visual features and boosted the performance of recent object discovery methods. However, the key challenge remains in extracting meaningful information from these features to discover objects without imposing restrictive assumptions about what constitutes an object. DINOv1 [4] produced attention maps that more faithfully reflect spatial object structure, making it better suited to our objective. Therefore, although DINOv2 [22] offers state-of-the-art representation learning, after several tests, we opted for DINOv1 to preserve the spatial fidelity necessary for unsupervised object discovery (see Table 3).

This is mainly due to the spatial inconsistency introduced by the registers in DINOv2, which interfere with the interpretability and usefulness of attention maps and feature activations for downstream localization tasks. Since registers have no spatial correspondence, they tend to absorb attention in ways that degrade object saliency estimation and boundary precision. This problem was later addressed in Vision Transformers Need Registers [8] by implementing a patch that mitigated the spatial inconsistency caused by registers. We compared our results using DINOv1, and DINOv2 with and without registers, as shown in Table 3. The iteration in which we introduced overlap (from version 0.6 to v0.8) proved to be critical in terms of accuracy. This is because objects with significant depth were being split across two different layers, and as a result, were either not discovered or detected as two separate objects. Introducing an overlap of 20–30% led to a substantial improvement in performance. This parameter will be explored in future work to make it dynamic, rather than fixed as it is currently.

Table 3. Ablation study of the DADO framework on the VOC07 dataset.

Version	Description	CorLOC
v0.1	Dinov2	53.74
v0.1	Dinov1	61.62
v0.2	Dinov1 and Depth	69.64
v0.4	Dinov1 and Depth, with weights	69.72
v0.6	Dinov1 and Depth-isolated layers	72.70
v0.8	Dinov1 and Depth-isolated layers with 30% overlap	78.30
v1.0	Dinov1 and Depth-isolated, overlaped and dynamic bins	78.32

Discussions. One of the major challenges in object discovery is the absence of labels, which also implies a lack of semantics. Although the features obtained from models like DINO are of high quality, the lack of semantic information complicates the distinction of individual objects. They can appear in images as independent and isolated entities, as parts of larger composite objects, adjacent to others with no space in between, in front of or behind, leading to occlusions at different scales. Figure 2 shows examples of these situations and how DADO

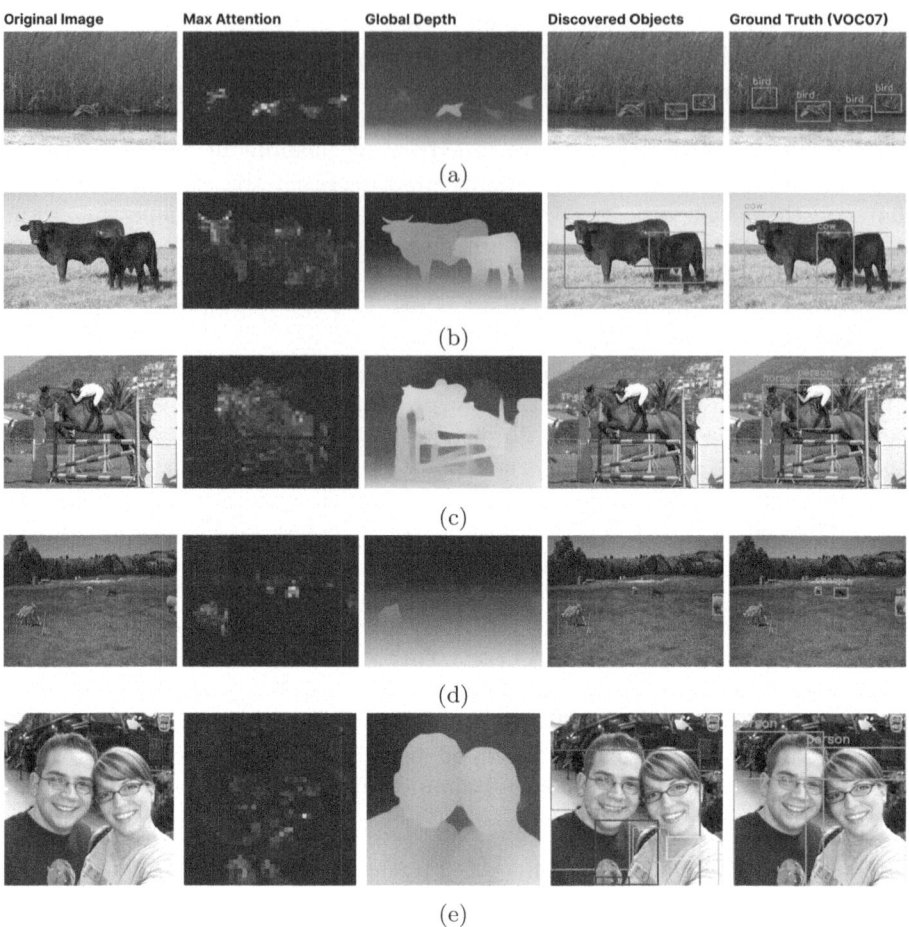

Fig. 2. Outputs of DADO. (a) Independent and isolated objects are effectively discovered by both attention mechanisms and depth cues. (b) Objects positioned in front of or behind others can be accurately separated using depth layers; in such cases, attention provides limited additional information. (c) Composite objects, such as the horse and rider, are very difficult to separate when they lie on the same plane–this represents the main weakness of our model. (d) Pascal VOC ground truth does not include the 'goat' class, but contains 'sheep'. DADO finds instances of both objects. (e) Separating two objects that are adjacent and on the same plane is particularly challenging for our model.

leverages attention and depth to resolve them. Note that our method successfully discovers both large and small objects, whether they are in the foreground or further in the background, and is also capable of separating them when they are overlapping (as in the case of the cows). However, it may also produce false positives in some situations, as shown in the fifth row.

Our approach, which incorporates depth layers, allows us to address some of these challenges, particularly occlusions. The core idea is that by leveraging depth information, we can achieve better separation of objects. Depth layers can help identify object boundaries and contours. This leads to more accurate object representation in the image, which ultimately improves feature quality and model performance.

Our findings reveal that object-centric images tend to produce attention maps of sufficient quality to detect unseen objects. In contrast, complex scenes containing important non-centric objects typically yield noisy and less discriminative attention maps. This noise arises because attention mechanisms prioritize the most salient visual features, which in cluttered environments may belong to background textures or secondary objects. We observed that in such challenging cases, depth representations can significantly improve localization performance. Depth cues provide spatial information that helps distinguish foreground objects from the background, even when visual attention alone is ambiguous. Separating the image into distinct depth planes and applying attention to each plane allows for the individualization of overlapping objects.

By employing dynamic weights that adapt based on either attention features or depth layer representations, we further improve model quality. A higher number of objects in an image typically results in a more dispersed attention map, making object identification harder. Dynamic weights allow the model to adjust its focus on attention or depth.

These findings suggest that combining depth information with attention-based discovery methods could be a promising direction for improving unsupervised object discovery in complex real-world scenes.

Limitations and Future Work. Our method has difficulty separating adjacent objects that lie in the same plane and lack visible gaps. Tokenization-based approaches, such as TokenCut and MOST, handle such cases more effectively. Additionally, low-level, non-semantic techniques, such as edge detection, superpixel grouping, and optimization-based methods such as graph cuts or watershed algorithms, can provide useful boundary cues without requiring semantic information. In future work, we aim to incorporate such strategies to improve performance in scenes with tightly packed objects.

5   Conclusions

In this paper, we introduce a novel unsupervised object discovery method in images based on inferred attention and depth features. Our model achieves state-of-the-art results and is highly adaptable to a wide range of visual scenes and object categories. These findings highlight the potential of combining mid-level cues for robust object discovery without supervision and pave the way for future extensions in image understanding and instance segmentation.

References

1. Alexe, B., Deselaers, T., Ferrari, V.: What is an object? In: IEEE Conference on Computer Vision and Pattern Recognition (2010). https://doi.org/10.1109/CVPR.2010.5540226
2. Biederman, I.: Recognition-by-components: a theory of human image understanding. Psychol. Rev. (1987). https://doi.org/10.1037/0033-295X.94.2.115
3. Caron, M., Misra, I., at.el.: Unsupervised Learning of Visual Features by Contrasting Cluster Assignments (2021). http://arxiv.org/abs/2006.09882
4. Caron, M., et al.: Emerging Properties in Self-Supervised Vision Transformers (2021). http://arxiv.org/abs/2104.14294
5. Chen, T., Kornblith, S., Norouzi, M., Hinton, G.: A Simple Framework for Contrastive Learning of Visual Representations (2020). http://arxiv.org/abs/2002.05709
6. Chen, Y., Li, W., Chen, X., Gool, L.V.: Learning semantic segmentation from synthetic data: a geometrically guided input-output adaptation approach (2019). https://arxiv.org/abs/1812.05040
7. Cho, M., Kwak, S., Schmid, C., Ponce, J.: Unsupervised object discovery and localization in the wild: part-based matching with bottom-up region proposals (2015). https://arxiv.org/abs/1501.06170
8. Darcet, T., Oquab, M., Mairal, J., Bojanowski, P.: Vision. Transformers Need Registers (2023). http://arxiv.org/abs/2309.16588
9. Deselaers, T., Alexe, B., Ferrari, V.: Localizing objects while learning their appearance. In: Daniilidis, K., Maragos, P., Paragios, N. (eds.) European Conference on Computer Vision (2010)
10. Everingham, M., Van Gool, L., Williams, C.K.I., Winn, J., Zisserman, A.: The PASCAL Visual Object Classes Challenge 2007 (VOC2007) Results. http://www.pascal-network.org/challenges/VOC/voc2007/workshop/index.html
11. Everingham, M., Van Gool, L., Williams, C.K.I., Winn, J., Zisserman, A.: The PASCAL Visual Object Classes Challenge 2012 (VOC2012) Results. http://www.pascal-network.org/challenges/VOC/voc2012/workshop/index.html
12. Grill, J.B., et al.: Bootstrap your own latent: a new approach to self-supervised. Learning (2020). http://arxiv.org/abs/2006.07733
13. He, K., Chen, X., Xie, S., Li, Y., Dollár, P., Girshick, R.: Masked autoencoders are scalable vision learners. In: IEEE Conference on Computer Vision and Pattern Recognition (2022)
14. He, K., Fan, H., Wu, Y., Xie, S., Girshick, R.: Momentum contrast for unsupervised visual representation learning. In: IEEE Conference on Computer Vision and Pattern Recognition (2020)
15. Hoyer, L., Dai, D., Chen, Y., Koring, A., Saha, S., Van Gool, L.: Three ways to improve semantic segmentation with self-supervised depth estimation. In: IEEE Conference on Computer Vision and Pattern Recognition (2021)
16. Hoyer, L., Dai, D., Wang, Q., Chen, Y., Van Gool, L.: Improving semi-supervised and domain-adaptive semantic segmentation with self-supervised depth estimation. Int. J. Comput. Vision 131(8) (2023)
17. Kirillov, A., et al.: Segment anything. In: IEEE International Conference on Computer Vision (2023)
18. Lee, Y.J., Grauman, K.: Learning the easy things first: self-paced visual category discovery. In: IEEE Conference on Computer Vision and Pattern Recognition (2011)

19. Marr, D.: Vision: A computational investigation into the human representation and processing of visual information. MIT Press (2010)
20. Mehrani, P., Tsotsos, J.K.: Self-attention in vision transformers performs perceptual grouping, not attention (2023). https://arxiv.org/abs/2303.01542
21. Naseer, M., Ranasinghe, K., Khan, S., Hayat, M., Khan, F.S., Yang, M.H.: Intriguing properties of vision transformers (2021). https://arxiv.org/abs/2105.10497
22. Oquab, M., et al.: DINOv2: Learning Robust Visual Features without Supervision (2023). http://arxiv.org/abs/2304.07193
23. Radford, A., et al.: Learning transferable visual models from natural language supervision (2021). https://arxiv.org/abs/2103.00020
24. Raghu, M., Unterthiner, T., Kornblith, S., Zhang, C., Dosovitskiy, A.: Do vision transformers see like convolutional neural networks? Adv. Neural Inf. Process. Syst. (2021)
25. Rambhatla, S.S., Misra, I., Chellappa, R., Shrivastava, A.: MOST: multiple object localization with self-supervised transformers for object discovery (2023). http://arxiv.org/abs/2304.05387
26. Ranftl, R., Bochkovskiy, A., Koltun, V.: Vision Transformers for Dense Prediction (2021). https://doi.org/10.48550/arXiv.2103.13413, http://arxiv.org/abs/2103.13413, arXiv:2103.13413 [cs]
27. Safadoust, S., Güney, F.: Multi-object discovery by low-dimensional object motion. In: IEEE International Conference on Computer Vision (2023)
28. Siméoni, O., et al.: Localizing objects with self-supervised transformers and no labels (2021). http://arxiv.org/abs/2109.14279
29. Siméoni, O., Zablocki, E., Gidaris, S., Puy, G., Pérez, P.: Unsupervised object localization in the era of self-supervised vits: a survey (2023). http://arxiv.org/abs/2310.12904
30. Sivic, J., Russell, B., Efros, A., Zisserman, A., Freeman, W.: Discovering objects and their location in images. In: International Conference on Computer Vision (2005). http://ieeexplore.ieee.org/document/1541280/
31. Trivedy, V., Almalki, A., Latecki, L.J.: Learning object focused attention (2025). https://arxiv.org/abs/2504.08166
32. Vandenhende, S., Georgoulis, S., Van Gansbeke, W., Proesmans, M., Dai, D., Van Gool, L.: Multi-task learning for dense prediction tasks: a survey. IEEE Trans. Pattern Anal. Mach. Intell. (2021). https://doi.org/10.1109/tpami.2021.3054719
33. Villa-Vásquez, J.F., Pedersoli, M.: Unsupervised object discovery: a comprehensive survey and unified taxonomy (2024). https://doi.org/10.48550/arXiv.2411.00868
34. Vo, H.V., Pérez, P., Ponce, J.: Toward unsupervised, multi-object discovery in large-scale image collections. In: European Conference on Computer Vision (2020)
35. Vo, V.H., Sizikova, E., Schmid, C., Pérez, P., Ponce, J.: Large-scale unsupervised object discovery. Adv. Neural Inf. Process. Syst. (2021)
36. Wang, Y., et al.: Tokencut: segmenting objects in images and videos with self-supervised transformer and normalized cut (2023). https://arxiv.org/abs/2209.00383

Ozone Concentration Estimation from Infrared Images Using Extinction Coefficient

Alexandra Duminil[1](✉) ⓘ, Jean-Philippe Tarel[1] ⓘ, and Jean Dumoulin[2] ⓘ

[1] University Gustave Eiffel, COSYS-PICS-L, 77454 Marne-la-Vallée, France
{alexandra.duminil,jean-philippe.tarel}@univ-eiffel.fr
[2] University Gustave Eiffel, Inria, COSYS-SII, I4S Team, 44344 Bouguenais, France

Abstract. Air quality assessment requires concentration measurement of various polluting gases, typically requiring multiple expensive sensors, each dedicated to a specific pollutant. Therefore, there is a need for cost-effective sensors capable of detecting one or more pollutants, with less reliability, such as camera-based sensors, but enabling denser sampling. In this paper, we investigate how the extinction coefficients estimated from an infrared camera may be useful for predicting ground-level ozone concentration. In addition to these coefficients, we show how weather and pollution measures collected from stations near the studied area are useful, with different machine learning methods, to better predict ozone concentration. The performance of the models is validated through a comprehensive evaluation using MAE, RMSE and R-Squared metrics. Parameter selection methods are also used to study the impact of different meteorological parameters and other pollutant concentrations on the prediction of ozone concentration.

Keywords: Pollution · Ozone prediction · Infrared images · Visibility

1 Introduction

Ambient ozone (O_3) is a naturally occurring gas in the atmosphere, formed through the chemical reaction of primary pollutants (nitrogen dioxide (NO2), volatile organic compounds (VOCs)) under ultraviolet radiation. However, it poses toxic risks to living organisms. Under persistent anticyclonic conditions, characterized by dry, sunny weather and low wind activity, it tends to gradually accumulate in the atmosphere [1]. As a result, implementing O_3 monitoring and prevention systems could significantly benefit public health. Monitoring air quality typically relies on measuring the concentration of various polluting gases, which is usually done using multiple expensive sensors, each dedicated to a specific pollutant. Therefore, there is a growing need for low-cost effective solutions, such as camera-based sensors, capable of detecting one or more pollutants, enabling more extensive and denser sampling. Several models have been proposed in the literature to estimate the concentration of pollutants and IQAs in particular, including statistical, deterministic, physically-based, and Machine Learning (ML)-based models. According to Kumar et al. [9], ML models have been shown to be more efficient than statistical and deterministic models. In particular, SVR [18], Random Forest [2], K-means, multiple linear regression, and MLP models are widely used. Liang et al. [10] investigated the performance of six ML classifiers in predicting Taiwan's IQA based on 11 years of data. The authors reported, in this study, that Adaptive Boosting (AdaBoost) and Stacking Ensemble were the most suitable, but prediction performance seems to vary between geographical regions. Several studies [12,17] have shown that the performance of prediction models is improved by adding meteorological data such as temperature, humidity, wind speed and direction. In [8], the linear regression model, which is a simple model, is not effective in taking into account local variations. Juarez et al. [7] proposed a comparison analysis of O_3 prediction in Delhi, India, using eight different ML methods. The study, carried out both seasonally and annually, showed improvements in $O3$ prediction, particularly during winter. Deep CNNs are not widely used for pollutant prediction but have been shown to be suitable for non-linear processes related to air quality [4].

Some studies proposed image processing-based methods, which focused on particulate matter concentration predictions [3,11,14,16]. Part of them aimed to analyze images by focusing on sky regions and estimating visual features such as the dark channel prior [11,16], or the color histogram [3]. Mohan et al. [14] proposed using pre-trained CNN methods to directly classify outdoor RGB images with different levels of pollution.

While previous studies have proposed image-based methods to predict particulate matter concentrations from RGB images, our work aims to go a step further by using infrared images to help estimate ozone concentrations in Champs-sur-Marne, a town around twenty kilometers east from Paris, France. Specifically, our work stands out by investigating the interest in using extinction coefficients k calculated from infrared camera images, a specific feature, jointly with other meteorological and pollution measures. To the best of our knowledge, this approach has not yet been investigated in the literature. We also studied the interest of different input measures with also their lagged values and the interest of using a parameter selection method such as SHAP. Different ML models are used to perform predictions including SVR, random forest, boosting, MLP and CNN.

2   Materials and Methods

2.1   Data

The data are collected from our system located on the roof of the University, at Champs-sur-Marne in France. It is composed of two RGB cameras, an infrared camera and a visibilimeter, which allow us to obtain meteorological data such as temperature, background luminance, cumulative water sum (cws) and the visibility distance. Pollution data, which are used to obtain the ground truth, are collected from the Airparif database [1] until 2018. Currently, the dataset comprises a single scene, as obtaining data over a long period of time using a specific sensor system is challenging. However, even though the dataset only contains a single scene, the system takes pictures continuously, enabling data to be obtained under different meteorological conditions.

The dataset used in this study consists of a data series of 18500 samples collected between 2018 and 2023. Data for 2020 were excluded due to a lower sample count compared to other years. To avoid imbalances, the number of samples for each remaining year is approximately equal. Then, missing and NaN values led to the removal of the data line.

2.2   Method

Air quality depends primarily on the intensity of pollutant emissions, which are influenced by various factors, particularly meteorological conditions. These conditions also determine the dispersion or accumulation of pollutants in the atmosphere. Wind and rain, for instance, promote the dispersion, mixing, and removal of pollutants. In contrast, persistent anticyclonic conditions, characterized by dry, sunny weather and the absence of wind at ground level, can lead to gradual accumulation of pollutants. Consequently, the summer season is characterized by hot and sunny days that favor the formation of ozone. Figure 1, shows the O_3 concentration over several years. A low concentration of O_3 is noticeable in autumn and winter, while it increases in spring and summer.

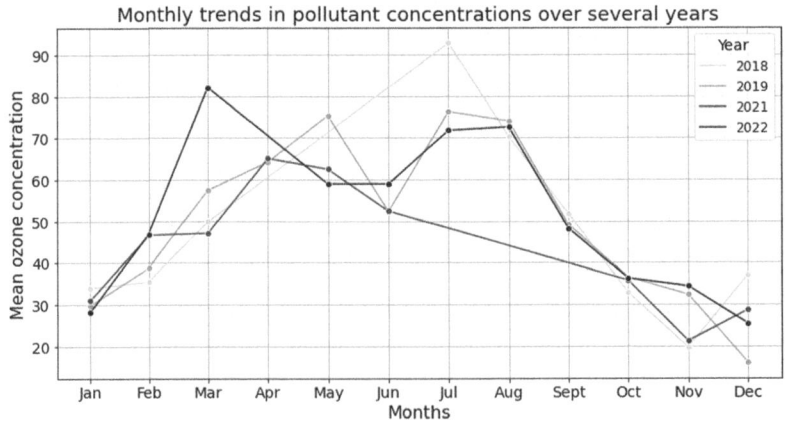

Fig. 1. Lines plot describing the monthly trends in O_3 concentrations over several years.

In addition to the meteorological data, we aim to exploit the infrared images provided by the camera. O_3 absorbs and emits in the infrared range and is strongly influenced by meteorological conditions, including sunshine and high temperatures. Figure 2 presents cropped images from the infrared camera with associated O_3 concentration. The right image shows higher contrast, likely due to seasonal temperature differences. The dark area, where the horizon is barely visible, may result from atmospheric absorption, where ozone would be involved. Infrared imagery can be used to visualize thermal variations in the atmosphere, leading indirectly to the location of ozone.

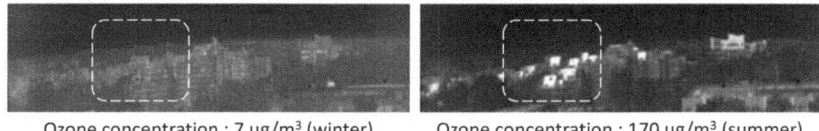

Ozone concentration : 7 µg/m³ (winter)   Ozone concentration : 170 µg/m³ (summer)

Fig. 2. Infrared camera images of areas of interest with associated O_3 concentration.

Extinction coefficients k. To achieve this, we rely on the use of extinction coefficients, which have been used in previous works to estimate the concentration of particulate matter in the atmosphere [11]. The visible veil, caused by high concentrations of particulate matter, is due to the interaction of light with the particles in the air, called light scattering, which causes an attenuation of transmitted light. The Koschmieder law [13] is a physically based model of the apparent luminance of objects on sky background near the horizon:

$$I = I_0 e^{-kd} + I_s(1 - e^{-kd}) \qquad (1)$$

where I is the intensity of an object seen at distance d, I_0 is the clear intensity of this object, I_s is the intensity of the sky and k is the atmospheric extinction coefficient to estimate. Equation 1 assumes constant lighting conditions. In practice, the weather and the position of the Sun vary between the days and seasons. For this reason, assuming a cold sky, the hypothesis of neglecting the second term of I in Eq. 1 is explored, leading to a simpler model with only the first term:

$$I \approx I_0 e^{-kd} \qquad (2)$$

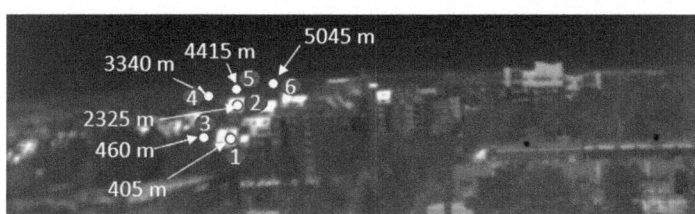

Fig. 3. Infrared image with distances in meters and points of interest with marks used for coefficient calculations.

Table 1. Label names of the estimated extinction coefficient k depending on the set of used point marks, with numbers as shown in Fig. 3.

Estimated k	Target object	Set of point marks
k_{optim}	vegetation	3,4,5
$k1_b$	buildings	1,2
$k2_v$	vegetation	3,4
$k3_v$	vegetation	3,5
$k4_v$	vegetation	3,6
$k5_v$	vegetation	4,6

Figure 3 presents the different points of interest with a number mark and displays their approximate distances to the camera in meters (m). Table 1 shows the label names of the estimated extinction coefficient k depending on the used point marks shown in Fig. 3. Points of interest were chosen on buildings (1 and 2) and on vegetation (3 to 6). The purpose is to investigate the optimum choice of the point marks set. Buildings tend to get warmer compared to vegetation. Thus, it could be relevant to analyze the impact of these choices on the calculation of the extinction coefficient. When considering the simpler model of Eq. 2, the k can be computed using only two image points at different known distances to simplify the unknown I_0 [5]. Therefore, k_n can be estimated as follows:

$$k_n = \frac{1}{(d_j - d_i)} log(\frac{I_i}{I_j}) \tag{3}$$

where I_i and I_j are the average intensities of a 5×5 area around points i and j, d_i and d_j are the distances between camera and pointed objects i and j (see Fig. 3). The k_n computations are independent of the global illumination conditions.

When considering the complete model of Eq. 1, the equation is with three unknowns: k, I_s and I_0, and thus three points at three different distances are requested to estimate k. These unknowns' values are obtained by least squares minimization. The k_{optim} can thus be computed from where the minimum of the following error is achieved:

$$
(I_3 - (I_{03} \, e^{-kd_3} + I_s(1 - e^{-kd_3}))^2 +
$$
$$
(I_4 - (I_{03} \, r_{34} \, e^{-kd_4} + I_s(1 - e^{-kd_4}))^2 +
$$
$$
(I_5 - (I_{03} \, r_{35} \, e^{-kd_5} + I_s(1 - e^{-kd_5}))^2 \tag{4}
$$

where r_{34} and r_{35} are the ratios of the mean intensities between points 3 and 4 and points 3 and 5, respectively. By comparing these different estimates, this will allow us to determine what is the best estimate of the extinction coefficient of O_3 concentration.

Data Analysis. Studies have shown that pollutant concentrations over time t depend on previous values [6]. To analyze these temporal dependencies, the Autocorrelation Function (ACF) and Partial Autocorrelation Function (PACF) are used. ACF captures the general correlation with past values, while PACF isolates the direct effect of each lag. The time increment in our case is one hour. As with pollutants, meteorological data and the k coefficients at a given time t are correlated with the two previous hours. These additional information are also used as input parameters. The suffixes $lag1$ and $lag2$ are used to differentiate time-shifted parameters in the rest of the paper.

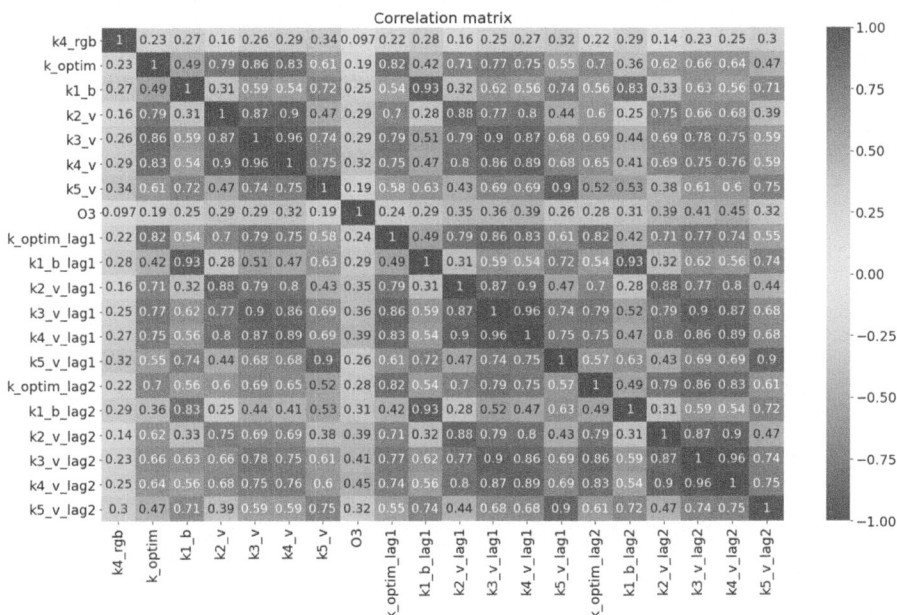

Fig. 4. Heatmap of the linear correlation matrix between O_3 and the different extinction coefficients.

Figure 4 presents the heatmap of the correlation matrix obtained from a set of data including the estimated k coefficients, and the Airparif data O_3 . The values turn red as positive correlations increase and blue for negative correlations. The weakest correlations with O_3 are k_{optim} and $k5_v$ followed by $k1_b$. A similar pattern was observed in the lagged versions, which tended to increase with lag. This suggests that these k values may not be optimal for estimating ozone concentration. For $k1_b$, the points of interest taken in the buildings may slightly distort the results due to their higher temperature. For $k5_v$, the distance between the two points is around 1705 m, corresponding to the shortest case. The best correlation values with O_3 are $k4_v$ and its lagged values. The points of interest were chosen in vegetation areas and at a greater distance than the other k values. To compare with the extinction coefficient obtained in the visible spectrum, the parameter $k4_{rgb}$ has been estimated similarly to $k4_v$ but on RGB images and is added at the first line of the correlation matrix. Its correlation value with O_3 is lower than 0.1, showing the interest of using infrared images instead of RGB. However, the correlation matrix captures only linear correlations. Thus, in addition, the mutual information (MI) between O_3 and the different extinction coefficients is computed to measure a non-linear link between them. The MI is presented in Fig. 5. The results confirm previous statements about the correlations of k_{optim}, $k5_v$ and $k1_b$ and their lagged values. In general, the coefficients with the highest MI scores are $k3_v$ and $k4_v$ with their lags.

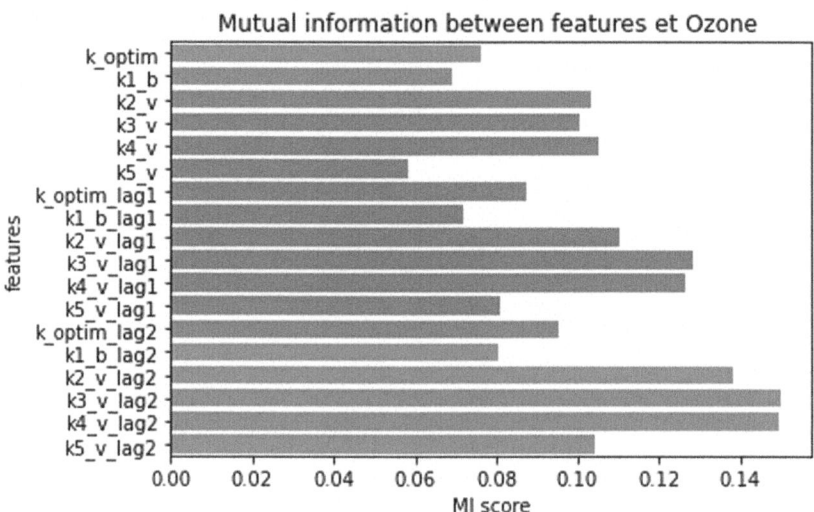

Fig. 5. Bar plot describing the mutual information between O_3 and the different extinction coefficients.

Meteorological Data. The use of meteorological data has been shown to be essential to predict or monitor the concentration of pollutants. Many studies have used data such as absolute humidity, air temperature, absolute pressure, and wind direction. In our study, our objective is to use the available meteorological data. The meteorological sensor is located on the station with the cameras on our acquisition site. The data retrieved are air temperature, background luminance, cumulative water sum (cws) and visibility distance. The background luminance corresponds to the ambient light and the cws is the total accumulated precipitation. The visibility distance may depend on particles in the air, rain or fog. Ozone concentration peaks appear in cycles, as shown in Fig. 1. The peaks occur at a certain time of year, mainly in summer and generally in the afternoon.

It thus seems relevant to add hours value, days value and seasons value as cyclical variables to the previous meteorological data vector. For example, the variable day is included as two variables: $sin_{day} = \frac{sin(2\pi \times day)}{7}$ and $cos_{day} = \frac{cos(2\pi \times day)}{7}$. The number 7 denotes the number of days in a week. This is performed similarly for hours and seasons.

The concentration of pollutant from Air-Quality stations close to Champs-Sur-Marne are also interesting to study. Figure 6 presents the MI with the different parameters previously selected and the O_3 concentration. We observed a strong link between $NO2$ and O_3. This can be due to the formation of O_3 by chemical reaction with primary pollutants, including $NO2$. Temperature and cws are also important features. The former is indeed essential for the formation of O_3, since its concentration increases with temperature. For the latter, precipitation favors the dispersion and leaching of pollutants. As a result, the O_3 concentration should decrease as cws increases. The importance of particulate matter PM2.5 and PM10 can be noticed. In fact, there are involved in slowing the fall of radical aerosols, which are one of the precursors of O_3 [15]. Some parameters, such as luminance, have more information in the lag2 version. Cyclic parameters appear less important, and MI between O_3 and the other data varies across features. Since some parameters may be redundant, it is essential to select the most predictive ones.

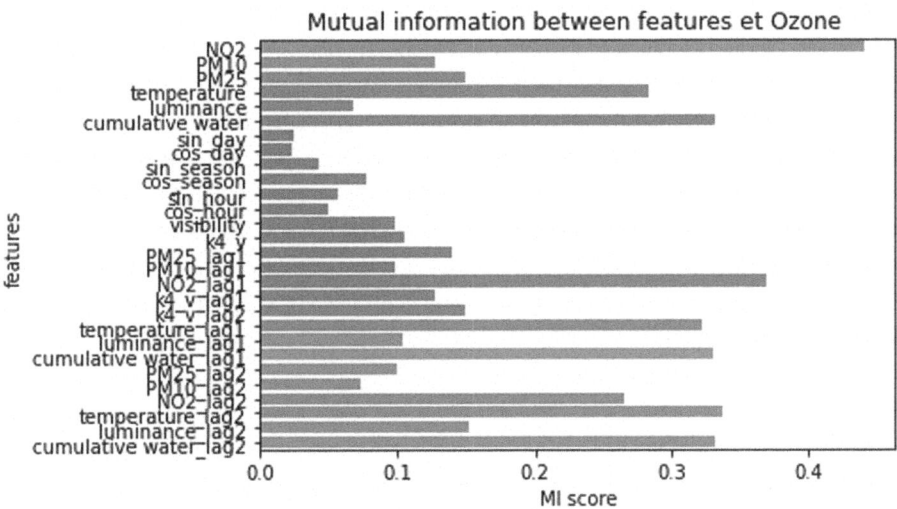

Fig. 6. Bar plot describing the mutual information between O_3 concentration and other selected parameters.

3 Results

The collected data serves to train different ML models in order to predict O_3 concentration. Inspired by the state-of-the-art, classical ML models are selected: Linear regression (LR), SVR, Random forest (RF), Xtreme gradient boosting (XB), as well as MLP and CNN. The CNN model consists of five layers of 1D convolution followed by two dense layers. The original data consists of a data series of approximately 18500 collected data from 2018 to 2023. The ML models are trained on data of size 11863 and 2966 for validation, including data from 2018 to 2023. The test sets $t1$ and $t2$ are based on data from 2022 and 2023, respectively. The data were split using odd months for training/validation and even months for testing, ensuring test independence while retaining data. Three performance metrics are used in this section: The Mean Absolute Error (MAE), Root Mean Squared Error (RMSE), and R-Squared (R^2), to measure the prediction errors for O_3 concentration prediction. Lower MAE and RMSE values and higher R^2 value demonstrate better performance.

Table 2. Prediction errors on validation (v) and test data ($t1$ and $t2$). The best values are in bold.

Methods	Without k parameters + lags						With k parameters + lags					
	LR	SVR	RF	XB	MLP	CNN	LR	SVR	RF	XB	MLP	CNN
MAE v	0.436	0.250	0.276	0.280	0.278	**0.188**	0.432	0.234	0.270	0.280	0.270	0.190
RMSE v	0.554	0.336	0.364	0.381	0.367	**0.255**	0.548	0.315	0.358	0.377	0.350	0.257
R^2 v	0.707	0.892	0.873	0.862	0.871	**0.938**	0.714	0.905	0.878	0.864	0.883	0.937
MAE $t1$	0.503	0.450	0.396	0.436	**0.412**	0.423	0.490	0.448	0.389	0.427	0.433	**0.399**
RMSE $t1$	0.615	**0.577**	0.504	0.554	**0.529**	0.544	0.599	0.581	0.496	0.545	0.557	**0.510**
R^2 $t1$	0.637	**0.681**	0.756	0.706	**0.732**	0.716	0.656	0.676	0.764	0.715	0.702	**0.751**
MAE $t2$	**0.473**	**0.398**	0.349	0.376	**0.384**	0.395	0.479	0.409	**0.342**	0.365	0.400	**0.375**
RMSE $t2$	**0.590**	**0.514**	0.453	0.485	**0.498**	0.502	0.598	0.534	**0.448**	0.472	0.522	**0.492**
R^2 $t2$	**0.696**	**0.769**	0.821	0.795	**0.783**	0.780	0.688	0.751	**0.825**	0.805	0.762	**0.788**

Table 2 presents the prediction errors for validation and tests with and without k parameters for different ML models. The general results on validation v and on tests $t1$ and $t2$ demonstrate better performance with the k parameters, compared to without. Specifically, the CNN, RF and Boosting models produce the best results on $t2$ and $t3$, which include the k parameters. Above all, RF appears to be the most suitable model for this type of task.

Based on these results, a SHAP analysis is applied on the CNN model to display the importance of each parameter, using the SHAP summary plot in Fig. 7. As we saw in the previous section, NO_2 has a significant impact on O_3 concentration. More specifically, in terms of SHAP values, higher NO_2 values tend to reduce O_3 levels. This means that high NO_2 levels are not required to

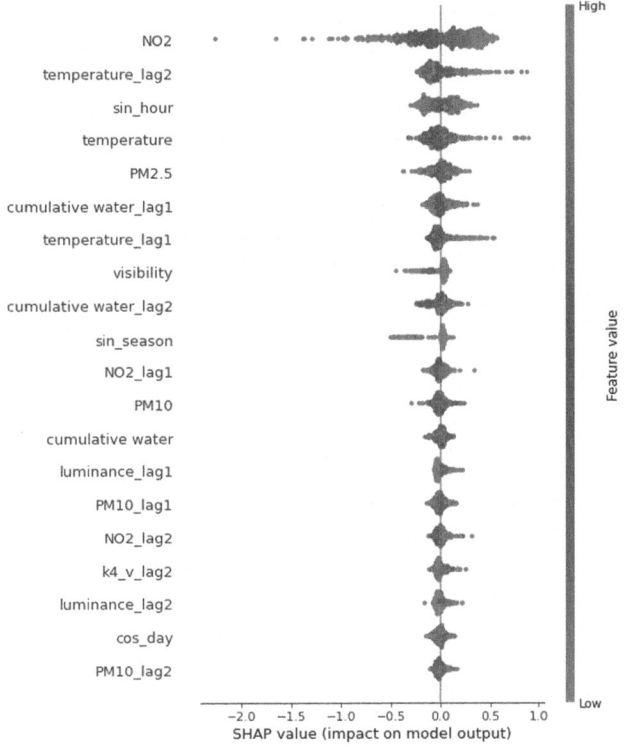

Fig. 7. Graph showing the importance of parameters using the SHAP method.

raise O_3, as they also correlates with meteorological factors such as temperature, especially at night and during the colder seasons. In addition, the plot shows that higher temperatures, luminance and k_4 values are linked to increased ozone concentrations, as indicated by their positive SHAP values. The visibility feature has negative values associated with a decrease in O_3 concentration. Finally, the others' values demonstrate their limited impact on O_3 concentration. In summary, the use of the SHAP method provides an overview of the impact of the various parameters on the learned model. However, the SHAP method appears to be relatively sensitive, since the results differ significantly depending on the types of model, the size of the datasets, and the number of input parameters.

4   Conclusion

The purpose of this work was to study the interest in using extinction coefficients computed from infrared images independently of their acquisition time to help predict O_3 concentration. The experimental study is conducted in Champs-sur-Marne, France, from a dataset collected quite regularly each hour from 2018 to 2023 under various different weather and traffic conditions. The first experiments are based on data from a single scene acquired under different weather conditions and throughout the day over several years. The preliminary results showed a significant correlation between these extinction coefficients and the O_3 concentration, making the use of camera data interesting for O_3 prediction. We were also able to measure the impact of other meteorological conditions that are relatively rarely studied in the literature, such as visibility or sky luminance. Then, we performed a comparison with several ML models and a CNN, which gave promising results on test sets. The perspective of collecting more data over the years to cover as many situations as possible and to improve the generalization properties of prediction models is important to consider. In future work, we also aim to extend the proposed method to more infrared scenes, improve the quantitative evaluation by adding advanced models, and to exploit the use of multi-spectral cameras.

Acknowledgment. Thanks to BRIGHTER project for funding. BRIGHTER has received funding from the Chips Joint Undertaking (JU) under grant agreement No 101096985. The JU receives support from the European Union's Horizon Europe research and innovation program and from France, Belgium, Portugal, Spain and Turkey.

References

1. Airparif: L'observatoire de la qualité de l'air en île-de-france (2023). https://www.airparif.fr/. Accessed 16 April 2025
2. Chen, G., et al.: Improving satellite-based estimation of surface ozone across China during 2008–2019 using iterative random forest model and high-resolution grid meteorological data. Sustain. Cities Soc. **69**, 102807 (2021)
3. Enggari, S., et al.: Identification of air pollution levels based on sky image using color histogram method. In: 2023 Sixth International Conference on Vocational Education and Electrical Engineering (ICVEE), pp. 77–83. IEEE (2023)
4. Eslami, E., Choi, Y., Lops, Y., Sayeed, A.: A real-time hourly ozone prediction system using deep convolutional neural network. Neural Comput. Appl. **32**, 8783–8797 (2020)
5. Hautiere, N., Aubert, D., Dumont, Ã., Tarel, J.P.: Experimental validation of dedicated methods to in-vehicle estimation of atmospheric visibility distance. IEEE Trans. Instrum. Meas. **57**(10), 2218–2225 (2008). https://doi.org/10.1109/TIM.2008.922096
6. Jairi, I., Ben-Othman, S., Canivet, L., Zgaya-Biau, H.: Enhancing air pollution prediction: a neural transfer learning approach across different air pollutants. Environ. Technol. Innov. **36**, 103793 (2024)
7. Juarez, E.K., Petersen, M.R.: A comparison of machine learning methods to forecast tropospheric ozone levels in Delhi. Atmosphere **13**(1), 46 (2021)
8. Jumin, E., et al.: Machine learning versus linear regression modelling approach for accurate ozone concentrations prediction. Eng. Appl. Comput. Fluid Mech. **14**, 713–725 (2020)

9. Kumar, K., Pande, B.: Air pollution prediction with machine learning: a case study of Indian cities. Int. J. Environ. Sci. Technol. **20**(5), 5333–5348 (2023)
10. Liang, Y.C., Maimury, Y., Chen, A.H.L., Juarez, J.R.C.: Machine learning-based prediction of air quality. Appl. Sci. **10**(24), 9151 (2020)
11. Liu, C., Tsow, F., Zou, Y., Tao, N.: Particle pollution estimation based on image analysis. PLoS ONE **11**(2), e0145955 (2016)
12. Madan, T., Sagar, S., Virmani, D.: Air quality prediction using machine learning algorithms–a review. In: 2020 2nd International Conference on Advances in Computing, Communication Control and Networking (ICACCCN), pp. 140–145. IEEE (2020)
13. Middleton, W.E.K.: Vision through the atmosphere. In: Geophysik II/Geophysics II, pp. 254–287. Springer (1957)
14. Mohan, A.S., Abraham, L.: Hybrid transfer learning approach using multiple pre-trained models for classification of outdoor images into AQI classes. In: 2022 IEEE 7th International Conference on Recent Advances and Innovations in Engineering (ICRAIE), vol. 7, pp. 283–288. IEEE (2022)
15. Qu, Y., et al.: Influence of atmospheric particulate matter on ozone in Nanjing, China: observational study and mechanistic analysis. Adv. Atmos. Sci. **35**, 1381–1395 (2018)
16. Samsami, M., Shojaee, N., Savar, S., Yazdi, M.: Classification of the air quality level based on analysis of the sky images. In: 2019 27th Iranian Conference on Electrical Engineering (ICEE), pp. 1492–1497. IEEE (2019)
17. Simu, S., et al.: Air pollution prediction using machine learning. In: 2020 IEEE Bombay Section Signature Conference (IBSSC), pp. 231–236. IEEE (2020)
18. Su, X., An, J., Zhang, Y., Zhu, P., Zhu, B.: Prediction of ozone hourly concentrations by support vector machine and kernel extreme learning machine using wavelet transformation and partial least squares methods. Atmos. Pollut. Res. **11**(6), 51–60 (2020)

Multi-modal Human-Robot Collaboration in Production Lines Through Speech Commands and Gestures

Vincenzo Carletti, Antonio Greco$^{(\boxtimes)}$, Domenico Longobardi,
Pierluigi Ritrovato, Alessia Saggese, and Mario Vento

University of Salerno, Salerno, Italy
{vcarletti,agreco,dlongobardi,pritrovato,asaggese,mvento}@unisa.it

Abstract. The integration of advanced artificial intelligence technologies in production lines enables an effective and safer collaboration between human workers and robotic platforms in Industry 5.0, where the focus is the reinforcement of the workers capability, improving their skills when cooperating with the robots. This novel collaboration paradigm requires a natural and efficient human-robot communication way in which speech command and gesture recognition emerge as fundamental components for enhancing collaboration and fostering adaptability in industrial environments. In this paper, we present an innovative multi-modal human-robot collaboration framework, based on speech command and gesture recognition, designed to meet the requirements of accuracy and real-time processing of an existing production line, used as test environment for the European FELICE project. The speech command recognition system performs voice activity detection and is able to reliably distinguish among a set of commands in a noisy industrial environments, by combining a Mel-spectrogram based representation for the voice with a speech recognition neural network based on a Conformer. As for the recognition of gestures, the task is performed using a one-stage detector based on MobileNetV3 SSD. The experiments, performed on datasets encompassing real, synthetic, and negative samples for speech commands as well as images acquired in a realistic use case for gesture recognition, have established the suitability of the proposed solution in challenging industrial settings.

Keywords: Industrial Robotics · Speech Recognition · Gesture Recognition · Deep Learning · Artificial Intelligence

1 Introduction

Industry 5.0 leverages artificial intelligence to improve human-robot collaboration, enhancing efficiency, safety, and decision-making [22]. A crucial aspect is making the interaction more natural by improving the ability of robots to understand human intentions [8]. Since speech and gestures are natural ways for humans to interact [34], their accurate and real-time recognition is crucial for effective collaboration [9]. Using these communication modalities not only increases efficiency but also bridges the gap between human cognitive abilities and robotic capabilities, even in disability cases [5].

Regarding the *FELICE: Flexible assembly manufacturing with human-robot collaboration and digital twin models* project, speech and gesture recognition are fundamental for orchestrating human-robot collaboration in manufacturing environments, aiming to make interaction more natural and improve the overall efficiency and ergonomics. As part of the project, a collaborative robot and an adaptive workstation are integrated into a production line. The system can receive commands in both English and Italian languages, a significant challenge as most speech recognition models are pre-trained only on the former [11]. The robot can take voice commands to handle tools or parts and perform navigation tasks, as well as gestures to start or stop an interaction, while a worker can use voice commands to adjust the workstation's configuration for ergonomic comfort [19,25]. Recognition tasks must be performed in real-time with an accuracy above 90%, and robust to the noisy industrial environment and worker movements. This requirement is challenging because state-of-the-art systems typically experience from 20% to 30% performance drop in such environments [20]. Moreover, while these systems often run on the cloud or powerful workstations [27], the target device required by the project is a low-power computing system [1] to be embedded in the robot and the workstation.

The complexity of facing such a problem on a production line is mostly related to background noise and workers that may not properly interact with automated systems [6]. For example, speech commands could be spoken from varying distances, at different speeds, or with the worker moving away from the system. Similar issues affect gestures, which can be performed at different distances and poses from the camera. These challenges can be solved more effectively through careful data acquisition and preparation (e.g., pre-processing and augmentation) rather than through changes to the overall system architecture.

Within this context, in this paper we describe our novel multi-modal human-robot collaboration framework based on speech command and gesture recognition designed to meet the aforementioned performance requirements in terms of accuracy, robustness and processing time in such a challenging industrial scenario over an embedded system with limited acceleration resources. We show that the proposed data acquisition, preparation, and augmentation allow the system to achieve promising performance in noisy and challenging industrial scenarios.

To the best of our knowledge, the proposed solution is the first end-to-end, multi-modal framework for human-robot collaboration in production lines. For speech command recognition, a lightweight end-to-end neural network was trained using dynamic data augmentation to ensure robustness against industrial noise and speech variations. Combined with a carefully designed Conformer-based architecture, Mel-spectrogram representation, and a voice activity detection module, the system achieved an error rate below 10%, meeting the required accuracy. For gesture recognition, a lightweight hand detector and gesture classifier was trained on real-use-case data. Robustness to occlusions and varying image quality was achieved through tailored data augmentation, and false positives were reduced by validating predictions over a sliding temporal window. This approach achieved 90% accuracy at a 2-meter distance, fulfilling the system requirement.

In the successive sections we delve into the technical details of our solution, providing a comprehensive analysis of the related works (see Sect. 2), describing the system architecture and the underlying algorithms (see Sect. 3), and discussing experimental results (see Sect. 4). Finally, we demonstrate the effectiveness of the proposed solution in improving human-robot collaboration within industrial production lines.

2 Related Works

Spoken Language Understanding (SLU) [23] is a widespread approach for speech command recognition where a speaker can utter non-predefined sentences that may or may not contain a command [36]. Although its popularity, SLU is not feasible for the previously described use case where a computationally efficient system is required to process short, pre-defined commands. The approach is designed to treat long commands, making it expensive in terms of training data and inference time [27], indeed. Then it is prone to propagation of errors from transcription mistakes [30], hence, a mistake in the text transcription may negatively affect command recognition. Finally, SLU does not utilize the original prosody of the command, which is useful for classification [23].

For these reasons, we adopted an *end-to-end model* [4,13,35], where a single neural network directly maps audio to a command, reducing data requirements and inference time [24]. End-to-end systems include meta-learning-based models [7,28], pre-trained models [16], and models trained from scratch [24,27]. For this work, models trained from scratch were chosen as they provide the best trade-off between accuracy and efficiency. A noteworthy example is MatchboxNet [24] that achieves over 90% accuracy on the Google Speech Command dataset.

For the sake of clarity, we consider an open word setting, where the system has to handle words that are not related to commands also; therefore, it is relevant to face a *reject option* [15]. The easiest approach in this case is to add negative samples in the dataset that are related to the rejection class [26].

It is important to point out that the problem of industrial noise overlapped to the speech commands is not often investigated in the literature, while in our use case is a crucial issue to address. Recent papers [17,20] demonstrate that intent recognition systems applied in noisy environments experience a drop in accuracy from 20% to 30%. We propose a solution that is robust to industrial noise conditions, with a performance drop lower than 10% in a real scenario.

Looking at the gesture recognition task, several works in the literature propose dynamic gestures for the interaction with the robot. According to them, a gesture is characterized by a temporal sequence of poses; therefore, to face the problem, the proposal is to use models capable of taking into account the evolution of the movement over time. Although their effectiveness in test scenarios, they have been demonstrated to be unfeasible in complex industrial environments [3].

In our use case, we have a finite set of simple commands (see Sect. 4.1), therefore we can focus on static gesture recognition methods. These methods can be divided into one-stage (end-to-end) or two-stage approaches [33]. Among them, two-stage methods can achieve higher accuracy because their separate networks for detection and classification can be pre-trained on very large datasets [14]. However, one-stage approaches, such as YOLO [29] and SSD [21], offer significantly lower memory usage and faster inference time at the cost of a potential slight decrease in accuracy. We focus on the latter, according to the efficiency and real-time requirements of our setup.

3 Proposed Solution

3.1 Hardware Platform

Considering the setup proposed in the FELICE project, the adaptive workstation has been equipped only with the microphone, as it is not possible to define gestures capable of reproducing the variation in height and inclination; while, the robot is equipped with a microphone and a camera, because some commands can also be given through hand gestures.

The microphone selected for both the systems (one for each) is the ReSpeaker Mic Array V2.0, a device optimized for speech signal that allows for identifying the direction of arrival (DOA, even when the speaker is far from the microphone) and suppressing background noise (environmental and reverberation). Furthermore, the computation for these functionalities is decentralized, since the DOA algorithm combined with beamforming, noise suppression, and acoustic echo cancellation are performed directly on board the microphone.

The camera used on the robot is an Intel RealSense Depth Camera D435, which provides a wide field of view and both RGB and depth images. It is recommended for robotics and fast-moving applications, with a range up to 10 m. The computing platform selected for both systems (one for each) is a NVIDIA Jetson Xavier NX board, a low-power device equipped with an embedded GPU with 384 cores (48 Tensor Cores, 21 TOPS), 8 GB of memory shared between the CPU and the GPU and 16 GB of storage. The neural networks designed for speech command and gesture recognition are executed in parallel on both the devices and have been optimized to run in real-time on limited hardware resources.

The underlying software platform is based on the Robot Operating System (ROS), a middleware widely adopted in robotics applications that makes our software system independent of the hardware and allows for smoother integration with the other software modules used in the project.

3.2 Speech Command Recognition

According to the typical processing pipeline [31], the realized speech command recognition method involves two steps: Voice Activity Detection (VAD) and speech command recognition. The former is required to process only audio samples with voice, to reduce false positives and the computational resources required for speech command classification.

Voice Activity Detection. The ReSpeaker Mic Array V2.0 already has a VAD implemented, but it is proprietary and can only be configured and not re-adapted for the specific purpose. Therefore, we used WebRTC VAD [38], a VAD specifically developed to run in real-time on devices with reduced computational capabilities; a feature that is essential since the module has to run together with the speech command classification and the gesture recognition models. Moreover, the underlying algorithm is extremely accurate in detecting the voice also in noisy environments; thus, it is very suitable to industrial environments where the signal-to-noise ratio (SNR) may be very low. The WebRTC VAD can be tuned using two parameters: the *sampling rate* and the *aggressiveness mode*. The aggressiveness is an integer from 0 (high sensitivity, higher false positive rate) to 3 (low sensitivity, higher false negative rate). We configured the sampling rate to 16 kHz to match our audio data and set the aggressiveness to 1. This setting provides a suitable balance, as any false positives can be filtered by the classification algorithm's reject option, ensuring that speech commands are not missed due to loud industrial noise.

Voice Recognition. For the voice command recognition task, we first represent the audio signal using a Mel-spectrogram, which is a visual representation of the short-term power spectrum of sound. This is obtained by applying a linear cosine transform to the logarithm of the power spectrum, mapped onto a nonlinear Mel scale. The Mel scale is designed to mimic the human auditory system's non-linear perception of sound, making it particularly well-suited for recognizing phonemes in human speech.

We trained and compared two models suitable for an embedded device: ResNet-8 [32], a small convolutional model for keyword spotting, and Conformer [12], a convolution-augmented transformer that combines self-attention and convolutions to capture both global and local correlations. Both models were trained using the same setup, which included an Adam optimizer with a weight decay of 10^{-4}, a *Reduce LR on Plateau* learning rate scheduler (from 10^{-2} and arrives at 10^{-5}), and an early stopping strategy with a patience of 8 to prevent overfitting.

To obtain a system more robust in industrial environments, the samples used to train and validate the neural networks have been randomly corrupted with noise that is typical in industrial environments, where the robot and the adaptive workstation must operate. In particular, a total of 2,635 (negative) noise samples have been collected, that can be divided into two groups. The first group includes 1,604 samples acquired in a production line during a typical working day, representing real-world industrial noise. These samples are collected at different distances and from different directions of arrival. The second group includes 1,031 samples sourced from the FreeSound dataset [10] representing sounds like drills, fans, and hammer. This noise was added to the speech samples collected for the project with a random SNR between 0 and 25 dB, a pre-processing strategy that allows the model to learn features robust to challenging industrial conditions.

3.3 Gesture Recognition

As introduced in Sect. 2, to meet the project requirements, we have considered a one-stage detector to perform the gesture recognition task. The system uses a Single Shot Detector (SSD), namely a MobileNetV3 SSD [18,21], to detect the hand and recognize the gesture simultaneously. The network, optimized to achieve a trade-off between accuracy and computational requirements, takes 320×320 images as input and provides a bounding box and a gesture category.

We started from a model pre-trained on ImageNet, and then we fine-tuned it on the dataset described in Sect. 4.1. An SGD optimizer was used with an initial learning rate of 0.001, a momentum of 0.9, and a weight decay of 0.0005. The learning rate is reduced by a factor of 0.1 if validation accuracy does not improve for 5 epochs, and an early stopping mechanism with a patience of 10 epochs prevents overfitting. To enhance robustness, data augmentation strategies like random crop and subsampling were applied to simulate real-world occlusions and distance variations.

Finally, to make the system more robust, the outputs of the classification module are evaluated through a sliding window that averages the predictions on the same hand. The windowing allows to obtain a more stable classification and reduces the errors. This operation is implemented with an overlap tracking module that follows the hands and manages a window for each of them.

4 Experimental Analysis

In this section we give detailed information about the datasets adopted for the experiments and the results of the validation for speech command and gesture recognition.

4.1 Dataset

Speech Command Recognition. According to the project requirements, the system uses two sets of speech commands in English and Italian that are reported in Table 1.

A custom dataset composed of real, synthetic, and negative samples was created for this work. Real speech commands were recorded from 218 English and 222 Italian speakers using a Telegram bot. This campaign collected 14,960 balanced real samples, totaling more than 12 h of recordings. Additionally, 5,904

Table 1. Speech commands required to control the adaptive workstation and instruct the robot to perform specific tasks.

	English	Italian
Workstation	Increase the height	Più alto
	Decrease the height	Più basso
	Increase the inclination	Più inclinato
	Decrease the inclination	Meno inclinato
Robot	Start	Start
	Release	Rilascia
	Bring me the gun screwdriver	Portami l'avvitatore elettrico
	Take the gun screwdriver	Prendi l'avvitatore elettrico
	Bring me the elbow screwdriver	Portami l'avvitatore a gomito
	Take the elbow screwdriver	Prendi l'avvitatore a gomito
	Bring me the windows control panel	Portami la mostrina comandi
	Take the windows control panel	Prendi la mostrina comandi
	Go	Libero
	Stop	Stop

balanced synthetic samples were collected using web services like Amazon Polly and Google Cloud, totaling around 5 h of recordings. To train the system to recognize non-command speech, 29,630 negative samples were collected, consisting of 20,083 samples from the Google Speech Commands dataset [37] and 9,551 from Mozilla Common Voices [2]. The dataset was split into training (70%), validation (15%), and test (15%) sets using a stratified strategy to maintain balance, ensuring that speakers in each set are unique. According to the procedure described in Sect. 3.2, the test set was artificially corrupted with noise to have fixed SNR values (0, 5, 10, 15, 20, 25), allowing for an evaluation of the system's resilience.

To validate the system's performance in a real industrial scenario, a separate dataset of 1,612 real, noisy samples was collected. This balanced dataset, totaling around 80 min of recordings, was pronounced by 13 Italian workers (speaking both English and Italian) on the CRF plant production line during daily operations. The data was captured using two microphones at varying distances: the robot's onboard microphone at 1–2 m and a laptop's embedded microphone at 3–4 m.

Gesture Recognition. The robot is controlled by three gesture commands:

- *Start*: performed by raising the right or left thumb;
- *Stop*: performed by raising the entire right or left hand;
- *Go*: performed with a "shaka" sign using either the right or left hand, which sends the robot to its home position;

The gesture dataset was acquired by reproducing specific industrial operation conditions, with a camera placed at a height of 1.2 m to record 16 workers performing gestures at three different distances (1 m, 2 m, and 5 m). The acquisition, which includes a negative class called *no gesture*, was performed in both normal and low lighting conditions. The dataset consists of 1,162 five-second video clips recorded at 30 FPS (150 frames per clip) with a 640×480 resolution, totaling around 170,000 manually labelled frames. The dataset was then divided by subject into three sets: 12 subjects for the training set, 2 for the validation set, and 2 for the test set.

4.2   Results

Speech Command Recognition. In Table 2 we report the accuracy achieved by the proposed speech recognition system. The Conformer maintains an accuracy over 90% even with substantially noisy samples (up to 5dB); the accuracy slightly decreases only with samples with noise energy equal to signal energy (SNR = 0dB). In addition, the Conformer is more robust to noise with respect to a smaller model like ResNet, especially with very noisy samples. Regarding the method for Italian speech command recognition, whose accuracy is reported in the same Table 2, the results are even better than the corresponding English system. The accuracy obtained on the Italian speech commands is even more promising than the ones achieved on the English ones. The performance is between 98% and 100% up to 10 dB, slightly decreasing for lower SNRs. We can thus conclude that the results achieved both on English and Italian speech commands allow to meet the operating requirement in less than 30 ms for each speech command.

When tested in a real-world industrial scenario, the Conformer model achieved 91% accuracy for Italian speech commands and 82% for English commands. A qualitative analysis suggests this discrepancy is due to a data bias, as all workers who recorded the commands were native Italian speakers, leading to inconsistent English pronunciation. While these results represent a 6 to 9% point drop in accuracy compared to a controlled environment, this reduction is substantially smaller than the 20–30% performance decrease reported in similar experiments. This area warrants further investigation for future improvements.

Gesture Recognition. The gesture recognition method achieves an accuracy of 0.91 at a distance of 1 m and 0.90 at 2 m. The proposed approach is able to achieve an accuracy higher than 90% when the worker is at a distance less than or equal to 2 m (operating requirement); the accuracy is lower when the distance increases (0.72 at 5 m), but this is out of the goals. The results are definitely positive both in terms of accuracy and processing speed, considering that the system requires less than 70 ms for processing each frame.

Table 2. Accuracy achieved by ResNet and Conformer for both the languages at different SNRs.

	English Commands						
Method-SNR(dB)	0	5	10	15	20	25	Avg
ResNet	0.68	0.89	0.95	0.97	0.97	0.97	0.90
Conformer	0.84	0.90	0.94	0.94	0.94	0.94	0.91
	Italian Commands						
Method-SNR(dB)	0	5	10	15	20	25	Avg
ResNet	0.84	0.93	0.99	0.99	0.99	0.99	0.95
Conformer	0.90	0.95	0.98	1.00	1.00	1.00	0.97

5   Conclusions

In conclusion, we demonstrated the feasibility and effectiveness of real-time human-robot collaboration through speech commands and gestures within a challenging and noisy production line environment. The results demonstrate that the proposed solution achieves an accuracy rate higher than 90%, realizing a reliable communication between workers and robotics systems. The robustness of the proposed solution is particularly noteworthy given the inherent challenges posed by the noisy industrial setting. Despite the environmental noise and the dynamic conditions characteristic of a production line, our system exhibits a commendable ability to accurately interpret and respond to both speech commands and gestures. This resilience underscores the adaptability of the solution to the challenging conditions of industrial environments. Furthermore, the response time of less than 100 ms (30 ms for speech commands and 70 ms for gestures) demonstrates a rapid responsiveness and the real-time applicability of the proposed solution, essential attributes for human-robot collaboration in production scenarios. Finally, the implementation of this solution on an embedded device with limited resources opens interesting frontiers for practical deployment in industrial settings. The fact that our software operates effectively within the constraints of such devices, demonstrates its scalability and potential for integration across production lines without the need of expensive hardware upgrades.

References

1. Anh, N.T., Hu, Y., He, Q., Linh, T.T.N., Dung, H.T.K., Guang, C.: LIS-Net: an end-to-end light interior search network for speech command recognition. Comput. Speech Lang. **65**, 101131 (2021)
2. Ardila, R., et al.: Common voice: a massively-multilingual speech corpus. arXiv preprint: arXiv:1912.06670 (2019)
3. Bini, S., Greco, A., Saggese, A., Vento, M.: Benchmarking deep neural networks for gesture recognition on embedded devices. In: IEEE International Conference on Robot & Human Interactive Communication (RO-MAN). IEEE (2022)

4. De Simone, G., Greco, A., Rosa, F., Saggese, A., Vento, M.: Context-aware data augmentation for enhanced speech command recognition in industrial environments. Sci. Rep. **15**(1), 1–16 (2025)

5. Drolshagen, S., Pfingsthorn, M., Hein, A.: Context-aware robotic assistive system: robotic pointing gesture-based assistance for people with disabilities in sheltered workshops. Robotics **12**(5), 132 (2023)

6. Foggia, P., Greco, A., Roberto, A., Saggese, A., Vento, M.: Degramnet: effective audio analysis based on a fully learnable time-frequency representation. Neural Comput. Appl. **35**(27), 20207–20219 (2023)

7. Foggia, P., Greco, A., Roberto, A., Saggese, A., Vento, M.: Few-shot re-identification of the speaker by social robots. Auton. Robot. **47**(2), 181–192 (2023)

8. Foggia, P., Greco, A., Roberto, A., Saggese, A., Vento, M.: A social robot architecture for personalized real-time human-robot interaction. IEEE Internet Things J. (2023)

9. Foggia, P., Greco, A., Roberto, A., Saggese, A., Vento, M.: Identity, gender, age, and emotion recognition from speaker voice with multi-task deep networks for cognitive robotics. Cogn. Comput., 1–11 (2024)

10. Fonseca, E., et al.: Freesound datasets: a platform for the creation of open audio datasets. In: International Society for Music Information Retrieval Conference (ISMIR 2017), pp. 486–493, Suzhou, China (2017)

11. Greco, A., Roberto, A., Saggese, A., Vento, M.: DENet: a deep architecture for audio surveillance applications. Neural Comput. Appl., 1–12 (2021)

12. Gulati, A., et al.: Conformer: convolution-augmented transformer for speech recognition. Interspeech **2020**, 5036–5040 (2020)

13. Haghani, P., et al.: From audio to semantics: approaches to end-to-end spoken language understanding. In: 2018 IEEE Spoken Language Technology Workshop (SLT), pp. 720–726. IEEE (2018)

14. He, K., Girshick, R., Dollár, P.: Rethinking ImageNet pre-training. In: Proceedings of the IEEE/CVF International Conference on Computer Vision, pp. 4918–4927 (2019)

15. Hendrickx, K., Perini, L., Van der Plas, D., Meert, W., Davis, J.: Machine learning with a reject option: a survey. arXiv preprint: arXiv:2107.11277 (2021)

16. Hsu, W.N., Bolte, B., Tsai, Y.H.H., Lakhotia, K., Salakhutdinov, R., Mohamed, A.: HuBERT: self-supervised speech representation learning by masked prediction of hidden units. IEEE/ACM Trans. Audio, Speech, Lang. Process. **29**, 3451–3460 (2021)

17. Khalil, D., et al.: An automatic speaker clustering pipeline for the air traffic communication domain. Aerospace **10**(10), 876 (2023)

18. Koonce, B., Koonce, B.: MobileNetV3. convolutional neural networks with swift for TensorFlow: image recognition and dataset categorization, pp. 125–144 (2021)

19. Kostopoulos, N., Kalogeras, D., Pantazatos, D., Grammatikou, M., Maglaris, V.: SHAP interpretations of tree and neural network DNS classifiers for analyzing DGA family characteristics. IEEE Access **11**, 61144–61160 (2023). https://doi.org/10.1109/ACCESS.2023.3286313

20. Li, C., Park, J., Kim, H., Chrysostomou, D.: How can i help you? An intelligent virtual assistant for industrial robots. In: ACM/IEEE International Conference on Human-Robot Interaction, pp. 220–224 (2021)

21. Liu, W., et al.: SSD: single shot multibox detector. In: European Conference on Computer Vision, pp. 21–37. Springer (2016)

22. Liu, Y., Caldwell, G., Rittenbruch, M., Belek Fialho Teixeira, M., Burden, A., Guertler, M.: What affects human decision making in human–robot collaboration?: A scoping review. Robotics **13**(2), 30 (2024)

23. Lugosch, L., Ravanelli, M., Ignoto, P., Tomar, V.S., Bengio, Y.: Speech model pre-training for end-to-end spoken language understanding. arXiv preprint: arXiv:1904.03670 (2019)

24. Majumdar, S., Ginsburg, B.: MatchboxNet: 1D time-channel separable convolutional neural network architecture for speech commands recognition. In: Interspeech (2020)

25. Papadaki, A., Pateraki, M.: 6D object localization in car-assembly industrial environment. J. Imaging **9**(3) (2023).https://doi.org/10.3390/jimaging9030072

26. Parkhi, O.M., Vedaldi, A., Zisserman, A.: Deep face recognition. In: British Machine Vision Conference (BMVC), pp. 1–12. BMVA Press (2015)

27. Qian, Y., et al.: Speech-language pre-training for end-to-end spoken language understanding. In: International Conference on Acoustics, Speech and Signal Processing (ICASSP), pp. 7458–7462. IEEE (2021)

28. Ravanelli, M., et al.: Multi-task self-supervised learning for robust speech recognition. In: International Conference on Acoustics, Speech and Signal Processing (ICASSP), pp. 6989–6993. IEEE (2020)

29. Redmon, J., Divvala, S., Girshick, R., Farhadi, A.: You only look once: unified, real-time object detection. In: IEEE Conference on Computer Vision and Pattern Recognition, pp. 779–788 (2016)

30. Shon, S., et al.: SLUE: new benchmark tasks for spoken language understanding evaluation on natural speech. In: ICASSP 2022-2022 IEEE International Conference on Acoustics, Speech and Signal Processing (ICASSP), pp. 7927–7931. IEEE (2022)

31. Tandel, N.H., Prajapati, H.B., Dabhi, V.K.: Voice recognition and voice comparison using machine learning techniques: a survey. In: International Conference on Advanced Computing and Communication Systems, pp. 459–465. IEEE (2020)

32. Tang, R., Lin, J.: Deep residual learning for small-footprint keyword spotting. In: IEEE International Conference on Acoustics, Speech and Signal Processing (ICASSP), pp. 5484–5488 (2018). https://doi.org/10.1109/ICASSP.2018.8462688

33. Tian, Z., Shen, C., Chen, H., He, T.: FCOS: fully convolutional one-stage object detection. In: IEEE/CVF International Conference on Computer Vision (ICCV) (2019)

34. Valner, R., Kruusamäe, K., Pryor, M.: TeMoto: intuitive multi-range telerobotic system with natural gestural and verbal instruction interface. Robotics **7**(1), 9 (2018)

35. Vygon, R., Mikhaylovskiy, N.: Learning efficient representations for keyword spotting with triplet loss. In: International Conference on Speech and Computer, pp. 773–785. Springer (2021)

36. Wang, D., Wang, X., Lv, S.: An overview of end-to-end automatic speech recognition. Symmetry **11**(8), 1018 (2019)

37. Warden, P.: Speech commands: a dataset for limited-vocabulary speech recognition. arXiv e-prints, pp. arXiv–1804 (2018)

38. Wiseman, J.: WebRTC VAD. https://pypi.org/project/webrtcvad/

Emerging Methods and Vision Applications

Emotion Recognition in Contemporary Dance Performances Using Laban Movement Analysis

Muhammad Turab[(✉)], Philippe Colantoni[iD], Damien Muselet[iD],
and Alain Trémeau[iD]

Laboratoire Hubert Curien - UMR 5516, Saint-Etienne, France
muhammad.turab.muslim.bajeer@etu.univ-st-etienne.fr,
{philippe.colantoni,damien.muselet,alain.tremeau}@univ-st-etienne.fr

Abstract. This paper presents a novel framework for emotion recognition in contemporary dance by improving existing Laban Movement Analysis (LMA) feature descriptors and introducing robust, novel descriptors that capture both quantitative and qualitative aspects of the movement. Our approach extracts expressive characteristics from 3D keypoints data of professional dancers performing contemporary dance under various emotional states, and train multiple classifiers, including Random Forests and Support Vector Machines. Additionally, provide in-depth explanation of features and their impact on model predictions using explainable machine learning methods. Overall, our study improves emotion recognition in contemporary dance and offers promising applications in performance analysis, dance training, and humanâĂŞcomputer interaction with highest accuracy of 96.85%.

Keywords: Laban Movement Analysis · Emotion Recognition · Explainable AI · 3D Body Pose Estimation

1 Introduction

In recent years, human emotion recognition has become an important research direction in the field of motion analysis and human-computer interaction. Emotion recognition involves the analysis of various emotional data by extracting features that describe the emotion and their relationship to classify the emotional state. It has been widely used in many fields including emotions in facial expressions [11,25], speech recognition [16], text [2] and psychological signals [15,22]. Emotion analysis is not limited to these fields, but it can expand to any field that involves movements. For instance, dance is a form of body movements that is used to express various emotions. Dance has been an integral part of human history, traditions, and cultures to express, tell stories, and convey emotions. Over the years, dance styles have evolved greatly reflecting cultural, social, and artistic changes. One of the new styles that has emerged is contemporary

M. Castrillón-Santana et al. (Eds.): CAIP 2025, LNCS 15622, pp. 317–328, 2026.
https://doi.org/10.1007/978-3-032-05060-1_27

dance that combines elements of jazz, ballet, and modern dance. It originates from a desire to break away from the rigid structures of traditional ballet and modern dance. Unlike traditional dance, it allows dancers to explore novel forms of movement and emotional expression without setting strict rules. Due to the free-form structure of the contemporary dance style, it becomes challenging to recognize emotions.

For such challenges, Laban Movement analysis (LMA) [17] provides a comprehensive framework for the evaluation and interpretation of human body movements. It is divided into four main components including Body, Effort, Shape, and Space. These components describe the structural, geometrical, and dynamic properties of motion. Body and Space components describe how the human body moves, either within the body or in relation with the 3D space surrounding the body. The Shape component describes the shape morphology of the body during the motion, whereas the Effort component focuses on the qualitative aspects of the movement in terms of dynamics, energy, and intent. LMA is more qualitative than quantitative, and a few studies have tried to propose feature descriptors that quantify the qualities of LMA to some extent. However, there are still a few challenges including capturing temporal dynamics of movement, and clear explanation of feature descriptors. Many current feature descriptors for movement analysis may not fully capture the dynamic characteristics without temporal dynamics. To address these challenges, we propose robust feature descriptors with temporal dynamics to capture the dynamic nuances of LMA qualities with their explanation on model predictions and individual performance. The main contributions of this study are as follows:

– We propose a sliding-window based approach for feature extraction to add the temporal context and capture the short-term and long-term patterns, and evolution of the movement.
– We propose complementary feature descriptors for the four LMA components that capture rich representations of movements.
– We perform a comprehensive evaluation on a contemporary dance dataset using Machine Learning (ML) methods, and analyze the interpretability of our descriptors with eXplainable AI methods.

2   Related Work

Emotion recognition in dance movements has been an active field due to its applications in human-computer interaction [9], performance analysis [26], and evaluation [6]. Understanding how dancers express emotions through movement provides valuable insights into the expressive behavior of the performer. To extract expressive behavior from movement, LMA has become a popular framework as it provides a structured language to describe both the quantitative and qualitative aspects of human motion. In study [1], LMA is used to recognize and analyze expressive gestures, including dancing, moving, waving, pointing, and stopping, with four emotions happy, angry, sad and neutral. Feature evaluation was performed using ML methods and subjective evaluation. The results show that the features are 87. 03% accurate in gesture classification. [27] used Kinect to track dancers and developed models for action and emotion recognition with 88.34% and 98.95% accuracy on skeleton data. Features like eigenvalue speed and skeleton pair distance help distinguish emotions. LMA was used in human action recognition in [23], with a two-stage approach in which the first feature descriptors are defined using three components of LMA, including body, space, and shape. The results were obtained on various datasets using Dynamic Time Warping (DTW) to measure similarity in actions. Several studies have focused on the classification of dance emotions using LMA-based feature descriptors. A method is proposed to extract the motion qualities from dance performances for indexing and analysis in [4]. LMA-based features were used to investigate the correlation of features with dancers expressed emotional state. The results show low inter-feature correlation between dancers and their emotional state. [3] investigated similarities of various emotional states using extracted LMA extracted features. The results show that the features can partially extract LMA components that can be used for the dance classification from emotion.

For automatic feature extraction and better results, Deep learning (DL) has been used. [29] applied a hybrid deep learning method based on Convolutional Neural Networks (CNN) and Long Short-Term Memory (LSTM) to effectively identify emotions. For body structure, spatial orientation, and force effect LMA-based feature descriptors were used to add emotional changes during movement. Their approach successfully recognizes emotions in dance movements with high accuracy.

3   Methodology

To improve emotion recognition in contemporary dance, our study is divided into six main parts: 1) Data pre-processing, 2) 3D human body pose estimation, 3) LMA feature extraction, 4) Multiclass classification, 5) Evaluation, and 6) Explainability of feature descriptors as shown in Fig. 1.

3.1   Data Pre-processing

We use contemporary dance performance videos from the Dance Motion capture dataset provided by the University of Cyprus [5]. This dataset contains performances of 5 dancers expressing 12 emotions including afraid, angry, annoyed, bored, excited, happy, miserable, pleased, relaxed, sad, satisfied and tired. The dataset has a few limitations, such as: 1) inconsistent video duration, 2) limited number of dancers and performances, 3) low frame rate, and 4) a fixed camera angle for a single perspective.

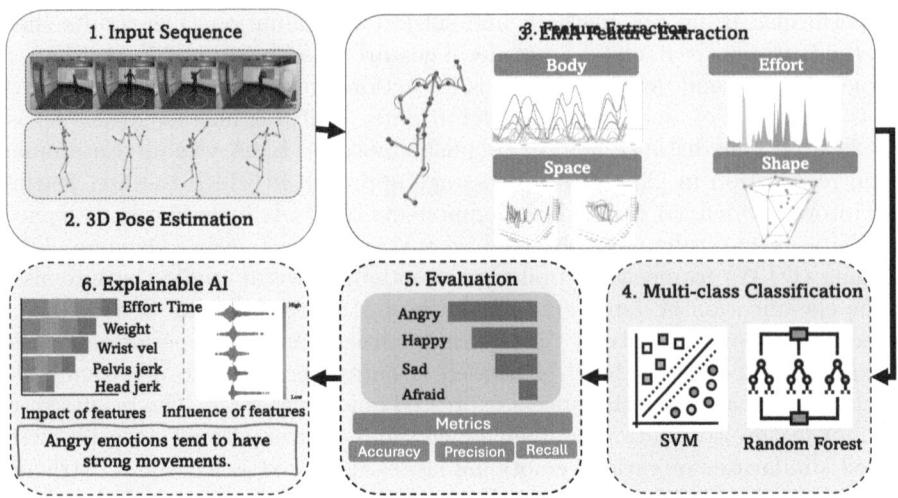

Fig. 1. Overview of the proposed method for Emotion Recognition in Contemporary Dance Performances.

3.2   3D Human Body Pose Estimation

To obtain the joints keypoint data from dance videos, we use Neural Localizer Fields (NLF) [24] for 3D body pose estimation. We compare NLF with other state-of-the-art (SOTA) methods including Mediapipe [19] and OpenPose [12] and choose NLF for several reasons including: 1) It is an advanced model trained on massive data to estimate 2D/3D points from single image, 2) the dance performances have many frames where some body parts are occluded or missing in some cases when dancer is very close to the camera and NLF performs really well in such cases outperforming existing SOTA methods. 3) Dancers back-and-forth movement introduces challenges in feature extraction due to shifting camera perspectives. NLF effectively addresses shifting camera perspective by providing the 3D keypoints in absolute camera-space.

3.3   LMA Feature Extraction

Laban Movement Analysis is a language and a structured framework to describe, and interpret human body movements. It categorizes human motion into four main components: BODY, EFFORT, SHAPE and SPACE. **Body** component describes the structural and physical characteristics of human body and is responsible for joints connectivity, movement and their influence on each other. For joint connectivity, we calculate the Euclidean distance and angles between the hands, shoulders, pelvis, knees, and ankles. For movement, we propose a new feature descriptor to describe the joint that initiates movement in a sequence, as shown in Eq. 1.

$$\text{Initiation}(t) = \frac{P_j(t+w) - P_j(t)}{\Delta t} > \tau \tag{1}$$

Here $P_j(t)$ denotes the position of joint j at time t, Δt is the time interval, w is the short time-window to add temporal dynamics, and τ is a data-driven threshold calculated using standard-deviation of the entire sequence. **Effort** component describes the intention, dynamic qualities of movement, and energy used during movement. It is associated with the change in emotion or mood, hence it is useful for motion expressivity and emotion description. We mainly focus on upper and lower body parts, such as hands, feet, head, and pelvis, as these joints are involved most in emotion expression. The Effort component has four factors: Space, Weight, Time, and Flow. Effort Space describes the attention of movement in space, and it can be direct where the movement is focused on single direction, and indirect where the movement is focused in multi-directions. For this we calculate the distance of each joint for a short time-window to the total distance covered by that joint using Eq. 2.

$$\text{Space}_j(T) = \frac{\sum_{i=1}^{T} \| P_j(t_i) - P_j(t_i - w) \|}{\| P_j(T) - P_j(t_1) \|} \tag{2}$$

Total space is calculated by multiplying the weights α_j of selected joints $\mathcal{J}$ to their space factor using Eq. 3. Joint weights are used to give more importance to extremities, as defined in [13].

$$\text{Space}(T) = \sum_{j \in \mathcal{J}} \alpha_j \, \text{Space}_j(T) \tag{3}$$

Effort Weight describes how powerful or strong a movement is, it can be strong or light. For this we calculate the kinetic energy of selected joints $\mathcal{J}$ using Eq. 4. Here $v_j(t_i)$ denotes the velocity of a joint j at time t_i.

$$Weight(t) = \sum_{j \in \mathcal{J}} E_j(t_i) = \sum_{j \in \mathcal{J}} \frac{1}{2} \alpha_j v_j(t_i)^2 \tag{4}$$

Effort Time captures a sense of urgency. It tells how quick or sustained a movement is executed. For this we calculate the acceleration of selected joints over a sliding time-window as shown in Eq. 5.

$$\text{Time}(T) = \sum_{j \in \mathcal{J}} \alpha_j \, \text{Time}_j(T) \tag{5}$$

where $\text{Time}_j(T) = \frac{1}{T} \sum_{i=1}^{T} a_j(t_i)$, and a_j denotes the acceleration of a joint j. All these features are inspired from [18]. **Space** component describes the relationship of movement with space. For this we calculate trajectory of whole sequence, curvature, and propose a new feature spatial dispersion that describes how dancer is using kinesphere or it's personal space. It is calculated as distance of upper body parts to torso, pelvis for lower body parts. total path covered, and total distance covered. **Shape** component describes how the body is changing shape during the movement. For this we calculate the volume of the body using ConvexHull algorithm implemented in Python SciPy package [28].

Once we quantify all LMA components, we obtain a descriptor vector composed of 54 features which is used as input to the classification models.

3.4   Multi-class Emotion Classification

For emotion classification we use ML methods including Support Vector Machines (SVM) [14], and Random Forest (RF) [10], since they are already used for emotion classification in [1,3,29]. Firstly, we use 3 fold cross-validation to divide the dataset into 3 sets including training, testing, and validation to overcome overfitting and data dependency. For both SVM and RF the best hyperparameter values were found using GridSearch [8], and the best parameter values were selected based on the validation accuracy.

3.5   Model Explainability Using SHAP

Machine learning models are often considered black-box systems, which means that the process by which they arrive at specific predictions remains unclear. It is challenging to determine the factors that influence their results. In order to improve our understanding for emotion classification in contemporary dance performances, we use SHapley additive explanations (SHAP) [21], which is a game theory-based method to explain the predictions of ML models by assigning importance values to each feature for a given prediction. It tells the contribution of each feature to a specific prediction. For SVM we applied KernelExplainer, and for Random Forest we applied TreeExplainer [20].

4   Results

In this section, we present the experimental results obtained using proposed LMA feature descriptors. For evaluation accuracy, precision, and recall metrics are used to assess the performance of our method. Next, we summarize the key quantitative results, highlighting how our approach compares with other methods. Table 1 shows the results of individual emotional states, It can be observed that RF demonstrates consistent performance across all classes and metrics. Similarly, the SVM shows balanced performance but is lower than that of RF. Our method maintains consistent performance across all emotional states, unlike [3], where some classes show high accuracy while others perform lower, clearly indicating that the features do not capture overlapping emotions.

Table 2 presents the comparison of different methods on the Dance Motion Capture dataset when compared to the proposed method. It is evident that the proposed method outperforms all other methods in terms of accuracy. RF achieves a higher accuracy of 96.85% compared to SVM's 93.90%.

Table 1. Classification results of RF and SVM for each emotion using sliding window of 25 consecutive frames.

Emotion	RF			SVM		
	Precision (%)	Recall (%)	F1-score (%)	Precision (%)	Recall (%)	F1-score (%)
Afraid	98.46	100.00	99.22	92.57	97.27	94.86
Angry	94.71	96.59	95.64	91.34	92.90	92.11
Annoyed	98.95	93.56	96.18	95.38	92.08	93.70
Bored	97.62	98.80	98.20	94.74	93.98	94.35
Excited	91.71	97.07	94.32	92.57	92.82	92.70
Happy	97.18	92.81	94.95	93.43	93.71	93.57
Miserable	98.27	98.61	98.44	92.96	91.67	92.31
Pleased	97.04	92.58	94.76	96.27	91.17	93.65
Relaxed	95.81	99.38	97.56	95.99	96.58	96.28
Sad	97.71	97.99	97.85	91.69	94.84	93.24
Satisfied	99.07	99.07	99.07	97.60	94.86	96.21
Tired	99.57	99.13	99.35	95.28	96.10	95.69
Average	**97.39**	**96.21**	**96.73**	**94.24**	**93.77**	**93.98**

Table 2. Comparison of emotion classification with state-of-the-art ML methods on Dance Motion Dataset [5] using sliding window of 25.

Study	Method	Accuracy (%)
[29]	Decision Tree	94.48
[29]	Random Forest	92.95
[3]	SVM	89.96
[3]	Random Forest	91.75
[7]	SVM	81.70
Ours	SVM	93.90
Ours	Random Forest	**96.85**

Ground Truth: Excited　**Ground Truth: Sad**　**Ground Truth: Angry**
Prediction: Excited　**Prediction: Sad**　**Prediction: Angry**

Fig. 2. Ground truth and model predictions for three unseen dance performances. Images from [5].

In Fig. 2, we show the predictions of our method for unseen movement sequences, where it successfully recognizes emotions in dance performances.

Figure 3a and Fig. 3b show the confusion matrices for both models. It can be observed that the SVM model is confused between classes such as happy, excited, and pleased, with each showing a high misclassification rate of 21 samples. There is also confusion among angry, excited, annoyed, relaxed, and sad, as these emotion styles exhibit similar movements. On the other hand, the RF model performs exceptionally well, with no confusion higher than 9 between angry and bored.

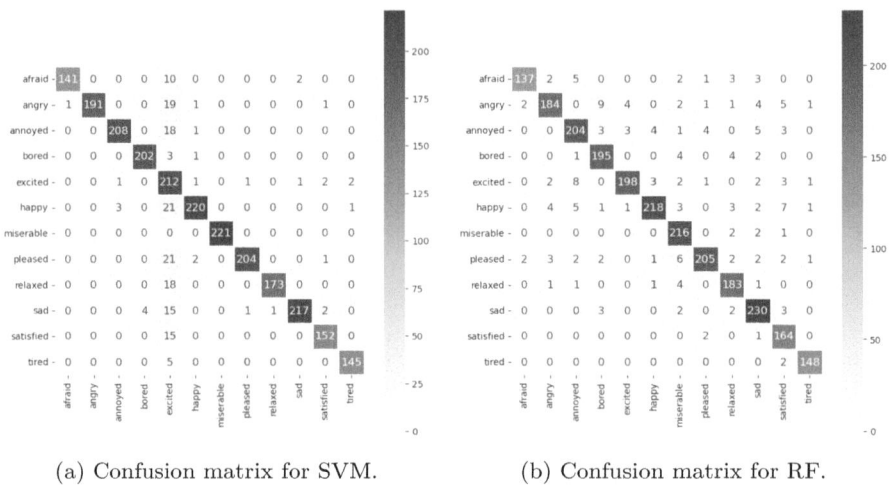

(a) Confusion matrix for SVM.   (b) Confusion matrix for RF.

Fig. 3. Confusion matrices showing the classification performance of SVM and RF models.

Our sliding window approach improves the overall accuracy of emotion classification, as shown in Fig. 4. It can be observed that a smaller window size results in lower accuracy, while increasing the window size improves the accuracy for both RF and SVM. In our case, the optimal window size is 25–30 for a balanced performance, and after 30 the improvement is negligible. Although it is evident that a larger window improves accuracy, it is also important to note that it may suppress nuanced emotional details. Therefore, selecting an optimal window size is crucial.

Currently, our method achieves promising results for emotion classification in contemporary dance. However, there remains a challenge regarding interpretability and explainability of the models, which has not been explored in existing research on emotion classification. We aim to investigate questions such as which features significantly affect the ability of models to classify angry movements from those that are happy or excited, as they look very similar visually. To answer such questions, we apply various method that explain the contributions of features individually or as a whole. The SHAP summary plot in Fig. 5

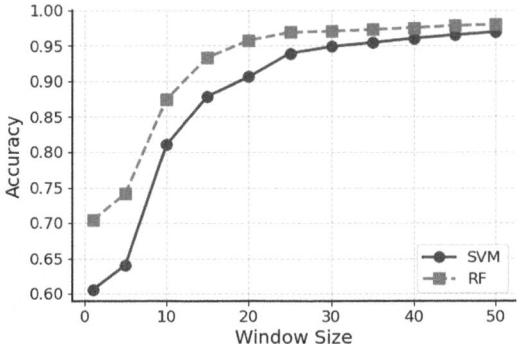

Fig. 4. Impact of sliding window size on accuracy.

shows most influential features in predicting emotions from movement data. Body volume has the highest impact, indicating that overall body expansion or contraction plays a key role in emotional expression. LMA Effort time has high impact as well which is essential to distinguish angry movements from sad. Overall, both global body features and localized joint behavior impact model's predictions.

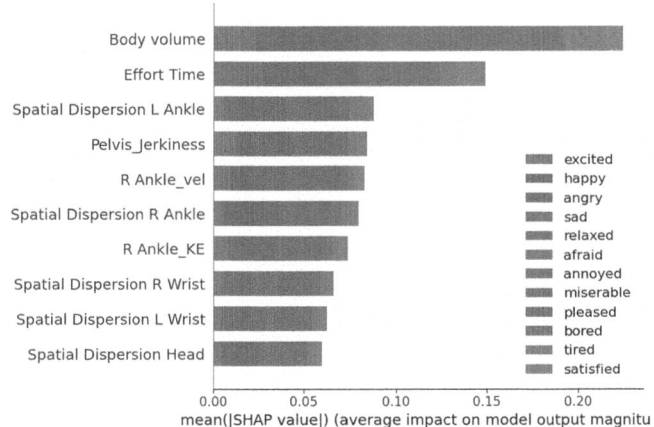

Fig. 5. Impact of top 10 features on models predictions.

Impact of key movement features on emotion predictions are illustrated in Fig. 6 and Fig. 7. The horizontal axis represents the SHAP value, where positive values push the prediction towards a target class and negative values push it away. In both plots, features such as *Body volume*, *Effort Time*, and *Pelvis Jerkiness* show strong positive influence, indicating that emotions associated with expansive or intense movements (e.g., anger or excitement) are driven by higher body expansion, energetic motion, and movement irregularities. Conversely, lower values in these features tend to align with sustained movements (e.g., sadness or tiredness), reflecting more contracted posture, reduced movement dispersion, and lower kinetic energy.

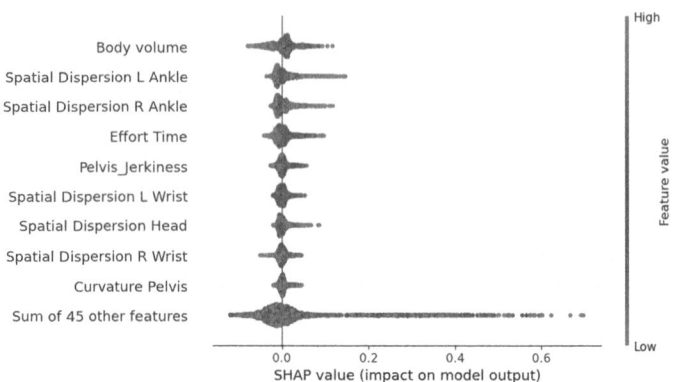

Fig. 6. Impact of features on the model's predictions for dance movements expressing anger.

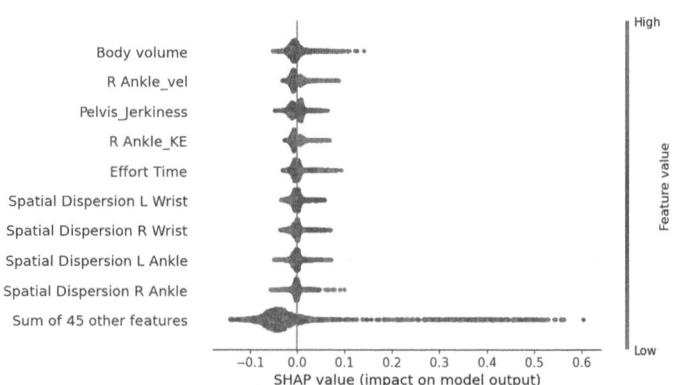

Fig. 7. Impact of features on the model's predictions for dance movements expressing sadness.

5   Conclusion

In this study, we improve existing Laban Movement Analysis (LMA) feature descriptors by adding the temporal dynamics with sliding window approach, introduce novel robust descriptors that capture both quantitative and qualitative aspects of emotions in dance. With a comparative analysis, we show that the proposed approach achieves higher accuracy compared to existing methods. Furthermore, we provide interpretations and explanations of the feature descriptors to better understand black-box models and their predictions. Overall, our study improves the understanding and recognition of dance emotions using improved LMA based feature descriptors, and contribute to more effective and transparent emotion and motion analysis in dance performances. Future work includes extending the framework to dance style classification.

Acknowledgments. This study was carried out as part of the PREMIERE project "Performing Arts in a new Era", https://premiere-project.eu/, funded by HORIZON-CL2-2021-HERITAGE-000201-04 (grant number 101061303 - PREMIERE).

References

1. Ajili, I., Ramezanpanah, Z., Mallem, M., Didier, J.Y.: Expressive motions recognition and analysis with learning and statistical methods. Multimedia Tools Appl. **78**, 16575–16600 (2019)
2. Alswaidan, N., Menai, M.E.B.: A survey of state-of-the-art approaches for emotion recognition in text. Knowl. Inf. Syst. **62**(8), 2937–2987 (2020). https://doi.org/10.1007/s10115-020-01449-0
3. Aristidou, A., Charalambous, P., Chrysanthou, Y.: Emotion analysis and classification: understanding the performers' emotions using the LMA entities. In: Computer Graphics Forum. vol. 34, pp. 262–276. Wiley Online Library (2015)
4. Aristidou, A., Chrysanthou, Y.: Feature extraction for human motion indexing of acted dance performances. In: 2014 International Conference on Computer Graphics Theory and Applications (GRAPP), pp. 1–11. IEEE (2014)
5. Aristidou, A., Shamir, A., Chrysanthou, Y.: Digital dance ethnography: organizing large dance collections. J. Comput. Cult. Herit. **12**(4) (2019). https://doi.org/10.1145/3344383
6. Aristidou, A., Stavrakis, E., Charalambous, P., Chrysanthou, Y., Himona, S.L.: Folk dance evaluation using laban movement analysis. J. Comput. Cult. Heritage (JOCCH) (2015)
7. Bai, J., Dai, R., Dai, J., Pan, J.: Emodescriptor: a hybrid feature for emotional classification in dance movements. Comput. Anim. Virtual Worlds **32**(6), e1996 (2021)
8. Bergstra, J., Bengio, Y.: Random search for hyper-parameter optimization. J. Mach. Learn. Res. (2012)
9. Brave, S., Nass, C.: Emotion in human-computer interaction. In: The Human-computer Interaction Handbook, pp. 103–118. CRC Press (2007)
10. Breiman, L.: Random Forest. Mach. Learn. **45**, 5–32 (2001)
11. Canal, F.Z., et al.: A survey on facial emotion recognition techniques: a state-of-the-art literature review. Inf. Sci. **582**, 593–617 (2022)
12. Cao, Z., Hidalgo, G., Simon, T., Wei, S.E., Sheikh, Y.: Openpose: Realtime multi-person 2D pose estimation using part affinity fields. IEEE Trans. Pattern Anal. Mach. Intell. (2019)

13. Contributors, M.: Openmmlab pose estimation toolbox and benchmark (2020). https://github.com/open-mmlab/mmpose
14. Cortes, C., Vapnik, V.: Support-Vector Netw. Mach. Learn. **20**, 273–297 (1995)
15. Egger, M., Ley, M., Hanke, S.: Emotion recognition from physiological signal analysis: a review. Electr. Notes Theor. Comput. Sci. **343**, 35–55 (2019)
16. Khalil, R.A., Jones, E., Babar, M.I., Jan, T., Zafar, M.H., Alhussain, T.: Speech emotion recognition using deep learning techniques: a review. IEEE Access **7**, 117327–117345 (2019)
17. von Laban, R.: The mastery of movement on the stage. MacDonald Evans (1950)
18. Larboulette, C., Gibet, S.: A review of computable expressive descriptors of human motion. In: Proceedings of the 2nd International Workshop on Movement and Computing (2015)
19. Lugaresi, C., et al.: Mediapipe: A framework for building perception pipelines (2019). arXiv preprint arXiv:1906.08172
20. Lundberg, S.M., et al.: From local explanations to global understanding with explainable AI for trees. Nat. Mach. Intell. **2**(1), 2522–5839 (2020)
21. Lundberg, S.M., Lee, S.I.: A unified approach to interpreting model predictions. In: Guyon, I., et al. (eds.) Advances in Neural Information Processing Systems 30, pp. 4765–4774. Curran Associates, Inc. (2017). http://papers.nips.cc/paper/7062-a-unified-approach-to-interpreting-model-predictions.pdf
22. Raj, K., Mileo, A.: Towards understanding graph neural networks: functional-semantic activation mapping. In: International Conference on Neural-Symbolic Learning and Reasoning. Springer (2024)
23. Ramezanpanah, Z., Mallem, M., Davesne, F.: Human action recognition using laban movement analysis and dynamic time warping. Proc. Comput. Sci. **176**, 390–399 (2020)
24. Sárándi, I., Pons-Moll, G.: Neural localizer fields for continuous 3D human pose and shape estimation (2024). arXiv preprint arXiv:2407.07532
25. Singh, A., Raj, K., Meghwar, T., Roy, A.M.: Efficient paddy grain quality assessment approach utilizing affordable sensors. AI (2024)
26. Sun, Q., Wu, X.: A deep learning-based approach for emotional analysis of sports dance. PeerJ Comput. Sci. (2023)
27. Tan, G., Wang, J.: Research on the mechanism of emotion expression in dance based on machine learning models. J. Combin. Math. Combin, Comput (2025)
28. Virtanen, P., et al.: Scipy 1.0: fundamental algorithms for scientific computing in python. Nat. Methods (2020)
29. Wang, S., Li, J., Cao, T., Wang, H., Tu, P., Li, Y.: Dance emotion recognition based on laban motion analysis using convolutional neural network and long short-term memory. IEEE Access **8**, 124928–124938 (2020)

Forecasting Sea Surface Temperature from Satellite Images with Graph Neural Networks

Giovanny A. Cuervo-Londoño[1] , Javier Sánchez[2(✉)] ,
and Ángel Rodríguez-Santana[1]

[1] Instituto Universitario de Investigación en Acuicultura Sostenible y Ecosistemas Marinos (ECOAQUA), Las Palmas de Gran Canaria, Spain
giovanny.cuervo101@alu.ulpgc.es, angel.santana@ulpgc.es
[2] Instituto Universitario de Cibernética, Empresas y Sociedad (IUCES), University of Las Palmas de Gran Canaria, 35017 Las Palmas de Gran Canaria, Spain
jsanchez@ulpgc.es

Abstract. Predicting the evolution of sea surface temperature (SST) is essential for applications in weather forecasting, maritime transport, and fisheries. Traditional ocean forecasting methods rely on physics-based numerical models, which face challenges such as data gaps, assimilation difficulties, and computational inefficiencies. Recent advances in Graph Neural Networks (GNNs) have shown promise in improving prediction accuracy and efficiency. In this work, we adapt a GNN model, initially designed for atmospheric forecasting, to oceanographic applications. We focus on the Canary Islands and the northwest African shore regions characterized by strong mesoscale dynamics. Our approach introduces a spatially masked loss function to address ocean-specific challenges like spatial discontinuities and observational data sparsity. We train our model using the L4 SST satellite images dataset from Copernicus Marine Service and compare its performance with state-of-the-art ConvLSTM-based models. Our results indicate that the adapted GNN model effectively captures mesoscale structures and outperforms ConvLSTM in both computational efficiency and accuracy. These findings suggest that graph-based deep learning approaches can overcome key limitations of current oceanographic models and provide a more flexible and scalable solution for forecasting oceanographic variables from satellite images.

Keywords: Graph neural network · Deep learning · Remote sensing · Forecasting · Oceanography

1 Introduction

Forecasting the evolution of oceanographic data obtained from satellite images is necessary for tasks related to weather prediction, maritime transport, and the fishery industry [2], among others. It relies heavily on forecasting mesoscale processes due to their environmental and economic impacts. These processes, which give rise to distinct structures, also influence mean currents and transport key ocean properties. Despite its importance, predicting mesoscale processes remains a challenging task [10].

Traditional ocean forecasting systems rely on numerical models that use physics-based equations, although they have several limitations. Gaps in observational data and ongoing challenges with current data assimilation techniques make it difficult to get a complete picture of the ocean. Additionally, these models do not fully leverage historical data or take advantage of modern hardware like GPUs, making them less efficient.

Deep learning models have recently emerged in global atmospheric forecasting, considerably improving the performance of numerical systems. There have appeared numerous models for predicting the state of the atmosphere, such as Pangu Weather [3], GraphCast [9], Aurora [4], NeuralGCM [8], or Gencast [11], which rely on foundational models, graph neural architectures, or diffusion models. These have considerably reduced inference times and computational costs.

These models rely on high-quality data but struggle with inconsistent, sparse, or noisy datasets, which can limit their accuracy. They also suffer from spectral bias in the long run, leading to numerical instability or unrealistic predictions. While these models work well for atmospheric systems, applying them to ocean data is more challenging because of the atmospheric interference and landmasses. Nevertheless, some models have recently appeared for predicting the ocean dynamics, such as Xihe [13], designed for global predictions, or SeaCast [7] and OceanNet [5], with a focus on regional forecasting.

Forecasting SST in coastal regions like the Canary Islands and northwestern Africa is crucial due to its significant implications for marine ecology, weather prediction, and fisheries. While prior deep learning approaches have shown strong results in atmospheric forecasting, regional oceanographic forecasting remains relatively unexplored.

In this work, we adapt a graph neural network [9], initially designed for atmospheric systems, to oceanographic forecasting. To tailor our model for this task, we adapt the underlying graph to a subregional domain and introduce a spatially masked loss function optimization that targets oceanic regions and prevents land-related artifacts during training. This approach differs from conventional methods, which are typically applied before training. By restricting loss computation to oceanic zones, we explore whether it can mitigate spectral bias by forcing the model to learn in selected marine areas during training.

Our study focuses on the Canary Islands and the northwest African shore subregion, which presents several challenges due to the oceanic and coastal interactions [12]. Traditional methods struggle to represent these interactions due to the highly nonlinear processes caused by strong mesoscale, upwelling fronts, and complex eddy fields. We use the L4 satellite-derived SST dataset [6] from the Copernicus Marine Service (CMEMS) for the Atlantic Ocean around Iberia, Biscay, and Ireland (IBI), limited to the Canary Islands subdomain. This dataset comprises nearly 40 years of daily SST satellite images from 1982 to 2020, with a resolution of 5.55 km per pixel.

By comparing our model to a state-of-the-art model, such as ConvLSTM [14], we assess the benefits and drawbacks of GNNs and how they can better capture these unresolved dynamics and offer greater flexibility. Additionally, we evaluate the model's sensitivity to mesoscale processes of high-dynamical complexity areas, such as frontal zones near the islands and African capes. The experimental results show that our model surpasses ConvLSTMs in computational efficiency and accuracy. We evaluate its performance in short- and long-range predictions, emphasizing its spatial behavior in our study area.

Our work introduces three main contributions: i) it adapts a global GNN for subregional SST forecasting; ii) it introduces a spatially masked loss function to target only oceanic areas, avoiding land artifacts; iii) the experiments evaluate mesh resolution trade-offs concerning both accuracy and computational cost.

Section 2 explains the dataset and the study area of this work. Section 3 details the architecture of our graph neural network. The experimental results (Sect. 5) evaluate and compare the model's performance with ConvLSTM. Finally, Sect. 6 discusses the contributions and limitations of this work and proposes future research directions.

2   Dataset and Study Area

This study focuses on the Canary Islands and the Moroccan subregion, which extends from 21°S to 33°N; see Fig. 1. This region has strong upwelling due to wind and seabed shape, bringing cold water and nutrients to the surface. The capes push coastal currents offshore, forming filaments that move water into the ocean. Irregular seabed modifies currents, creating cyclonic eddies and boosting marine life.

We use the L4 SST dataset [6] from the Copernicus Marine Service (CMEMS) for the Atlantic Ocean around Iberia, Biscay, Ireland (IBI), and the northwestern European shelf domain; see Fig. 1 on the left. It comprises nearly 40 years of SST satellite images from 1982 to 2020, providing a gap-free daily SST estimate, using multiple satellite observations and ensuring temporal consistency. The image resolution is about 0.05° ($\approx 5.55\ km$ per pixel).

Our study area is a subdomain within the IBI region, ranging from 19.55° to 34.525° latitude and -20.97° to -5.975° longitude, covering an area of approximately $2,462,475\ km^2$; see Fig. 1 on the right. This region is represented by a grid of 300 × 300 cells. The temporal range of the data used spans from January 1, 1982, to December 31, 2020, corresponding to a total of 14,245 frames (daily images) and a storage size of 10.25 GB.

We preprocessed the dataset to fill in missing values by first generating a binary land-sea mask, smoothed using a Gaussian filter. Then, we replaced the missing values in the continent with the average SST to maintain data continuity. This information is used as boundary conditions for our method. Figure 2 shows several samples of the L4 dataset in 2020.

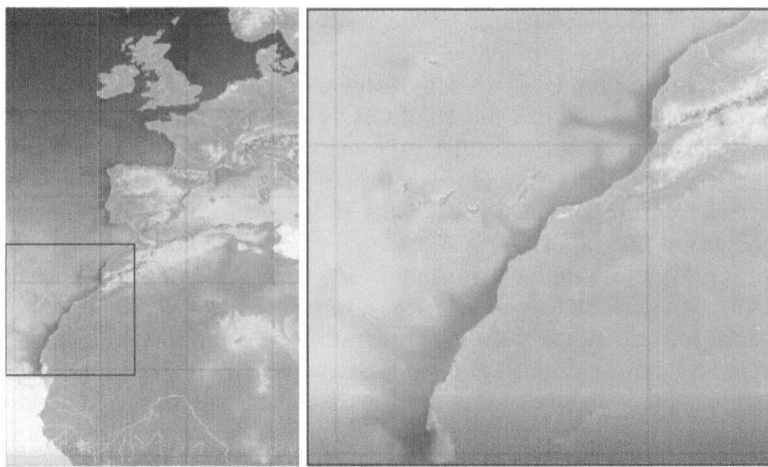

Fig. 1. Iberia, Biscay, and Ireland (IBI) region represented in the L4 dataset on the left and the subregion we use in this work on the right, comprising the Canary Islands, Madeira, and the northwestern African shore.

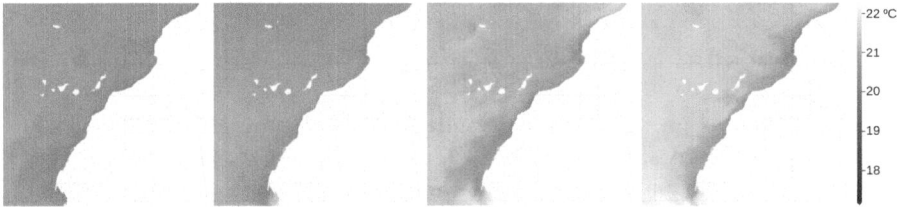

Fig. 2. Several samples of the L4 dataset. These images are examples of the SST on different days of 2020, corresponding to January, April, July, and October, respectively. Colors represent the sea surface temperature in °C, with bright colors representing higher temperatures. (Color figure online)

3   Forecasting with Graph Neural Networks

Graph Neural Networks (GNN) are an efficient mechanism to model dependencies based on the geometry of the problem. This section explains our adaptation of a GNN architecture, based on the GraphCast [9] model, for regional oceanographic forecasting. On the one hand, we limit the number of variables to the SST and replace the original graph with a planar mesh, which reduces the complexity of the model and increases the computational efficiency. On the other hand, we adjust the latent size of the features in each node to reduce memory requirements.

The input variable is a spatiotemporal volume, represented as $\mathbf{x} : \mathbb{R}^3 \to \mathbb{R}$. x_i^t is a scalar value, with i standing for node $v_i \in \mathcal{V}^g$, and t the time instant.

Our GNN is defined as an autoregressive model,

$$\hat{\mathbf{x}}^{t+1} = f(\mathbf{x}^t, \mathbf{x}^{t-1}), \tag{1}$$

where $\hat{\mathbf{x}}^{t+1}$ is estimated from two previous values and can be fed into the model to predict future states. The graph, $\mathcal{G}(\mathcal{V}^g, \mathcal{V}^m, \mathcal{E}^m, \mathcal{E}^{g2m}, \mathcal{E}^{m2g})$, is composed of grid nodes, $\mathcal{V}^g$, mesh nodes, $\mathcal{V}^m$, bidirectional edges connecting mesh nodes, $\mathcal{E}^m$, and directed edges from grid to mesh nodes, $\mathcal{E}^{g2m}$, and vice-versa, $\mathcal{E}^{m2g}$.

Grid nodes are defined as $\mathbf{v}_i^g = [x_i^t, x_i^{t-1}, \mathbf{f}_i^{t-1}, \mathbf{f}_i^t, \mathbf{f}_i^{t+1}, \mathbf{c}_i]$, i.e., a combination of current and past SST values, temporal forcings, and static properties depending on spatial coordinates. Temporal forcings depend on the local time of day and the year progress, and constants on a binary land-sea mask and the node position.

Mesh nodes are defined as $\mathbf{v}_i^m = [\cos(\phi_i), \sin(\lambda_i), \cos(\lambda_i)]$, with ϕ_i and λ_i the longitude and latitude of node i, respectively. Mesh edges are defined as $\mathbf{e}_{s,r}^m = [\text{distance}(s, r), \mathbf{p}_s - \mathbf{p}_r]$, i.e., the edge length and the difference between the spatial locations of the sender node, $mathbfp_s$, to the receiver node, $\mathbf{p}_r$. Unidirectional edges from the grid to the mesh, $\mathbf{e}_{s,r}^{g2m}$, and vice-versa, $\mathbf{e}_{s,r}^{m2g}$, are similarly defined. A mesh representation is shown in Fig. 3.

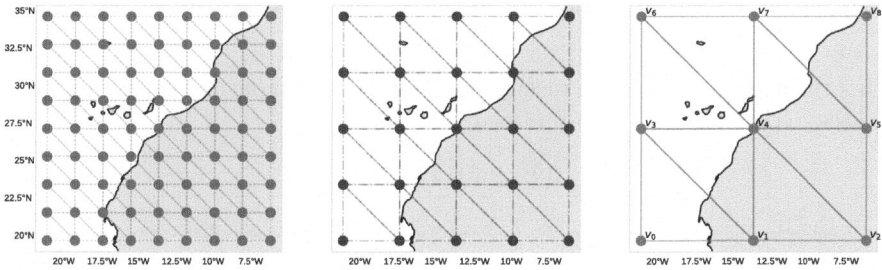

Fig. 3. The mesh nodes are organized into three distinct resolution levels: the finest level on the left (M^2, green) consists of 56 nodes, the intermediate level in the middle (M^1, blue) includes 16 nodes, and the coarsest level on the right (M^0, red) comprises 9 nodes. (Color figure online)

The architecture of the GNN is composed of an *encoder* that converts the input data from the grid to the mesh, a *processor* that comprises multiple message-passing layers, and a *decoder* that converts the output data from the mesh to the grid; see Fig. 4. The *encoder* embeds the variables, $\mathbf{v}_i^g$, $\mathbf{v}_i^m$, $\mathbf{e}_{s,r}^m$, $\mathbf{e}_{s,r}^{g2m}$ and $\mathbf{e}_{s,r}^{m2g}$, into the latent space as

$$\tilde{\mathbf{v}}_i^g = \text{MLP}_{\mathcal{V}^g}(\mathbf{v}_i^g); \quad \tilde{\mathbf{v}}_i^m = \text{MLP}_{\mathcal{V}^m}(\mathbf{v}_i^m) \tag{2}$$

$$\tilde{\mathbf{e}}_{s,r}^{g2m} = \text{MLP}_{\mathcal{E}^{g2m}}(\mathbf{e}_{s,r}^{g2m}); \quad \tilde{\mathbf{e}}_{s,r}^{m2g} = \text{MLP}_{\mathcal{E}^{m2g}}(\mathbf{e}_{s,r}^{m2g}); \quad \tilde{\mathbf{e}}_{s,r}^m = \text{MLP}_{\mathcal{E}^m}(\mathbf{e}_{s,r}^m); \tag{3}$$

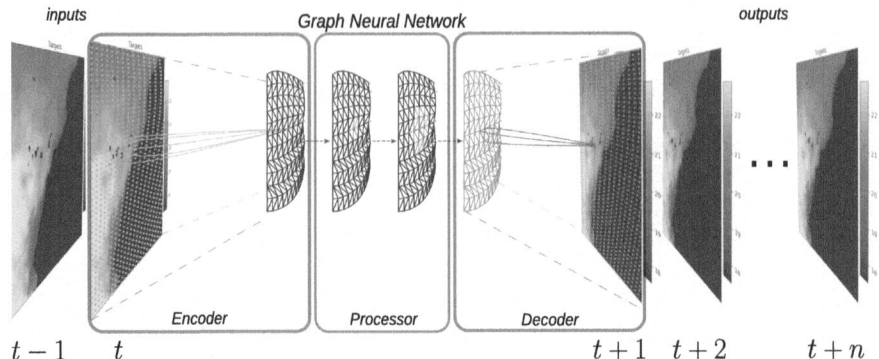

Fig. 4. Diagram of the *encoder-processor-decoder* architecture. The SST data is embedded in the mesh through the *encoder*, the *processor* transforms node features using multiple graph layers, and the *decoder* converts the internal mesh representation into a future SST forecast.

with MLP a multi-layer perceptron. The information is then transferred to the mesh using interaction networks (IN) [1], as follows:

$$\mathbf{d\tilde{e}}_{s,r}^{g2m} = \mathrm{MLP}_{\mathcal{E}^{g2m}}([\mathbf{\tilde{e}}_{s,r}^{g2m}, \mathbf{s}^g, \mathbf{r}^m]); \quad \mathbf{d\tilde{v}}_i^m = \mathrm{MLP}_{\mathcal{V}^m}([\mathbf{\tilde{v}}_i^m, \sum_{s \in \mathcal{V}^g; r=\tilde{\mathbf{v}}_i^m} \mathbf{\tilde{e}}_{s,r}^{g2m}]), \quad (4)$$

with s the sender node, r the receiver node, and [.] the concatenation of multiple features. Grid nodes are also updated as

$$\mathbf{d\tilde{v}}_i^g = \mathrm{MLP}_{\mathcal{V}^g}(\mathbf{\tilde{v}}_i^g), \tag{5}$$

In the final step of the *encoder*, we use residual connections as

$$\mathbf{\tilde{v}}_i^g \leftarrow \mathbf{\tilde{v}}_i^g + \mathbf{d\tilde{v}}_i^g; \quad \mathbf{\tilde{v}}_i^m \leftarrow \mathbf{\tilde{v}}_i^m + \mathbf{d\tilde{v}}_i^m; \quad \mathbf{\tilde{e}}_{s,r}^{g2m} \leftarrow \mathbf{\tilde{e}}_{s,r}^{g2m} + \mathbf{d\tilde{e}}_{s,r}^{g2m}. \tag{6}$$

The *processor* contains various layers with the same mesh structure, whose parameters are calculated through message-passing. The edge and node features are updated as

$$\mathbf{d\tilde{e}}_{s,r}^m = \mathrm{MLP}_{\mathcal{E}^m}([\mathbf{\tilde{e}}_{s,r}^m, \mathbf{s}^m, \mathbf{r}^m]); \quad \mathbf{d\tilde{v}}_i^m = \mathrm{MLP}_{\mathcal{V}^m}([\mathbf{\tilde{v}}_i^m, \sum_{s \in \mathcal{V}^m; r=\tilde{\mathbf{v}}_i^m} \mathbf{\tilde{e}}_{s,r}^m]), \quad (7)$$

with sender, s, and receiver, r, nodes in the mesh in this case, and they are updated with residual connections:

$$\mathbf{\tilde{v}}_i^m \leftarrow \mathbf{\tilde{v}}_i^m + \mathbf{d\tilde{v}}_i^m; \quad \mathbf{\tilde{e}}_{s,r}^m \leftarrow \mathbf{\tilde{e}}_{s,r}^m + \mathbf{d\tilde{e}}_{s,r}^m. \tag{8}$$

The *decoder* performs the inverse operation to the *encoder*. The edge and node features are calculated with the following expressions:

$$\mathbf{d\tilde{e}}_{s,r}^{m2g} = \mathrm{MLP}_{\mathcal{E}^{m2g}}([\mathbf{\tilde{e}}_{s,r}^{m2g}, \mathbf{s}^m, \mathbf{r}^g]); \quad \mathbf{d\tilde{v}}_i^g = \mathrm{MLP}_{\mathcal{V}^g}([\mathbf{\tilde{v}}_i^g, \sum_{s \in \mathcal{V}^g; r=\tilde{\mathbf{v}}_i^g} \mathbf{\tilde{de}}_{s,r}^{m2g}]). \quad (9)$$

A residual connection is used to update the information of the grid nodes coming from the embedding in the encoder $\tilde{\mathbf{v}}_i^g \leftarrow \tilde{\mathbf{v}}_i^g + \mathbf{d}\tilde{\mathbf{v}}_i^g$, and the output prediction is obtained with another MLP as $\hat{\mathbf{y}}^t = \mathrm{MLP}_{\mathcal{V}^g}(\tilde{\mathbf{v}}_i^g)$. Finally, the forecasting is obtained through a residual connection as $\hat{\mathbf{x}}^{t+1} = \mathbf{x}^t + \hat{\mathbf{y}}^t$. All input variables are normalized to zero mean and unit variance. The size of the output layer in the last MLP, corresponding to the *decoder*, is one for the prediction of the value of the SST at each node.

4   Model Configuration and Training Details

The dataset was split by years from 1982 to 2012 in the training set, from 2013 to 2016 in the validation set, and from 2017 to 2020 in the test set, with approximately a ratio of 80%-10%-10%. The training phase followed the methodology explained in [9]. It involved 150 epochs with a half-cosine learning rate decay, starting at 10^{-3}. The model was trained to predict one lead time using three-time-instant samples. We used the AdamW optimizer with $\beta_1 = 0.9$, $\beta_2 = 0.95$, $\epsilon = 10^{-8}$, and $\lambda = 0.1$. We also implemented gradient clipping when its norm was bigger than 32.

After an extensive hyperparameter search, we selected a configuration for the GNN model based on the model accuracy and our computational resources. We employed a three-level mesh as shown in Fig. 3. The processor module performed 6 message-passing steps within an 8-dimensional MLP latent space, with all MLPs containing a single hidden layer.

We used the Mean Square Error (MSE) to optimize the neural network, weighted by latitude to compensate for the spherical geometry of the grid cells. The loss function is given by:

$$\mathcal{L}_{MSE} = \frac{1}{|\mathbb{G}|} \sum_{v_i \in \mathbb{G}} w_\phi \left(\hat{x}_i^t - x_i^t \right)^2,$$

where $|\mathbb{G}|$ is the number of grid nodes, v is a node of the grid, and $w_\phi = \cos(\phi) / \left(\frac{1}{|\mathbb{G}|} \sum_{\mathbb{G}} \cos(\phi) \right)$ is a spatial varying weight that depends on the latitude ϕ. This weight decreases as we move towards the poles to account for smaller cell sizes.

We conducted experiments and hyperparameter search in a server equipped with 8 Quadro RTX 4000 GPUs (8 GB each). We used the Distributed Data-Parallel (DDP) approach for parallel training, with each GPU maintaining a copy of the model and processing independent batches, synchronizing gradients across devices through an all-reduce operation. This hybrid parallel-serial approach ensured efficient training while fully utilizing our hardware resources.

5   Results

In this section, we assess the performance of our model and compare it with state-of-the-art models. Table 1 shows the RMSE obtained for different hyper-parameters. We test the mesh resolution using three configurations (M^2, M^4, and M^6), various latent sizes, and several message passing steps in the processor. The batch size was 16 in all experiments. We observe that the RMSE is low for the three configurations; thus, it does not depend on the mesh resolution. The accuracy for M^2 and M^6 is similar, although the latter is more computationally demanding since the mesh contains exponentially more nodes. This is reasonable since the input data resolution is low, and M^2 seems to contain enough nodes to represent the input data.

Table 1. Comparison of the Root Mean Squared Error (RMSE) depending on the latent size, the number of resolution levels (M^2, M^4, and M^6), and the number of message-passing steps. Bold letters indicate the two most accurate solutions.

Configuration	Latent-Size	Message-Passing	RMSE ($^\circ C$)
GNN model (M^2)	2	2	0.0317
	4	6	0.0296
	8	7	**0.0294**
GNN model (M^4)	2	2	0.0323
	4	6	0.0301
	8	7	0.0299
GNN model (M^6)	2	2	0.0319
	4	6	**0.0293**

On the other hand, we observe an improvement when we increase the latent size and the number of message-passing steps. In this case, a latent size of eight seems the best value for an internal representation of the SST. In the work proposed in [9], the latent size was much bigger, according to the number of input variables. The difference between six and seven message-passing steps is low, so we retain the configuration with M^2, a latent size of eight, and six message-passing steps.

Figure 5 compares the predictions of our model with the ground truth values for five- to twenty-day forecasts. We observe that the error is low in short-time predictions and increases with longer predictions. As we observed in the 15- and 20-day forecasts, the method is very accurate in the deep sea and less accurate near the shore, where the temperature varies significantly.

Next, we compare our model with a state-of-the-art method based on the ConvLSTM architecture [14]. This model was trained with an input image of dimensions 256×256, and was optimized with the AdamW optimizer, a learning rate of 10^{-2}, a weight decay of 0.1, and 150 epochs.

Figure 6 shows the average RMSE of both models for 20-day forecasts. We calculated 20 predictions for each sample in the dataset and averaged the results for every lead time. The graph neural model maintains a lower error ratio in all lead times, providing more accurate solutions. On the other hand, Fig. 7 shows images of the spatial average of 5-, 10-, 15-, and 20-day forecasts for the whole

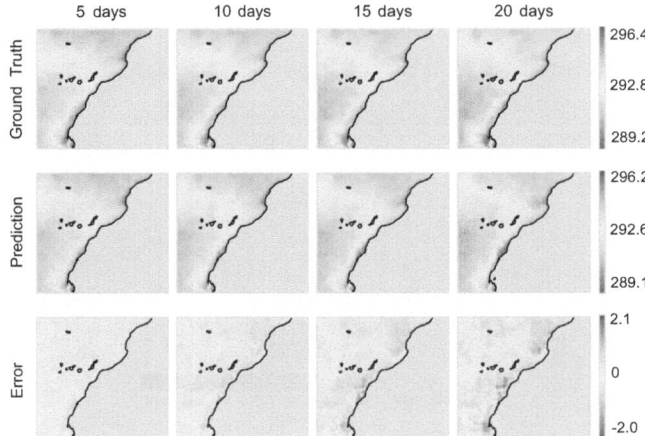

Fig. 5. Forecast comparison for several lead times. The initial date is April 5, 2020, with columns representing forecasts at 5-, 10-, 15-, and 20-day horizons, respectively. The top row shows the ground truth (GT) value for each day, the middle row shows model predictions for the same days, and the bottom row quantifies the error as the difference per pixel between the GT and the prediction.

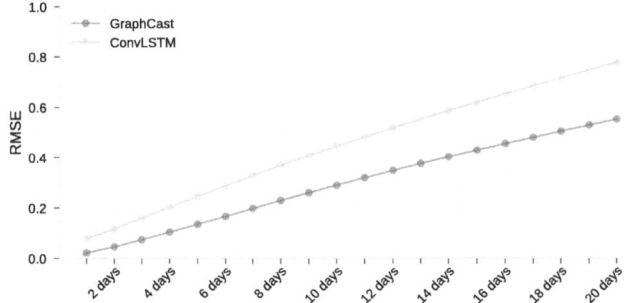

Fig. 6. Comparison of the RMSE values obtained by our model and the ConvLSTM model across a 20-day forecast horizon.

dataset. The graph model on the top clearly shows a lower average error in all the pixels of the images, meaning that the graph neural network can yield better predictions at any point in the domain. We also observe that the most significant errors are located near the coast.

An interesting observation is that, while the ConvLSTM produces smooth and continuous solutions, the graph neural network can estimate more discontinuous solutions. This allows the model to preserve small-scale variations in its predictions. We also note the presence of triangular shapes induced by the underlying mesh.

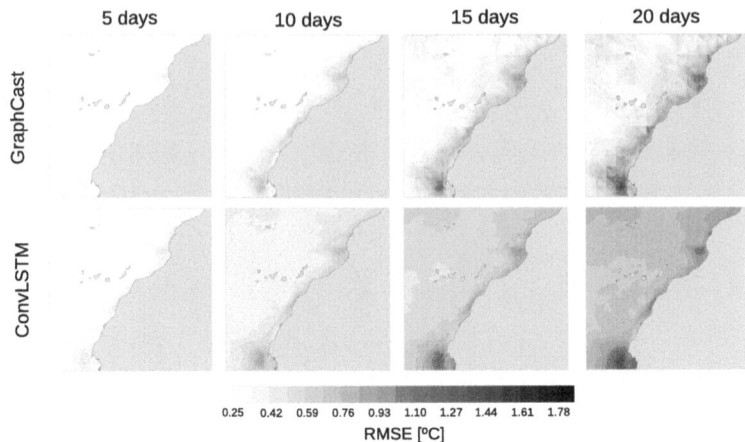

Fig. 7. Spatial distribution of the RMSE for 5-, 10-, 15-, and 20-day lead time forecasts. Results represent daily point-wise averages across the study domain. The graph model yields lower errors in most of the pixels. The highest errors are produced near the coast.

6   Conclusion

In this work, we adapted a GNN for global weather prediction to sub-regional oceanographic forecasting. We trained the model with the L4 dataset to analyze ocean dynamics in the Canary Islands and the northwest African shore. The spatial distribution of errors indicates that the model produces higher errors near the coast. These areas exhibit intense coastal interactions and complex bathymetric gradients. The lack of explicit atmospheric forcings, the bathymetry, and the coarse resolution of satellite data limit its ability to predict the SST in these regions. Our model consistently outperformed ConvLSTM in short- and long-range ocean forecasting, better capturing fine-scale ocean structures.

Some possible solutions to the triangular artifacts include refining the mesh, increasing connectivity between grid nodes, or using convolutional layers in the decoder. Addressing these limitations will be crucial for improving the model. Future work will also compare with other state-of-the-art methods.

Disclosure of Interests. The authors have no competing interests to declare that are relevant to the content of this article.

References

1. Battaglia, P.W., Pascanu, R., Lai, M., Rezende, D., Kavukcuoglu, K.: Interaction networks for learning about objects, relations and physics. Tech. rep., Cornell University (2016). https://doi.org/10.48550/arXiv.1612.00222
2. Bell, M.J., Lefèbvre, M., Traon, P.Y.L., Smith, N., Wilmer-Becker, K.: GODAE: the global ocean data assimilation experiment. Oceanography **22**(3), 14–21 (2009). https://doi.org/10.5670/oceanog.2009.62

3. Bi, K., Xie, L., Zhang, H., Chen, X., Gu, X., Tian, Q.: Accurate medium-range global weather forecasting with 3D neural networks. Nature **619**(7970), 533–538 (2023). https://doi.org/10.1038/s41586-023-06185-3

4. Bodnar, C., et al.: A foundation model of the atmosphere. Tech. rep., Cornell University (2024). https://doi.org/10.48550/arXiv.2405.13063

5. Chattopadhyay, A., Gray, M., Wu, T., Lowe, A.B., He, R.: OceanNet: a principled neural operator-based digital twin for regional oceans. Sci. Rep. **14**(1), 21181 (2024). https://doi.org/10.1038/s41598-024-72145-0

6. CMEMS: European north west Shelf/Iberia Biscay Irish seas - high resolution L4 sea surface temperature reprocessed (2024). https://doi.org/10.48670/moi-00153

7. Holmberg, D., Clementi, E., Roos, T.: Regional ocean forecasting with hierarchical graph neural networks. arXiv e-prints: arXiv:2410.11807 (2024)

8. Kochkov, D., et al.: Neural general circulation models for weather and climate. Nature **632**, 1060–1066 (2023). https://doi.org/10.1038/s41586-024-07744-y

9. Lam, R., et al.: Learning skillful medium-range global weather forecasting. Science **382**(6677), 1416–1421 (2023). https://doi.org/10.1126/science.adi2336

10. Mourre, B., et al.: Assessment of high-resolution regional ocean prediction systems using multi-platform observations: illustrations in the Western Mediterranean sea. New Front. Oper. Oceanogr. (2018). https://doi.org/10.17125/gov2018.ch24

11. Price, I., et al.: Probabilistic weather forecasting with machine learning. Nature **637**, 84–90 (2024). https://doi.org/10.1038/s41586-024-08252-9

12. Sangrà, P., et al.: The Canary eddy corridor: a major pathway for long-lived eddies in the subtropical North Atlantic. Deep Sea Res. Part I: Oceanogr. Res. Pap. **56**(12), 2100–2114 (2009). https://doi.org/10.1016/j.dsr.2009.08.008

13. Wang, X., et al.: XiHe: a data-driven model for global ocean eddy-resolving forecasting. Tech. Rep. arXiv:2402.02995, Cornell University (2024)

14. Yang, J., Zhang, T., Zhang, J., Lin, X., Wang, H., Feng, T.: A ConvLSTM nearshore water level prediction model with integrated attention mechanism. Front. Mar. Sci. **11** (2024). https://doi.org/10.3389/fmars.2024.1470320

PLRF-NMS: A Piecewise Linear Rational Function in Non-Maximum Suppression

Ivar Persson[1]([⊠]) , Håkan Ardö[2] , and Mikael Nilsson[1]

[1] Division of Computer Vision and Machine Learning, Centre for Mathematical
Sciences, Lund University, Lund, Sweden
{ivar.persson,mikael.nilsson}@math.lth.se
[2] Spiideo, Malmö, Sweden
hakan.ardo@spiideo.com

Abstract. Activation functions are fundamental components in neural
networks, enabling non-linear transformations essential for tasks like signal processing, control systems, image analysis, economics, and robotics.
They play a crucial role in facilitating processes such as noise reduction,
segmentation, and decision-making across various applications. Splines
offer an alternative approach to traditional activation functions (e.g.,
ReLU or Sigmoid), providing flexibility and adaptability to enhance function approximation. In this work a specific spline, the Piecewise Linear
Fractional Function (PLRF), is introduced and proposed as a re-scoring
mechanism for soft Non-Maximum Suppression (NMS) in object detection pipelines. The PLRF is parametrized with up to four hyperparameters within the range $(0, 1)$ and the paper presents two black-box optimization techniques, GridFib and HybridNM, to refine hyperparameters.
Experimental results on two different datasets indicate that the PLRF
achieves higher scores compared to Greedy-NMS and Soft-NMS methods.
Furthermore, the number of function evaluations needed with the proposed optimization methods reduces computational evaluations needed
relative to the Bayesian optimization technique commonly used in this
context.

Keywords: Non-Maximum Suppression · Intersection over Union ·
Object Detection

1 Introduction

Activation functions are widely used in fields beyond neural networks [15],
including signal processing, control systems, image analysis, economics, and
robotics, enabling efficient non-linear transformations across domains. In this
context, splines [2,4,19,26] offer smooth, flexible function approximations, typically realised as piecewise polynomials such as cubic- or B-splines [4].

Activation functions, or spline variants, find specific application in post-
processing object detection outputs by merging overlapping detections. Object
detection, a central task in computer vision, generates candidate detections,

© The Author(s), under exclusive license to Springer Nature Switzerland AG 2026
M. Castrillón-Santana et al. (Eds.): CAIP 2025, LNCS 15622, pp. 340–350, 2026.
https://doi.org/10.1007/978-3-032-05060-1_29

typically axis-aligned bounding boxes, paired with classification scores. Due to frequent overlapping outputs with correlated scores, Non-Maximum Suppression (NMS) [6] is widely used to refine results. NMS prunes detections through a re-scoring function based on mutual overlaps, analogous to an activation function. It remains a core component in widely adopted object detection systems, such as different generations of YOLO [20], Mask-RCNN [10] and Swin Transformer [17].

The standard approach for Non-Maximum Suppression (NMS) in object detection remains Greedy-NMS, introduced by Dalal and Triggs [6], which removes detections exceeding a predefined overlap threshold relative to a higher-scoring candidate. Greedy-NMS has been widely adopted across major detection frameworks [7,10,20]. Several variants have been proposed to improve its perfor-mance, notably Soft-NMS [3], which decays scores rather than removing detec-tions. Other approaches include Fitness NMS [25], convolutional and learned NMS methods [11,12], IoU-Net [13], and recent speed-focused NMS variants [22].

In this work, we revisit Non-Maximum Suppression (NMS) and Soft-NMS [3] for object detection, and extend the concept of Intersection over Union (IoU) from bounding boxes to circles for ground-plane-based person detection. We introduce the Piecewise Linear Rational Function (PLRF), a re-scoring func-tion with four tunable hyperparameters, for use within Soft-NMS, denoted as PLRF-NMS. We further propose black-box optimization strategies to fine-tune the PLRF parameters. For multiple hyperparameters, a combination of grid search and the Nelder-Mead method [18] and for single-parameter tuning, a two-level grid search followed by Fibonacci search [8,14,23]. The effective-ness of PLRF-NMS is demonstrated both for object detection on the COCO dataset [16] and ground-plane-based person detection on the Spiideo SoccerNet SynLoc dataset [1].

2   Non-Maximum Suppression

Non-Maximum Suppression (NMS) operates on a set of N detections $\mathcal{D} = \{d_1, \ldots, d_N\}$ with corresponding scores $\mathcal{S} = \{s_1, \ldots, s_N\}$, aiming to produce a refined set of M detections $\mathcal{E} = \{e_1, \ldots, e_M\}$ and associated scores $\mathcal{T} = \{t_1, \ldots, t_M\}$, reducing false positives through overlap pruning. Detection scores are assumed non-negative ($s_i \geq 0$), where zero indicates absence of a valid detec-tion. Each detection shape d_i may represent a bounding box, parametrised by $(x_{i1}, y_{i1}, x_{i2}, y_{i2})$, or a circle, parametrised by (x_i, y_i, r_i). Other parametrised shapes can also be considered. Given a method to compute Intersection over Union (IoU) between two detections d_k and d_i, one can for example use the greedy re-scoring function

$$f(\text{IoU}(d_k, d_i)) = \begin{cases} 1, & \text{IoU}(d_k, d_i) < \tau_{\text{IoU}} \\ 0, & \text{IoU}(d_k, d_i) \geq \tau_{\text{IoU}} \end{cases} \quad (1)$$

An update to all $i = 1, 2, \ldots, N$ detection scores as $s_i \leftarrow s_i f(\text{IoU}(d_k, d_i))$ can effectively remove all detections that overlaps d_k with an IoU greater than the

---

Algorithm 1: The pseudo code for Non-Maximum Suppression (NMS) with a general re-scoring function f.

Input   : List of detection shapes $\mathcal{D} = \{d_1, d_2, \ldots, d_N\}$, corresponding detection scores $\mathcal{S} = \{s_1, s_2, \ldots, s_N\}$, a detection threshold $\tau_{detection}$ and hyperparameters related to the the re-scoring function (e.g. τ_{IoU} in Greedy-NMS, see Eq. (1)).

Output: Filtered resulting M detection shapes in list $\mathcal{E}$ and corresponding detection scores $\mathcal{T}$

begin
    $\mathcal{R} \leftarrow \{\}$
    $j \leftarrow 0$
    while v_{flag} **do**
        $k \leftarrow \operatorname{argmax} \mathcal{S}$
        if $s_k < \tau_{detection}$ **then**
         | **break**
        end
        $j \leftarrow j + 1$
        $t_j \leftarrow s_k$
        $\mathcal{E} \leftarrow \mathcal{E} \cup d_k$
        $\mathcal{T} \leftarrow \mathcal{T} \cup t_j$
        $s_k \leftarrow 0$
        for d_i **in** $\mathcal{D}$ **do**
         | $s_i \leftarrow s_i f(\text{IoU}(d_k, d_i))$
        end
    end
    Return $\mathcal{E}, \mathcal{T}$
end

---

threshold τ_{IoU} by setting detection scores to zero. Given a re-scoring function, see e.g. Eq. (1), a chosen score threshold of accepting a detection $\tau_{\text{detection}}$ and a hyperparameter τ_{IoU} related to the a threshold on IoU in the re-scoring function, see τ_{IoU} in Eq. (1), a Greedy-NMS algorithm can be formed, see Algorithm 1.

While a Greedy-NMS can be formed using Eq. (1) as the re-scoring function, other re-scoring functions were proposed in Soft-NMS [3]. Essentially, instead of using Eq. (1) in Algorithm 1, a soft activation function

$$f(\text{IoU}(d_k, d_i)) = e^{-\frac{\text{IoU}(d_k, d_i)^2}{\sigma}} \tag{2}$$

with the hyperparameter σ was the proposed re-scoring function in Soft-NMS [3]. When there is no overlap, i.e. IoU $= 0$ the value of Eq. (2) will be 1, however, the value for complete overlap, i.e. IoU $= 1$ will depend on the value of σ. This Soft-NMS will still be a greedy algorithm that does not find the globally optimal re-scoring of detection boxes, but it generalizes NMS, with greedy NMS being a special case using a binary weighting function (Eq. (1)). The Soft-NMS paper [3] explicitly mentions that other functions with more parameters could also be considered, which is what we will explore in this paper.

Intersection over Union, $\text{IoU}(d_k, d_i)$, plays a crucial role in NMS, where it helps eliminate redundant detected objects, often bounding boxes, by selecting the most confident detections. We will explore the default bounding boxes but also how to utilize IoU from circles.

2.1 Intersection over Union

The most commonly occurring form in object detection is the bounding box, it can be found in many works [6,7,10,20], but a short revisit is made here for completeness. Detections in form of bounding boxes can be used to calculate the IoU. The IoU is simply defined as the fraction between the intersection between the boxes and their union.

Although less common, circular representations for object detection have been studied [28,29]. We adopt this approach for ground-plane person detection, where, from an overhead view, a person's footprint is approximately circular, mitigating the directional bias of rectangles. To compute IoU between circles, the distance between centers is evaluated: if it exceeds the sum of radii, IoU is zero; if it is within the absolute difference, the IoU is the ratio of areas. Otherwise, the intersection is computed by summing sector areas and subtracting the associated triangular segments.

3   Piecewise Linear Rational Function

This section will propose and explore a four parameter function for use as a re-scoring function in a Soft-NMS variant. Here the input x is considered in range $[0, 1]$, or given previous sections, $x = 1 - \text{IoU}(d_k, d_i)$. Note that we want to keep the detection score if there is no overlap and remove it (set it to zero) if IoU is one. By using $1 - \text{IoU}$, the activation function can be formulated as monotonic increasing rather than decreasing. In the present setup, we require the activation function $y = f(x)$ to satisfy the following properties:

- The function operate on the domain $x \in [0, 1]$, since $1 - \text{IoU}$ is in this range.
- The function has a codomain of $y \in [0, 1]$, i.e. keep detection score unchanged (y=1), remove detection score (y=0) or partially suppress the detection.
- The function passes $(x, y) = (0, 0)$, since complete overlap ($x = 0$) should remove the detection ($y = 0$).
- The function passes $(x, y) = (1, 1)$, since zero IoU ($x = 1$) should keep the detection unchanged ($y = 1$).

In addition to those properties above, which are directly relatable to the NMS context, we introduce some additional design idea around forming the function:

- It is desirable to realize this function as a rational function only using addition, subtraction, multiplication, and division for simplicity and speed.
- One point is defined (p_x, p_y), where $p_x, p_y \in [0, 1]$, that the function passes through and acts as the break-point between two function pieces.

– The function should interpolate between different forms, similar to Soft-NMS [3], ranging from a linear function (Soft-NMS linear variant [3]) at one end to a step function, or Heaviside function, corresponding to standard NMS at the other.
– The function consist of two almost independent parts, with the only connection being the break-point (p_x, p_y).
– The interpolation between a linear and step function should form some form of smooth step function, or a sigmoid-like function, with a single interpolation hyperparameter for each part, $\alpha_l \in [0, 1]$ for $x < p_x$ and $\alpha_u \in [0, 1]$ for $x > p_x$. The α subscript l indicating lower and u the upper part of the function.

To satisfy the outlined design criteria, we employ a piecewise linear rational function to interpolate between linear and step-function behaviour. The function is parametrised by four hyperparameters; p_x, p_y, which define the breakpoint, and α_l, α_u, which control the shape in the respective upper and lower regions. The function is composed of two segments: one between $(0,0)$ and (p_x, p_y), and the other between (p_x, p_y) and $(1, 1)$, with separate interpolation characteristics. The proposed activation function (or re-scoring function in the NMS context) that adheres to the desired properties listed above is formed using

$$f_{\text{part}}(z, \alpha) = \frac{\alpha z - z}{2\alpha z - \alpha - 1} \tag{3}$$

which is used in this proposed Piecewise Linear Rational Function (PLRF)

$$f(x) = \begin{cases} f_{\text{part}}(\frac{x}{p_x}, \alpha_l)p_y, & x < p_x \\ 1 - f_{\text{part}}(\frac{1-x}{1-p_x}, \alpha_u)(1 - p_y), & \text{else} \end{cases} . \tag{4}$$

Please note that this has four hyperparameters $p_x, p_y, \alpha_l, \alpha_u \in (0, 1)$, hence it is a convenient bounded parameter space.

We consider the default values for the four parameters to all be equal to 0.5, see Fig. 1A. Variants of the PLRF that fix some of the parameters can also be considered. For example, a one variable version we will consider is if $p_x = p_y = 0.5$ and the same interpolation factor is used, i.e. $\alpha = \alpha_l = \alpha_u$, see examples in Fig. 1B. We will denote this setup PLRF_α and the four parameter setup PLRF. More examples of non-symmetric parameter setups with two trainable parameters are in Fig. 1C and D, three parameters or all four parameters in Fig. 1E and F. The choice of reducing the number of tunable parameters can be user-defined depending on the optimization task. Note that the PLRF function is continuous but may not be differentiable. One important setup of the PLRF, used later in the paper, is to consider the case of setting $p_x = \tau_{\text{IoU}}$ with $\alpha_u = \alpha_l = 0.999, p_y = 0.5$. This is a close approximation to the original Greedy-NMS weight function, Eq. (1).

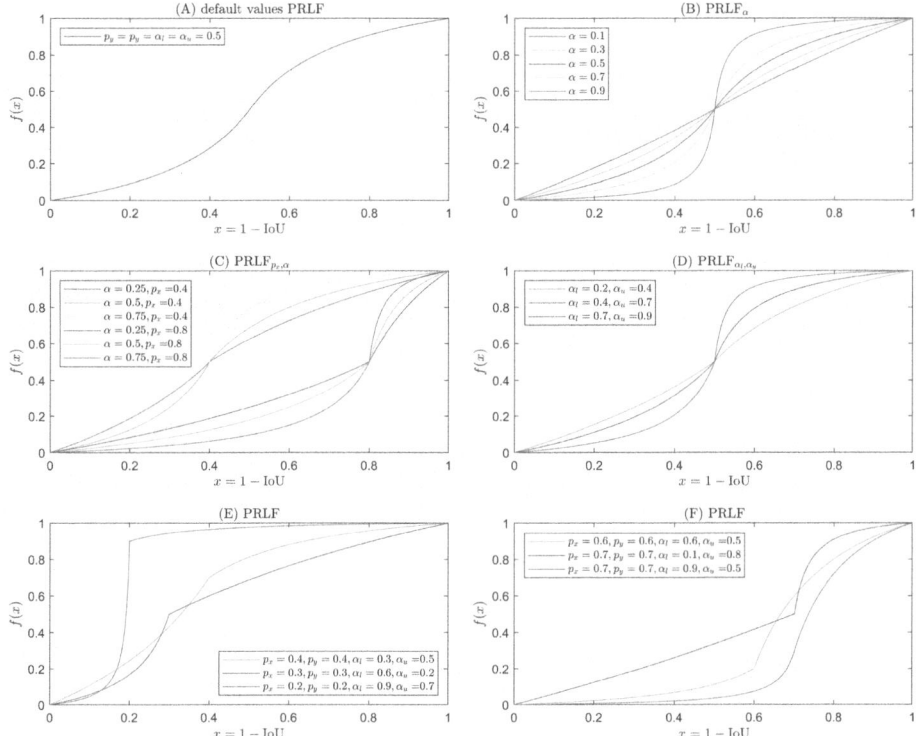

Fig. 1. Examples of the Piecewise Linear Rational Function (PLRF) with different parameter choices.

4   Black-Box Optimization of PLRF Parameters

Various black-box Optimization choices could be considered for finding suitable hyperparameters for the PLRF if one considers it used for object detection. In the following subchapters we will describe the black-box methods we use in this work, including the Neural Pipeline Search (NePS) library [24]. But, in order to possibly speed up the optimization, and also with aim to find a more optimal parameter setup, we will describe our own optimization variants for one or multiple parameters, denoted GridFib and HybridNM, respectively. These are inspired by our knowledge by the hyperparameter space for the PLRF and the object detection context, and the idea is that they will be more effective than black-box models with no prior knowledge.

4.1   Neural Pipeline Search

One hyperparameter optimization strategy is Bayesian optimization. We use the Neural Pipline Search (NePS) library [24] to search for and optimize the parameters in the given parameter space.

4.2   GridFib

We propose a method, GridFib, for single hyperparameter tuning (e.g. for PLRF_α). This is a combination between a grid search on two levels and Fibonacci search [8,14,23]. The parameter space for the PLRF_α in one dimension is the open interval $(0, 1)$. The procedure involves an initial grid search with nine function evaluations, followed by a finer grid search with eight additional function evaluations and finally we assume the function is locally unimodal and perform a Fibonacci search, see Algorithm 2. The unimodal property may not hold, so we track the minimum across all Fibonacci search evaluations and select it instead of assuming the last point is optimal.

Algorithm 2: GridFib: Hybrid Grid and Fibonacci Search

Input: Function $f(x)$, number of Fibonacci iterations N_{fib} and an small positive value ϵ
Output: Estimated optimal point x_{opt}.

begin
    $S_1 \leftarrow \{0.1, 0.2, \ldots, 0.9\}$
    $x_1^* \leftarrow \text{argmin}_{x \in S_1} f(x)$
    $S_2 \leftarrow \{x_1^* - 0.08, x_1^* - 0.06, x_1^* - 0.04, x_1^* - 0.02, x_1^* + 0.02, x_1^* + 0.04, x_1^* + 0.06, x_1^* + 0.08\}$
    $x_2^* \leftarrow \text{argmin}_{x \in S_2} f(x)$
    $a \leftarrow x_2^* - 0.02 + \epsilon$
    $b \leftarrow x_2^* + 0.02 - \epsilon$
    Setup for Fibonacci search in range $[a, b]$
    for $k \leftarrow 1$ **to** N_{fib} **do**
       Perform Fibonacci search step and evaluate $f(x_k)$
       Store x_k in S_3
    end
    $x_{\text{opt}} \leftarrow \text{argmin}_{x_k \in S_3} f(x_k)$
    return x_{opt}
end

Algorithm 3: Hybrid Nelder-Mead Optimization

Input: Function $f(\mathbf{x})$ including penalty if $\mathbf{x} \notin (0, 1)$, Nelder-Mead stopping criteria ϵ_{NM}
            and number of Fibonacci iterations N_{fib}
Output: Optimal point $\mathbf{x}_{\text{opt}}$.

begin
    $x_{opt} = \text{GridFib search on PLRF}_{p_x}$ with $\alpha_1 = \alpha_u = 0.99$ and $p_y = 0.5$
    $\alpha_{opt} = \text{GridFib search on PLRF}_\alpha$ with $p_x = x_{opt}$ and $p_y = 0.5$
    $\mathbf{x}_0 \leftarrow (x_{opt}, 0.5, \alpha_{opt}, \alpha_{opt})$
    $k \leftarrow 0$ and store $(x_k, f(x_k))$
    Perform Nelder-Mead optimization step to get $\mathbf{x}_1$
    $k \leftarrow 1$ and store $(x_k, f(x_k))$
    while $|f(\mathbf{x}_k) - f(\mathbf{x}_{k-1})| > \epsilon_{NM}$ **do**
       Perform Nelder-Mead optimization step and $k \leftarrow k + 1$
       Store $(x_k, f(x_k))$
    end
    $\mathbf{x}_{\text{opt}} \leftarrow \text{argmin}_{\mathbf{x}_k} f(\mathbf{x}_k)$
    return $\mathbf{x}_{\text{opt}}$
end

4.3   HybridNM

In the multi parameter case we use the Nelder-Mead method [18], with strategic initialization. The same principle applies to optimization with two, three, or four hyperparameters, but we focus on the four-parameter case in this discussion (i.e. for PLRF). Generally, the initialization of Nelder-Mead could be crucial [27]. For this reason we consider two steps of the GridFib algorithm before the Nelder-Mead optimization, to find a suitable initialization. The initial GridFib sweep finds a suitable p_x for $\alpha_l = \alpha_u = 0.999$, i.e. a greedy-NMS approximation. Next, given the p_x, $\alpha = \alpha_l = \alpha_u$ is estimated to give the initial point for Nelder-Mead. An added penalty is added to the function during Nelder-Mead if the $(0, 1)$ bounds are not met for all parameters. Given the initial GridFib sweeps we can with certainty know that any solution from HybridNM will be as good as, or better than Greedy-NMS.

5   Experiments and Results

We conduct two different experiments, one using the object detection dataset COCO [16] and the other experiment on the Spiideo SoccerNet SynLoc [1] dataset. On the COCO dataset we run detections with MMDetection [5] with NMS turned off in order for us to apply different NMS methods, including our PLRF-NMS.

Table 1. Results on COCO dataset with different detection algorithms, mAP metric reported for all test-setups. Greedy-NMS results are from using network and hyperparameters from MMDetection [5]. Soft-NMS and PLRF$_\alpha$ were optimized with GridFib and PLRF was optimized with HybridNM

	Greedy-NMS [6]	Soft-NMS [3]	PLRF$_\alpha$	PLRF
YOLOX-S [9]	39.9	39.9	40.0	40.3
Faster-RCNN [21]	37.2	38.1	37.9	38.2

We notice that PLRF achieves a higher mAP score for both detection algorithms, while PLRF$_\alpha$ performs on par with Soft-NMS. Please note that the mAP values in Table 1 may be different from the original articles, given that the parameters from MMDetection [5] are used. The numerical values may be different for different versions of the MMDetection model.

For the second experiment, we apply NMS to detections produced on the Spiideo SoccerNet SynLoc dataset. Each soccer player is modelled as a circle with a 1 m radius on the ground plane. Evaluation follows the F1-score and Frame Accuracy metrics, as defined in [1], where Frame Accuracy measures the proportion of images with perfect predictions, i.e., no false positives or negatives and detections within 0.48 m.

Table 2. Results when optimizing with F1-score as optimizing factor. The columns depicts number of tunable parameters, number of function evaluations, and also the F1-score and Frame accuracy.

	# params	# f-eval	F1	Frame acc
NePS Soft-NMS	1	50	0.955	0.555
NePS PLRF$_\alpha$	1	50	0.959	0.560
NePS PLRF	4	500	0.959	0.557
GridFib PLRF$_\alpha$	1	28	0.959	0.559
HybridNM PLRF	4	192	0.959	0.560

The optimization is done using both our own implemented GridFib and HybridNM, but also NePS. For GridFib and HybridNM we use $N_{fib} = 11$ and $\epsilon_{NM} = 0.0001$. We use NePS to optimize the σ hyperparameter for Soft-NMS, see Eq. (2), α in PLRF$_\alpha$ and all four parameters in PLRF. Then we use GridFib on PLRF$_\alpha$ and HybridNM on the four hyperparameter PLRF. Results can be found in Table 2.

When comparing the results in Table 2 we can see that PLRF$_\alpha$ has a slightly higher F1-sore than Soft-NMS in these tests, showing the viability of the proposed rational function in one parameter. Comparing function evaluations we can also see that utilizing GridFib achieves similar results with fewer function evaluations than Bayesian optimization. When optimizing on the F1-score we note that the four hyperparameter PLRF performs on par with the single parameter PLRF$_\alpha$. The HybridNM PLRF does the optimization in fewer function evaluations than NePS.

6   Conclusions

We have reviewed NMS, Soft-NMS, and IoU for bounding boxes and circles, proposing the Piecewise Linear Rational Function (PLRF) with up to four hyperparameters. We introduced two black-box optimization schemes, GridFib and HybridNM, for PLRF tuning.

PLRF was tested on both the COCO dataset and Spiideo SoccerNet SynLoc, where results indicate greater or on-par performance. PLRF$_\alpha$ (and PLRF) has greater performance than Soft-NMS in F1-score on Spiideo SoccerNet SynLoc, while GridFib and HybridNM achieves these results with fewer evaluations than NePS. The four-parameter PLRF matches the single-parameter PLRF$_\alpha$ on Spiideo SoccerNet SynLoc, and HybridNM optimizes more efficiently than NePS. When considering COCO, PLRF$_\alpha$ does performs on par with greedy-NMS, suggesting that multiple parameters are needed in some cases. With an F1-score of 0.959, performance may be near a soft ceiling on the Spiideo dataset for the used detection algorithm.

In summary, our results show that modest performance improvements can be achieved over Soft-NMS by refining the choice of soft thresholding functions, including the use of additional parameters such as the PLRF proposed in this study. Furthermore, we demonstrate that it is possible to design a task specific black box optimization method for parameter findings, HybridNM, which achieves greater computational efficiency, measured in terms of fewer function evaluations, compared to general approaches like NePS.

References

1. Ardö, H., et al.: Spiideo soccernet synloc - single frame world coordinate athlete detection and localization with synthetic data. In: In Proceedings of the 20th International Joint Conference on Computer Vision, Imaging and Computer Graphics Theory and Applications - Volume 2: VISAPP, pp. 278–285. INSTICC, SciTePress (2025)
2. Bartels, R.H., Beatty, J.C., Barsky, B.A.: An Introduction to Splines for Use in Computer Graphics & Geometric Modeling. Morgan Kaufmann Publishers Inc., San Francisco (1987)
3. Bodla, N., Singh, B., Chellappa, R., Davis, L.S.: Soft-NMS — improving object detection with one line of code. In: 2017 IEEE International Conference on Computer Vision (ICCV), pp. 5562–5570 (2017). https://doi.org/10.1109/ICCV.2017.593
4. de Boor, C.: A Practical Guide to Splines. Springer, revised edn. (2001)
5. Chen, K., et al.: MMDetection: open MMLab detection toolbox and benchmark (2019)
6. Dalal, N., Triggs, B.: Histograms of oriented gradients for human detection. In: 2005 IEEE Computer Society Conference on Computer Vision and Pattern Recognition (CVPR'05), vol. 1, pp. 886–893 (2005)
7. Felzenszwalb, P.F., Girshick, R.B., McAllester, D., Ramanan, D.: Object detection with discriminatively trained part-based models. IEEE Trans. Pattern Anal. Mach. Intell. 32(9), 1627–1645 (2010)
8. Ferguson, D.E.: Fibonaccian searching. Commun. ACM 3(12), 648 (1960). https://doi.org/10.1145/367487.367496
9. Ge, Z., Liu, S., Wang, F., Li, Z., Sun, J.: YOLOX: exceeding yolo series in 2021 (2021)
10. He, K., Gkioxari, G., Dollár, P., Girshick, R.: Mask R-CNN. In: 2017 IEEE International Conference on Computer Vision (ICCV), pp. 2980–2988 (2017)
11. Hosang, J., Benenson, R., Schiele, B.: A convnet for non-maximum suppression. In: Rosenhahn, B., Andres, B. (eds.) Pattern Recognition, pp. 192–204. Springer International Publishing, Cham (2016)
12. Hosang, J., Benenson, R., Schiele, B.: Learning non-maximum suppression. In: 2017 IEEE Conference on Computer Vision and Pattern Recognition (CVPR), pp. 6469–6477 (2017). https://doi.org/10.1109/CVPR.2017.685
13. Jiang, B., Luo, R., Mao, J., Xiao, T., Jiang, Y.: Acquisition of localization confidence for accurate object detection. In: Ferrari, V., Hebert, M., Sminchisescu, C., Weiss, Y. (eds.) Computer Vision - ECCV 2018, pp. 816–832. Springer International Publishing, Cham (2018)
14. Kiefer, J.: Sequential minimax search for a maximum. Proc. Am. Math. Soc. 4, 502–506 (1953)
15. Kunc, V., Kléma, J.: Three decades of activations: a comprehensive survey of 400 activation functions for neural networks (2024)

16. Lin, T., et al.: Microsoft COCO: common objects in context. CoRR **abs/1405.0312** (2014)
17. Liu, Z., et al.: Swin transformer: hierarchical vision transformer using shifted windows. In: 2021 IEEE/CVF International Conference on Computer Vision (ICCV), pp. 9992–10002 (2021)
18. Nelder, J.A., Mead, R.: A simplex method for function minimization. Comput. J. **7**(4), 308–313 (1965)
19. Perperoglou, A., Sauerbrei, W., Abrahamowicz, M., Schmid, M.: A review of spline function procedures in R. BMC Med. Res. Methodol. **19** (03 2019)
20. Redmon, J., Divvala, S., Girshick, R., Farhadi, A.: You only look once: unified, real-time object detection. In: 2016 IEEE Conference on Computer Vision and Pattern Recognition (CVPR), pp. 779–788. IEEE Computer Society, Los Alamitos (2016)
21. Ren, S., He, K., Girshick, R., Sun, J.: Faster R-CNN: towards real-time object detection with region proposal networks. In: Cortes, C., Lawrence, N., Lee, D., Sugiyama, M., Garnett, R. (eds.) Advances in Neural Information Processing Systems, vol. 28. Curran Associates, Inc. (2015)
22. Si, K.S., et al.: Accelerating non-maximum suppression: a graph theory perspective. In: Globerson, A., Mackey, L., Belgrave, D., Fan, A., Paquet, U., Tomczak, J., Zhang, C. (eds.) Advances in Neural Information Processing Systems, vol. 37, pp. 121992–122028. Curran Associates, Inc. (2024)
23. Snider, A.D.: Fibonacci Search, pp. 1–11. Springer International Publishing, Cham (2023). https://doi.org/10.1007/978-3-031-29219-4_1
24. Stoll, D., et al.: Neural pipeline search (NePS) (2023). https://github.com/automl/neps
25. Tychsen-Smith, L., Petersson, L.: Improving object localization with fitness NMS and bounded IoU loss. In: 2018 IEEE/CVF Conference on Computer Vision and Pattern Recognition, pp. 6877–6885 (2018)
26. Wahba, G.: Spline Models for Observational Data. Society for Industrial and Applied Mathematics, Philadelphia (1990)
27. Wessing, S.: Proper initialization is crucial for the Nelder–Mead simplex search. Optim. Lett. **13**(4), 847–856 (2019)
28. Yang, X., Li, C., Ruan, R., Liu, L., Chao, W., Luo, B.: EllipseIoU: a general metric for aerial object detection. In: Pattern Recognition and Computer Vision: 5th Chinese Conference. PRCV 2022, Shenzhen, China, November 4–7 2022, Proceedings, Part III, pp. 537–550. Springer-Verlag, Berlin, Heidelberg (2022)
29. Zhang, H., Liang, P., Sun, Z., Song, B., Cheng, E.: CircleFormer: circular nuclei detection in whole slide images with circle queries and attention. In: Greenspan, H., et al. (eds.) Medical Image Computing and Computer Assisted Intervention - MICCAI 2023, pp. 493–502. Springer Nature Switzerland, Cham (2023)

R3ST: A Synthetic 3D Dataset with Realistic Trajectories

Simone Teglia[1]([⊠]), Claudia Melis Tonti[1], Francesco Pro[1], Leonardo Russo[1], Andrea Alfarano[2], Matteo Pentassuglia[3], and Irene Amerini[1]

[1] Sapienza University of Rome, Rome, Italy
{teglia,melistonti,pro,russo,amerini}@diag.uniroma1.it,
russo.2015563@studenti.uniroma1.it
[2] INSAIT, Sofia University, Sofia, Bulgaria
andrea.alfarano@insait.ai
[3] EURECOM, Biot, France
matteo.pentassuglia@eurecom.fr

Abstract. Datasets are essential to train and evaluate computer vision models used for traffic analysis and to enhance road safety. Existing real datasets fit real-world scenarios, capturing authentic road object behaviors, however, they typically lack precise ground-truth annotations. In contrast, synthetic datasets play a crucial role, allowing for the annotation of a large number of frames without additional costs or extra time. However, a general drawback of synthetic datasets is the lack of realistic vehicle motion, since trajectories are generated using AI models or rule-based systems. In this work, we introduce **R3ST** (Realistic 3D Synthetic Trajectories), a synthetic dataset that overcomes this limitation by generating a synthetic 3D environment and integrating real-world trajectories derived from SinD, a bird's-eye-view dataset recorded from drone footage. The proposed dataset closes the gap between synthetic data and realistic trajectories, advancing the research in trajectory forecasting of road vehicles, offering both accurate multimodal ground-truth annotations and authentic human-driven vehicle trajectories. We publicly release our dataset here (https://r3st-website.vercel.app/).

Keywords: Synthetic Dataset · Traffic Analysis · Trajectory Forecasting · Computer Vision

1 Introduction

Understanding and studying complex urban traffic scenarios is vital for traffic analysis, urban planning, and enhancing road safety. In particular, urban intersections poses significant challenges for Traffic Monitoring Systems (TMS) due to their complex layouts and dynamic traffic conditions, as they are the most accident-prone locations in urban environments. A significant hurdle in achieving this safety is the need for a comprehensive understanding of complex

traffic environments and the precise prediction of other traffic participants movements. The advancement of predicting the motion of road agents is therefore
vital for developing advanced traffic monitoring systems. The analysis of traffic dynamics requires extensive datasets that effectively integrate the behavior
of actual traffic participants with comprehensive annotations. Previous works
like [13,15,16,20] either failed to address the inherently sophisticated nature
of intersections, proposing datasets that contain real trajectories but lacking
the necessary realism of an intersection scenario, or presented HD-Maps based
datasets, that are not always suitable for intelligent TMS due to the absence of
such precise data in real time.

Fig. 1. The R3ST Dataset. It provides photorealistic synthetic images (first), depth
maps (second), instance segmentation masks (third), and object detection bounding
boxes (fourth), enabling diverse computer vision tasks. Each task is presented on a
different frame.

The scarcity of a comprehensive dataset reflecting real road agent behavior
poses serious challenges in creating precise and universally applicable motion
prediction models, thereby constraining the efficacy of intelligent traffic monitoring in practical situations. The behavior of drivers can noticeably deviate
from the expected patterns due to the intrinsic unpredictability of their intentions, bad habits, poor road design, or high traffic levels. Those factors make
simulation-based traffic models insufficient to completely capture the full spectrum of human responses to varying driving conditions, despite the considerable
advances of simulation environments. To address these limitations and bridge the
gap between realism and precise ground-truth annotations, we introduce R3ST
(Realistic 3D Synthetic Trajectories), a novel large-scale synthetic dataset that

uniquely integrates real, human-driven trajectories derived from actual aerial drone footage (the SinD dataset [28]) into a constructed digital twin simulation (Fig. 1). By leveraging real-world vehicle trajectories, R3ST captures authentic traffic behaviors while providing comprehensive multimodal annotations, such as instance segmentation, depth maps, and bounding boxes, facilitated by photorealistic rendering techniques and advanced annotation tools like Vision Blender [4].

Specifically, R3ST features diverse urban intersections modeled and rendered within realistic environments created with Blender [6]. Unlike conventional synthetic datasets relying solely on rule-based or AI-generated vehicle motion, R3ST's vehicles follow real-world annotated trajectories, reflecting genuine human decision-making dynamics. Furthermore, multiple camera angles, realistic sensor parameters, and high-quality rendering settings ensure that R3ST closely mimics actual road-camera images, significantly enhancing the dataset's applicability and value for diverse computer vision tasks. Additionally, we leverage R3ST to evaluate pre-trained deep-learning models focusing on critical tasks like vehicle detection, instance segmentation, and monocular depth estimation. Through the innovative fusion of realistic human-driven trajectories and precise synthetic data generation, R3ST represents an advancement over previous datasets, offering researchers optimal capabilities for evaluating and training autonomous driving models, trajectory forecasting algorithms, and traffic analysis tools. The dataset presented in this paper offers critical foundations for future research in Intelligent Transportation Systems and contribute directly to safer, smarter urban mobility.

2   Related Work

Datasets are essential for developing and optimizing Intelligent Traffic Monitoring Systems, particularly in complex urban scenarios such as intersections. These environments pose unique challenges due to the unpredictable behavior of road users, varying lighting conditions, and the need for real-time decision-making. To address these problems, a variety of datasets have been proposed, ranging from real-world to synthetic data. In this section, we review existing datasets used for road agents' perception and motion prediction, highlighting their strengths and weaknesses. To the best of our knowledge, no existing work has explored synthetic datasets that integrate real trajectories.

Real-World Datasets. Real-world datasets are collected from sensors like cameras, LiDAR, or GPS, that are placed in real-world traffic situations. Earlier works in the field of urban scene understanding like KITTI [10], CityScapes [7], SemanticKITTI [3], nuScenes [25], the Waymo Open Dataset [20] and Argoverse 2 [27] provided complex scenarios allowing tasks like object detection, semantic segmentation and trajectory prediction, but from an on-board Point Of View (POV) or relying on HD-Maps. UAVDT [9], highD [15], DroneVehicle [23], and SinD [28] instead contain images recorded by drones, giving a new point of

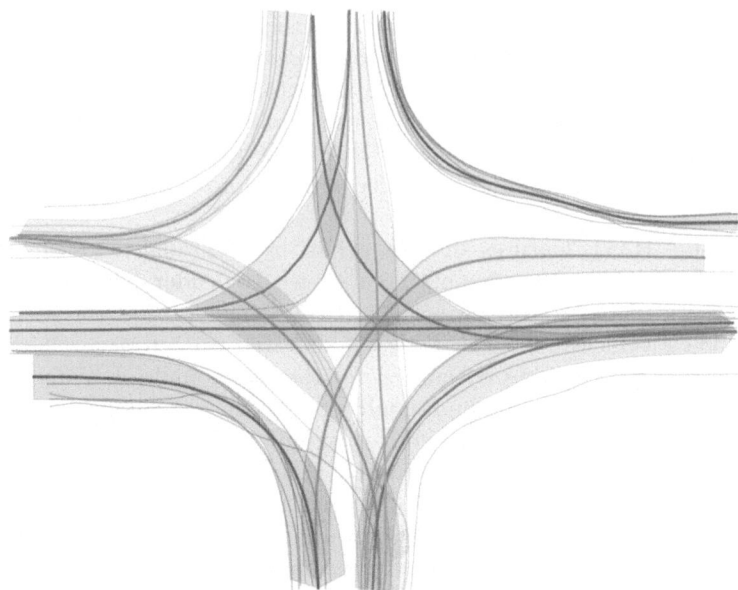

Fig. 2. Visualization of clustered vehicle trajectories in a R3ST crossroad. Each colored trajectory represents a cluster of similar paths, while the shaded regions indicate variance within each cluster.

view with respect to previous datasets, addressing object detection and object tracking. Notable effort have been made by TUMTraf [30], which contains 4.8k humanly annotated frames captured by two LiDAR and two cameras pointed at an intersection. However, the creation of a real-world dataset with precise annotations of vehicle trajectories remains an ongoing challenge, due to the difficulties and the high costs of acquiring large-scale, high-quality trajectory data.

Synthetic Datasets The problem of collecting real-world images and annotating them correctly, with an high expense of cost and time, has led to the creation of synthetic datasets. Synthetic datasets have the advantage of being fully controllable. They allow the generation of diverse traffic scenarios and perfectly annotate frames since the attributes of each object in the scene, like position, size, and movement, are known to the generating environment. This, without additional costs or extra time, ensures consistency and eliminates human errors in the annotation process. VIPER [18] and SYNTHIA [19] exploited game engines like Grand Theft Auto V or Unity to generate synthetic datasets. IDDA [2], SHIFT [22], AIODrive [26] and TUMTraf Synthetic [29] instead rely on CARLA [8], a popular simulation environment built on top of Unreal Engine 5, to create large-scale synthetic datasets for a variety of tasks, addressing the problem of the lack of various domains in a single dataset. Similarly, SynTraC [5] and Omni-IMOT [21] are among the CARLA-based datasets the most closely relate to our work, as they focus specifically on intersections. However, a significant limitation of synthetic datasets lies in the realism of vehicle movements. CARLA or game engines fail to recreate realistic trajectories since their movement relies on rule-based approaches that cannot model the unpredictability of human decision-making. This discrepancy poses important challenges in developing Intelligent TMS that are only trained on synthetic data, as models may struggle to generalize to real-world traffic scenarios.

Table 1. Comparison of size, nature and additional annotations of existing perception datasets. R3ST is the only dataset that leverages the advantages of synthetic data generation, combining it with realistic trajectories.

Dataset	Year	Size	Real/Synthetic	POV	Depth	Instance Segmentation	Realistic Trajectories
KITTI [10]	2012	41K frames	Real	On-Board	✗	✗	✗
nuScenes [25]	2019	40K frames	Real	On-Board	✗	✗	✗
SinD [28]	2019	7 hours	Real	Drone	✗	✗	✓
Argoverse2 [27]	2023	2M frames	Real	On-Board	✓	✗	✓
TUMTraf-I [30]	2023	4.8K frames	Real	Street Camera	✗	✗	✓
SYNTHIA [19]	2016	13.4K frames	Synthetic	Street Camera	✓	✓	✗
SynTraC [5]	2024	6 hours	Synthetic	Street Camera	✗	✗	✗
Omni-MOT [21]	2020	14M+ frames	Synthetic	Street Camera	✗	✗	✗
SHIFT [22]	2022	2.5M frames	Synthetic	On-Board	✗	✓	✗
R3ST	**2024**	80K+ frames	Synthetic	**Street Camera**	✓	✓	✓

3   The R3ST Dataset

Here we introduce the R3ST Dataset, a synthetic dataset that encompasses diverse urban intersections, multiple camera angles, and five vehicle types: cars, trucks, buses, motorcycles, and bicycles. This dataset has been created to leverage the benefits of synthetic data while avoiding the artificial driving behaviors typical of game engines and driving simulators like Unity or CARLA. Our dataset also provides rich annotations, including instance segmentation, depth information, and bounding boxes for object detection and object tracking. R3ST consists of photo-realistic rendering of two different intersections, each with four views, for more than 80K frames. In Table 1 we report the properties of R3ST compared with the most famous urban scene understanding datasets, highlighting that our dataset is the only one that combines the synthetic nature of the images with real trajectories.

3.1   Virtual World Generation

R3ST has been generated by rendering the virtual intersections created with Blender [6]. The motivation behind the choice of this 3D editor is the high level of freedom that it offers in creating any kind of virtual environment and the ease with which it allows the integration of external realistic trajectories. In fact, unlike typical synthetic datasets where vehicle motion is dictated by AI-driven or

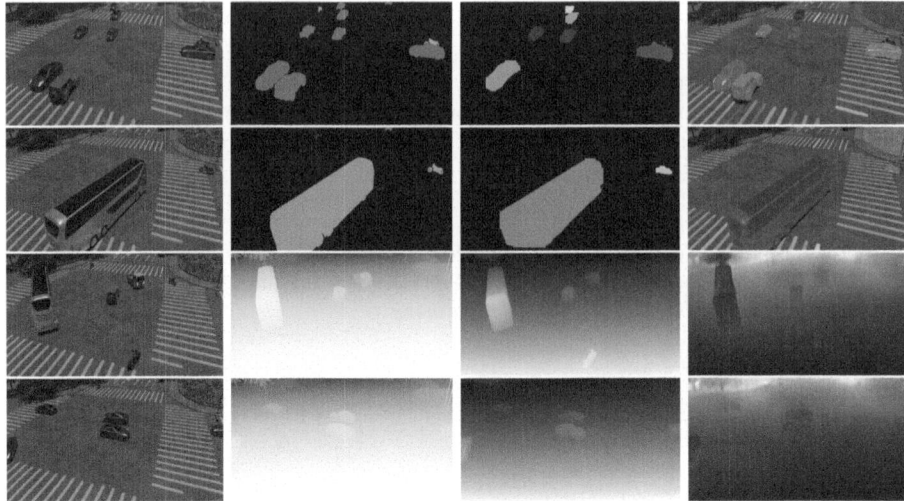

Fig. 3. Qualitative results for instance segmentation (top two rows) and monocular depth estimation (bottom two rows). The first column shows RGB frames, the second column contains ground truth annotations, while the third and fourth columns present model predictions. Instance segmentation results are obtained using YOLO-Seg [12] and SAM2 [17] online demo, while monocular depth estimation is performed with AnyDepth [11] and Pixelformer Large [1] pre-trained on KITTI [10].

rule-based algorithms, R3ST incorporates real-world vehicle trajectories derived from two of the four scenarios proposed by SinD [28], a bird's-eye-view dataset with precise annotation of vehicle positions extracted from real drone footage. As visible in Fig. 2, which shows the variance of clustered trajectories, this approach ensures that the motion patterns of vehicles in R3ST replicate the real-world traffic scenarios. For each trajectory of SinD, we associated the corresponding vehicle type. If the vehicle is a car we randomly selected among three different car meshes. To increase the realism further, we carefully selected materials for each mesh to achieve a visual fidelity comparable to real-world data. We modeled two of the four scenarios presented by SinD: an intersection in Tianjin city and an intersection in Chongqing city. These virtual environments were created from scratch in Blender, utilizing external libraries such as BlenderKit [24] to import free, realistic materials, 3D vehicle models, and building models. Additionally, we employed free add-ons like Sampling Tree Generator [14], which uses a procedural algorithm to generate vegetation. This algorithm dynamically creates trees based on user-defined parameters. Furthermore, to simulate images captured from road cameras, we positioned them on light poles, setting the vertical sensor at 22 mm, the field of view (FOV) to 35.3°, and the F-Stop to 2.8. In the reconstructed scenes, four cameras were placed, each framing one of the four directions of traffic at the respective intersections. Vehicle animations were then rendered from the perspective of each camera.

Multimodal Annotations. To enhance the usability of R3ST we leveraged Vision Blender [4] to compute additional multimodal annotations that can be used in a range of computer-vision applications. These per-frame annotations add information that goes beyond the scope of trajectory forecasting and are useful for different tasks like instance segmentation and monocular depth estimation (MDE). Additionally, we directly derive the 3D bounding box of each object in the scene from the Blender World Environment and project them on the image plane to get the 2D bounding boxes. We organize these annotations in YOLO format, where every object is encoded by its class label, the normalized center coordinates (cx, cy), along with the normalized width and height of the bounding box.

4 Experiments

We leverage multimodal annotations to demonstrate the applicability of real world tasks to our dataset. We aim to assess how pre-trained models perform on R3ST in tasks like Object Classification, for which we show the results in Table 2, and Monocular Depth Estimation and Instance Segmentation, for which we present results in Fig. 3.

To address object detection, we tested YOLO11-large directly on R3ST, but its performance was uneven, with good precision for cars but low on motorcycles, probably due to the poor quality of the mesh used. To mitigate this, we fine-tuned YOLO11-large on R3ST for 15 epochs to show how a model trained on real images can generalize fast on our dataset. In Table 2 we report the resulting mAP@50 and mAP@50–95.

Table 2. Object detection performance of YOLO11-large on the R3ST test set, composed of 900 images.

Class	mAP@50	mAP@50–95
Car	0.995	0.986
Van	0.978	0.925
Motorcycle	0.990	0.972
Bicycle	0.994	0.975
Overall	0.989	0.965

5 Conclusion and Future Works

In this paper, we introduced R3ST, the first synthetic dataset that integrates real-world vehicles trajectories, bridging the gap between realism and the flexibility of synthetic environments. We believe our dataset is valuable especially for trajectory forecasting, as well as other urban scene understanding tasks. Future works will focus on expanding the dataset with more scenes and increasing the realism of them by including different type of models for each vehicle type, as well as more challenging light conditions, to better address the domain shift problem.

Acknowledgments. This study has been partially supported by Italian Ministry of Enterprises and Made in Italy (Ministero delle Imprese e del Made in Italy - MIMIT) with the project PMDI under the agreements for innovation in the automotive sector D.M. 31/12/2021 and DD 10/10/2022.

Partially financed by the European Union - Next Generation EU, Mission 4 Component 1 CUP B53C23003540006. This research was partially carried out by Claudia Melis Tonti within the framework of the National Ph.D. Program funded by the Italian PNRR (Mission 4, Component 1), under Ministerial Decree No. 118/2023, in the Department of Computer, Control and Management Engineering 'Antonio Ruberti' - Sapienza University of Rome.

References

1. Agarwal, A., Arora, C.: Attention attention everywhere: monocular depth prediction with skip attention (2022). https://arxiv.org/abs/2210.09071
2. Alberti, E., Tavera, A., Masone, C., Caputo, B.: IDDA: a large-scale multi-domain dataset for autonomous driving. IEEE Robot. Autom. Lett. **5**(4), 5526–5533 (2020)
3. Behley, J., et al.: SemanticKITTI: a dataset for semantic scene understanding of LiDAR sequences. In: Proceedings of the IEEE/CVF International Conference on Computer Vision (ICCV) (2019)
4. Cartucho, J., Tukra, S., Li, Y., S. Elson, D., Giannarou, S.: VisionBlender: a tool to efficiently generate computer vision datasets for robotic surgery. In: Computer Methods in Biomechanics and Biomedical Engineering: Imaging & Visualization, pp. 1–8 (2020)
5. Chen, T., et al.: SynTrac: a synthetic dataset for traffic signal control from traffic monitoring cameras. arXiv preprint arXiv:2408.09588 (2024)
6. Community, B.O.: Blender - a 3D modelling and rendering package. Blender Foundation, Stichting Blender Foundation, Amsterdam (2018). http://www.blender.org
7. Cordts, M., et al.: The cityscapes dataset for semantic urban scene understanding. In: Proceedings of the IEEE Conference on Computer Vision and Pattern Recognition (CVPR) (2016)
8. Dosovitskiy, A., Ros, G., Codevilla, F., Lopez, A., Koltun, V.: CARLA: an open urban driving simulator (2017). https://arxiv.org/abs/1711.03938
9. Du, D., et al.: The unmanned aerial vehicle benchmark: object detection and tracking. In: Proceedings of the European Conference on Computer Vision (ECCV), pp. 370–386 (2018)
10. Geiger, A., Lenz, P., Stiller, C., Urtasun, R.: Vision meets robotics: the KITTI dataset. Int. J. Robot. Res. (IJRR) (2013)
11. He, X., Guo, D., Li, H., Li, R., Cui, Y., Zhang, C.: Distill any depth: distillation creates a stronger monocular depth estimator (2025). https://arxiv.org/abs/2502.19204

12. Jocher, G., Qiu, J.: Ultralytics YOLO11 (2024). https://github.com/ultralytics/ultralytics
13. Katariya, V., Noghre, G.A., Pazho, A.D., Tabkhi, H.: A POV-based highway vehicle trajectory dataset and prediction architecture. arXiv preprint arXiv:2303.06202 (2023)
14. Kegler, L.: Sampling tree generator - blender addon (2017)
15. Krajewski, R., Bock, J., Kloeker, L., Eckstein, L.: The highD dataset: a drone dataset of naturalistic vehicle trajectories on German highways for validation of highly automated driving systems. In: 2018 21st International Conference on Intelligent Transportation Systems (ITSC), pp. 2118–2125. IEEE (2018)
16. Moers, T., Vater, L., Krajewski, R., Bock, J., Zlocki, A., Eckstein, L.: The exiD dataset: a real-world trajectory dataset of highly interactive highway scenarios in Germany. In: 2022 IEEE Intelligent Vehicles Symposium (IV), pp. 958–964 (2022). https://doi.org/10.1109/IV51971.2022.9827305
17. Ravi, N., et al.: SAM 2: segment anything in images and videos (2024). https://arxiv.org/abs/2408.00714
18. Richter, S.R., Hayder, Z., Koltun, V.: Playing for benchmarks (2017). https://arxiv.org/abs/1709.07322
19. Ros, G., Sellart, L., Materzynska, J., Vazquez, D., Lopez, A.M.: The synthia dataset: a large collection of synthetic images for semantic segmentation of urban scenes. In: Proceedings of the IEEE Conference on Computer Vision and Pattern Recognition, pp. 3234–3243 (2016)
20. Sun, P., et al.: Scalability in perception for autonomous driving: Waymo open dataset. In: Proceedings of the IEEE/CVF Conference on Computer Vision and Pattern Recognition, pp. 2446–2454 (2020)
21. Sun, S., Akhtar, N., Song, X., Song, H., Mian, A., Shah, M.: Simultaneous detection and tracking with motion modelling for multiple object tracking. In: Vedaldi, A., Bischof, H., Brox, T., Frahm, J.-M. (eds.) ECCV 2020. LNCS, vol. 12369, pp. 626–643. Springer, Cham (2020). https://doi.org/10.1007/978-3-030-58586-0_37
22. Sun, T., Segu, M., Postels, J., Wang, Y., Van Gool, L., Schiele, B., Tombari, F., Yu, F.: SHIFT: a synthetic driving dataset for continuous multi-task domain adaptation. In: Proceedings of the IEEE/CVF Conference on Computer Vision and Pattern Recognition, pp. 21371–21382 (2022)
23. Sun, Y., Cao, B., Zhu, P., Hu, Q.: Drone-based RGB-infrared cross-modality vehicle detection via uncertainty-aware learning. IEEE Trans. Circuits Syst. Video Technol. 32(10), 6700–6713 (2022)
24. Tichý, J.: Blenderkit - online asset library for blender (2017). https://www.blenderkit.com
25. Vora, S., Lang, A.H., Helou, B., Beijbom, O.: PointPainting: sequential fusion for 3D object detection (2020). https://arxiv.org/abs/1911.10150
26. Weng, X., Man, Y., Cheng, D., Park, J., O'Toole, M., Kitani, K.: All-in-one drive: a large-scale comprehensive perception dataset with high-density long-range point clouds. arXiv (2020)
27. Wilson, B., et al.: Argoverse 2: next generation datasets for self-driving perception and forecasting (2023). https://arxiv.org/abs/2301.00493
28. Xu, Y., et al.: SIND: a drone dataset at signalized intersection in China. In: 2022 IEEE 25th International Conference on Intelligent Transportation Systems (ITSC), pp. 2471–2478. IEEE (2022)

29. Zhou, X., et al.: Warm-3D: a weakly-supervised sim2real domain adaptation framework for roadside monocular 3D object detection. In: 2024 IEEE 27th International Conference on Intelligent Transportation Systems (ITSC), pp. 3489–3496 (2024). https://doi.org/10.1109/ITSC58415.2024.10919929

30. Zimmer, W., Creß, C., Nguyen, H.T., Knoll, A.C.: TUMTraf intersection dataset: all you need for urban 3D camera-lidar roadside perception. In: 2023 IEEE 26th International Conference on Intelligent Transportation Systems (ITSC), pp. 1030–1037 (2023). https://doi.org/10.1109/ITSC57777.2023.10422289

Adaptive Meshes in Graph Neural Networks for Predicting Sea Surface Temperature Through Remote Sensing

José G. Reyes[1] , Giovanny A. Cuervo-Londoño[2] ,
and Javier Sánchez[1(✉)]

[1] IU Cibernética, Empresas y Sociedad (IUCES), University of Las Palmas de Gran
Canaria, Las Palmas de Gran Canaria, Spain
jose.reyes121@alu.ulpgc.es, jsanchez@ulpgc.es
[2] IU Investigación en Acuicultura Sostenible y Ecosistemas Marinos (ECOAQUA),
University of Las Palmas de Gran Canaria, Las Palmas de Gran Canaria, Spain
giovanny.cuervo101@alu.ulpgc.es

Abstract. Accurate sea surface temperature (SST) forecasting is key
for understanding marine and climatic dynamics, but remains challenging in high-variability regions such as coastal zones. Deep learning techniques have recently surpassed traditional numerical methods in computational efficiency and accuracy in prediction tasks. In particular,
graph neural networks (GNNs) have demonstrated outstanding performance in forecasting climate variables and are attracting interest for
modeling ocean dynamics. This work aims to adapt a GNN, originally
designed for atmospheric data, to predict the temperature at the ocean
surface. However, this type of neural network typically relies on regular
meshes, which struggle to capture nonlinear oceanographic processes.
Therefore, we propose to use a physically-informed mesh that adapts
node density based on the bathymetry of the sea, prioritizing coastal
areas. Our method integrates satellite-derived SST data with flexible
graph topologies by restructuring latent representations through physics-
aware graphs. The model is optimized with the L4 SST satellite images
dataset from the Copernicus Marine Service. The results demonstrate
that adaptive meshes reduce forecasting errors compared to regular grids,
particularly near the coast. This approach bridges geospatial data and
graph-based learning, showing that node allocation based on static forcings enhances model performance. The results highlight the potential of
geometric deep learning for operational oceanography, offering improved
interpretability and accuracy in complex geophysical systems based on
remote sensing.

Keywords: Graph Neural Networks · Adaptive Mesh · Deep
Learning · Sea Surface Temperature · Remote Sensing

M. Castrillón-Santana et al. (Eds.): CAIP 2025, LNCS 15622, pp. 361–372, 2026.
https://doi.org/10.1007/978-3-032-05060-1_31

1   Introduction

Forecasting SST is a fundamental task in physical oceanography, with broad applications ranging from climate modeling and marine ecosystem management to fisheries and extreme event prediction. Accurate SST predictions are especially significant in coastal regions, where dynamic processes such as upwelling, eddies, and thermal fronts exhibit complex and nonlinear behavior. While physically grounded, traditional numerical models often face limitations in computational cost and spatial resolution, particularly in high-variability areas where fine-grained predictions are needed.

In recent years, deep learning approaches, and specifically Graph Neural Networks (GNNs), have emerged as a promising alternative, offering flexible data representations and powerful capabilities to learn spatiotemporal patterns from remote sensing data [9,17].

GNNs have demonstrated remarkable progress in various geophysical forecasting tasks, including weather prediction, traffic flow estimation, and now increasingly in oceanography. These models operate on graph-structured data, enabling them to model irregular domains and capture long-range spatial dependencies more effectively than traditional convolutional architectures. Recent works such as GraphCast [8], SeaCast [6], and GNN-Surrogate [16] have shown that GNNs can outperform conventional models, especially when deployed on large-scale Earth system datasets. However, a common limitation of most existing GNN-based prediction models is their reliance on fixed, uniform spatial discretizations, which fail to capture the spatial heterogeneity of ocean processes, particularly near the coast [15].

Recent advancements in adaptive graph techniques have addressed these shortcomings: i) Adaptive Graph Learning: Models like AGLNM [17] dynamically update the adjacency matrix during training via a graph loss mechanism, uncovering hidden spatial dependencies to improve SST prediction accuracy. SD-LPGC [9] further constructs static and dynamic graphs to capture long- and short-term spatial patterns with personalized convolutional filters; ii) Hierarchical and Multiscale Mesh Adaptation: To balance resolution and computational cost, GNN-Surrogate [10,16] employs unstructured hierarchical meshes with adaptive resolutions, enabling efficient exploration of ocean simulation parameter spaces. Multiscale GNN architectures with Adaptive Mesh Refinement (AMR) [14] mimic multigrid solvers, locally refining nodes in regions that require high fidelity. iii) Attention Mechanisms and Memory Networks: GMSAN [15] uses multi-head self-attention to extract global dependencies and build adaptive graphs among multiple SST points. Graph Memory Neural Networks (GMNN) [10] integrate a graph encoder with temporal LSTM modules to handle irregular SST regions and missing-data zones caused by islands or coastlines.

Graph Convolution Networks (GCN) coupled with Computational Fluid Dynamics (CFD) have demonstrated the ability to predict previously unseen flow behaviors [4], and tools like MeshCNN [5] and PolyGen [12] have advanced direct mesh analysis and generation with high accuracy and efficiency.

Anisotropic remeshing techniques [11] refine meshes based on geometric complexity and velocity gradients, concentrating resolution where needed. Moreover, studies reveal an inverse relationship between node count in Graph Element Networks (GENs) [1] and RMSE, with strategic node placement in high-complexity regions offering superior accuracy under limited computational budgets.

This paper addresses these issues by introducing a novel, physically-informed, mesh strategy for GNN-based SST forecasting. Instead of relying on uniformly distributed nodes, we propose a graph construction paradigm that adapts the spatial resolution based on the bathymetry. Our approach increases node density in coastal regions, where SST dynamics are more complex, and reduces it in deep ocean areas with more stable thermal patterns. This leads to the development of adaptive meshes that enhance the representational power of GNNs without incurring additional computational costs.

Our contributions are threefold: i) we propose a new mesh construction method where node density is inversely related to ocean depth, guided by bathymetric data, ensuring more accurate representation of SST dynamics in coastal zones; ii) we embed these adaptive meshes within a bipartite graph neural network inspired by GraphCast and SeaCast, incorporating a multi-resolution structure for efficient spatial information propagation; iii) we conduct experiments on the Canary current coastal upwelling system using the L4 SST dataset from the Copernicus Marine Service. Results show that our bathymetry-based mesh improves the performance of uniform mesh baselines, particularly in some regions near the coast.

The paper is organized as follows: Sect. 2 introduces the dataset and the architecture of our graph neural network, including the design of the proposed adaptive meshes. Section 3 presents the experimental results, evaluating the impact of different mesh strategies. Finally, Sect. 4 discusses implications, limitations, and directions for future research.

2   Methods and Data

To address the inherent limitations of regular grids in predicting dynamic coastal phenomena, we developed a novel methodology centered on adapting the underlying graph topology guided by relevant physical forcings. Our starting point is an initial graph with a regular grid structure, upon which we apply two successive transformations designed to optimize spatial representativeness in those regions where ocean dynamics exhibit greater complexity.

2.1   High Resolution SST Dataset

We use the High Resolution Level 4 Reprocessed SST dataset provided by the Copernicus Marine Service [2]. Our analysis focuses on the region of the Canary Islands and the northwest African coast, which is included within the spatial extent of this product.

This dataset provides daily gap-free SST maps classified as a Level 4 product. The horizontal resolution of these maps is $0.05° \times 0.05°$, allowing for detailed analysis of temperature structures in our area of interest.

The temporal extent of this dataset spans from January 1, 1982, to December 31, 2023, providing us with an extensive time series for training and evaluating our forecasting models. The product represents a daily-mean SST field at 20 cm depth. The data used to generate this product comes from satellite observations of the European Space Agency Sea Surface Temperature Climate Change Initiative (ESA SST CCI) for the period 1982-2016 and from the Copernicus Climate Change Service (C3S) L3 product for the period 2017 to present. The processing level is 4, meaning it is a spatially complete analysis.

We selected the spatial extent from latitude $19.55°$ to $34.52°$ N and longitude $-20.97°$ to $-5.98°$ E, which includes the Canary Islands and the northwest African coast, the focus of our study. This area is a subset of the original product's full domain. This dataset is updated annually and is available in NetCDF-4 format. This high-resolution and reprocessed dataset from the Copernicus Marine Service provides a solid and reliable basis for investigating the forecasting of SST in our area of interest.

2.2   Architecture of the Graph Neural Network

Our graph neural network architecture draws inspiration from the groundbreaking GraphCast model [8], which achieved major advancements in machine learning-based weather forecasting. Building on GraphCast's conceptual framework, Neural-LAM [13] introduced a similar encode-process-decode structure, as shown in Fig. 1, but employed a distinct bipartite graph design. This approach was later adapted by SeaCast [6] for specialized oceanographic applications [3]. Our work builds on the bipartite graph structure proposed by Neural-LAM and SeaCast, adapting it to model the complex, non-linear dynamics of the ocean.

The model works in an autoregressive way, $\hat{\mathbf{x}}^{t+1} = f(\mathbf{x}^t, \mathbf{x}^{t-1})$, where a future SST state $\hat{\mathbf{x}}^{t+1}$ is estimated from two previous time instants $\mathbf{x}^t$ and $\mathbf{x}^{t-1}$, mapped from a latitude-longitude grid into a graph-based representation.

The bipartite graph is denoted as $\mathcal{G}$ ($\mathcal{V}^g$, $\mathcal{V}^m$, $\mathcal{E}^m$, $\mathcal{E}^{g2m}$, $\mathcal{E}^{m2g}$), comprising two distinct sets of nodes: Grid nodes ($\mathcal{V}^g$) that are arranged in a regular grid, directly corresponding to the spatial structure of the input SST data given in matrix format. They serve as the initial representation of data; Mesh nodes ($\mathcal{V}^m$) that are organized in a hierarchical multi-level graph with three levels of resolution (L_1, L_2, L_3). Each subsequent coarser level contains progressively fewer nodes and longer-range connections, allowing for the representation of multiscale oceanographic processes; see Fig. 2. The focus of our contribution lies in adapting this multi-level mesh component.

Edges within the mesh connect its nodes ($\mathcal{E}^m$), while edges $\mathcal{E}^{g2m}$ and $\mathcal{E}^{m2g}$ facilitate information exchange between the grid nodes and the mesh nodes.

The *encoder* transforms the input data from the grid into the mesh with information about the location of each node. The *processor* comprises several layers that facilitate efficient local and global information propagation through

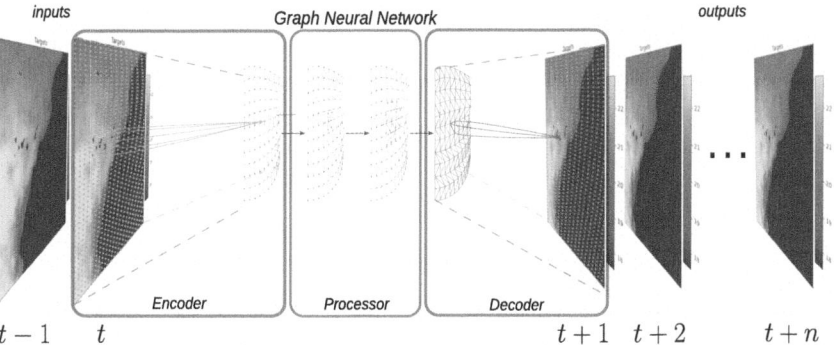

Fig. 1. The *encoder-processor-decoder* structure: The *encoder* embeds the information from the grid into the mesh; multiple message-passing steps are carried out in the layers of the *processor* to transform node features; the SST forecast is obtained at the *decoder*, which converts the mesh features into grid values. The grid, $\mathcal{V}^g$, is a regular map that contains the input SST data. The mesh, $\mathcal{V}^m$, processes the node features, sending information through the edges, $\mathcal{E}^m$. Grid pixels are sent to the mesh nodes through grid-to-mesh edges, $\mathcal{E}^{g2m}$, and back through mesh-to-grid edges, $\mathcal{E}^{m2g}$, represented in green lines.

message-passing mechanisms. Subsequently, the *decoder* projects the processed features back to the spatial grid, predicting the future state as a residual update. The output of the decoder can then be fed into the system in an autoregressive way to forecast the SST over multiple days.

The training process is organized in two phases: In the first one, batches of three consecutive time instant samples are organized randomly, and the parameters of the network are calculated using the AdamW optimizer during 150 epochs; In the second phase, the samples are increased to contain up to four output estimates, so that the network can predict multiple forecasts in the future, optimizing the parameters for each lead time. The dataset is split into four years for the training set, one year for validation, and one year for the test set. Although the dataset covers 40 years, we selected 6 years due to the complexity of the neural network and our computational resources, which is enough to assess the model's performance.

One benefit of this model is that it can produce skillful medium-range forecasts significantly faster than traditional systems, achieving a full 10-day global forecast in under one minute on a single TPU device.

2.3 Mesh Configurations

In this work, we investigate the impact of various mesh configurations, $\mathcal{V}^m$, on the predictive performance of our GNN, particularly in the context of coastal upwelling. To overcome the limitations of the static grid configurations, particularly in capturing the complex dynamics of the coast, we developed a physically informed node densification mesh. We start with an initial regular grid and apply successive transformations to optimize spatial representativeness.

We compare the mesh configurations, each utilizing the same number of nodes at each corresponding level $\{L_1 : 3548, L_2 : 394, L_3 : 45\}$ for a fair comparison across our hierarchical graph structure. Table 1 shows the statistics of the three meshes.

Table 1. Mesh Statistics for each level (L_1, L_2, L_3) and configuration (Uniform, Random, Bathymetry-based), representing the number of nodes and edges, and the number of connections (min, max, avg) between nodes. At L_1, all three meshes have 3568 nodes, but the Uniform mesh has significantly more edges. The number of nodes decreases to 394 (L_2) and 45 (L_3), with the U-mesh maintaining approximately 20% and 10% more edges than the R-mesh and B-mesh at these levels and a higher average degree.

Mesh size	Level 1			Level 2			Level 3		
	U-mesh	R-mesh	B-mesh	U-mesh	R-mesh	B-mesh	U-mesh	R-mesh	B-mesh
Nodes	3568	3568	3568	394	394	394	45	45	45
Edges	27578	21216	21192	2848	2298	2248	268	242	208
Min degree	4	6	4	4	6	4	4	6	4
Max degree	16	24	26	16	22	22	16	16	20
Avg degree	15.46	11.89	11.88	14.46	11.66	11.41	11.91	10.76	9.24

Uniform Grided Mesh (U-mesh): Following the approach in [6,13], this mesh is adapted to the geometry of the coastline, with nodes distributed regularly over the ocean domain and edges connecting the closest neighbors, forming potentially intersecting rectangular triangles (Fig. 2). This contrasts with more geometrically regular strategies employing non-intersecting equilateral triangles [7,8].

Random Mesh (R-mesh): In this configuration, we place the nodes randomly in the mesh following a uniform distribution and create edges with Delaunay triangulation. The resulting graph has fewer edges than the U-mesh due to the absence of intersecting triangles, allowing us to assess the importance of regular node distribution and connection density.

Batimetry-Based Mesh (B-Mesh): This strategy distributes the nodes according to the bathymetry of the study area. Recognizing the higher SST variability in shallow waters, we concentrate more nodes in these areas and fewer in deeper and stable waters. Node placement follows the inverse probability of depth, $p(x) = \frac{1}{B(x)+\epsilon}$, with $B(x)$ being the bathymetry at point x and ϵ a small constant. The probability distribution function is given by $F(x) = \frac{p(x)}{\sum p(x)}$.

The hierarchical organization of all mesh configurations (U-mesh, R-mesh, and B-mesh) follows the structure proposed in [13], with connections between nodes at the same level (L_i) and across adjacent levels (L_{i-1} and L_{i+1}). Coarser levels (L_{i+1}) feature fewer nodes and long-range connections for efficient information propagation, while finer levels (L_{i-1}) have a higher node density and shorter connections for modeling local interactions; see Fig. 2 and Table 1. Additionally, in our specific implementation of the bipartite graph, we have modified the connection strategy between $\mathcal{V}^g$ and $\mathcal{V}^m$ such that each $\mathcal{V}_i^g$ is consistently

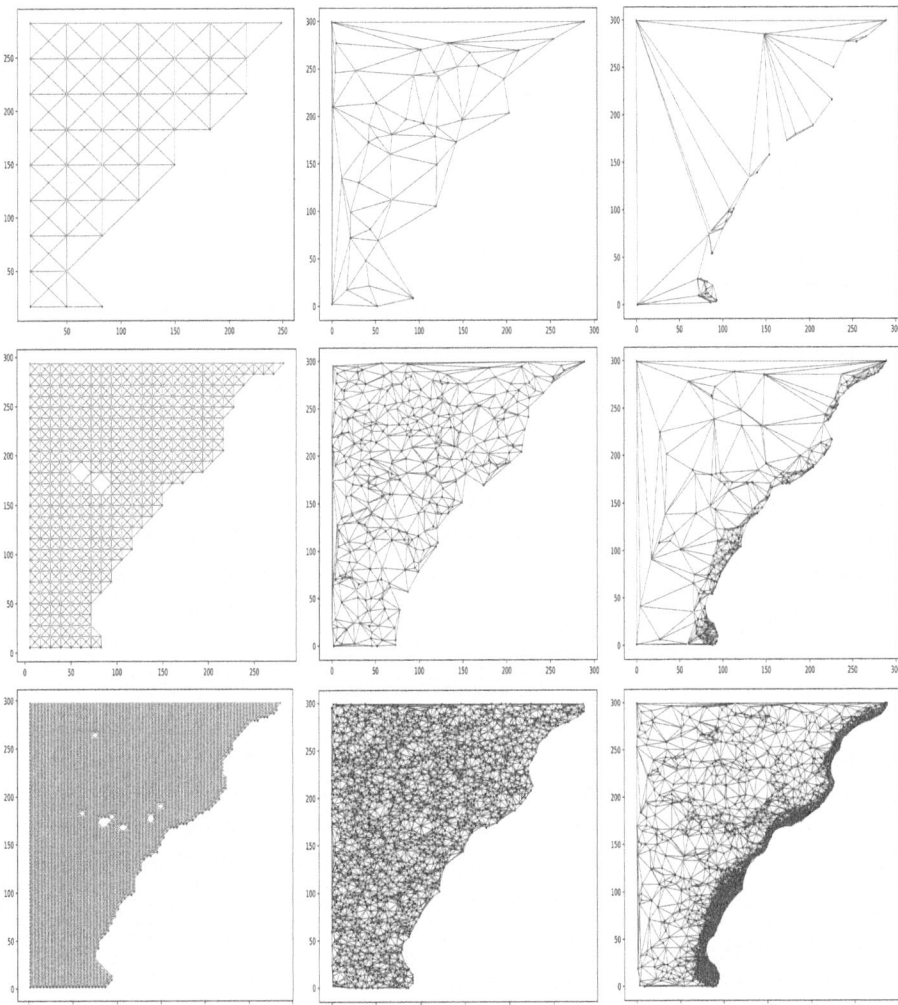

Fig. 2. Mesh configurations. The first column shows three levels of the uniform mesh (U-mesh); In the middle column, the random mesh (R-mesh); and the bathymetry-based mesh (B-mesh) on the right. The first row shows the coarsest scale, L_3, of each mesh, with fewer nodes and edges; the second row shows an intermediate scale, L_2; and the last row, the finest scale, L_1, with high density of nodes and edges.

connected to exactly three nodes in $\mathcal{V}^m$ at level L_1. This differs from the original connection scheme in [13] and ensures a consistent information flow between the two graph components.

These configurations will allow us to understand the influence of the distribution of nodes in the mesh and the number of edges between them. It will also allow us to study the accuracy improvement that results from the node density, especially in regions near the coast.

3   Results

In the experiments, the model was trained with the three mesh configurations, and we compared the RMSE score using the test dataset. A global assessment revealed that the random and bathymetry-based distributions did not yield a significant statistical reduction in RMSE over the entire domain. The mean RMSE remained virtually unchanged compared to the U-mesh baseline, indicating that the new meshes alone are insufficient to enhance overall accuracy. This is reasonable as the improvements are located in a small fraction of the domain.

However, the spatial analysis of the RMSE in Fig. 3 revealed a decrease in the upwelling area near the capes. In these locations, the RMSE improved by an average of 2.36% after applying the B-mesh. This localized improvement suggests that the concentration of nodes captures coastal processes more accurately.

Interestingly, when comparing the configurations separately by region, oceanic and coastal, with different total numbers of nodes each, we found that a sparse version, with approximately 75% fewer nodes in open waters, maintained the same level of accuracy without deteriorating the global performance. This suggests that the location of nodes is more relevant than their absolute quantity: removing nodes in areas with linear oceanic behavior does not penalize accuracy and reduces computational costs.

Although the RMSE was globally similar, the predictions made with the B-mesh showed greater spatial coherence and superior detail of mesoscalar filaments compared to the regular mesh, evidencing a more realistic resolution of upwelling structures (Fig. 4), with a moderate improvement near the capes.

These findings indicate that topological densification partially addresses the coast prediction problem; however, their global effectiveness requires a broader combination of forcings and optimization criteria to achieve consistent RMSE reductions across the entire ocean domain. The results reveal that the adaptive redistribution of nodes in geometric graphs applied to oceanographic prediction models offers limited improvements but provides valuable insights into the interaction between mesh topology, physical forcings, and local accuracy.

The localized reduction of RMSE in the capes was not only due to node density, but also to the coincidence between the adaptive mesh and a real physical pattern, underscoring that the alignment between graph topology and relevant forcings is crucial in areas where the forcing is the leading mechanism.

As anticipated in Sect. 2, densification based exclusively on bathymetry imposes a bias: it favors shallow areas without guaranteeing that these correspond to processes of interest. This bias was evident in regions where upwelling

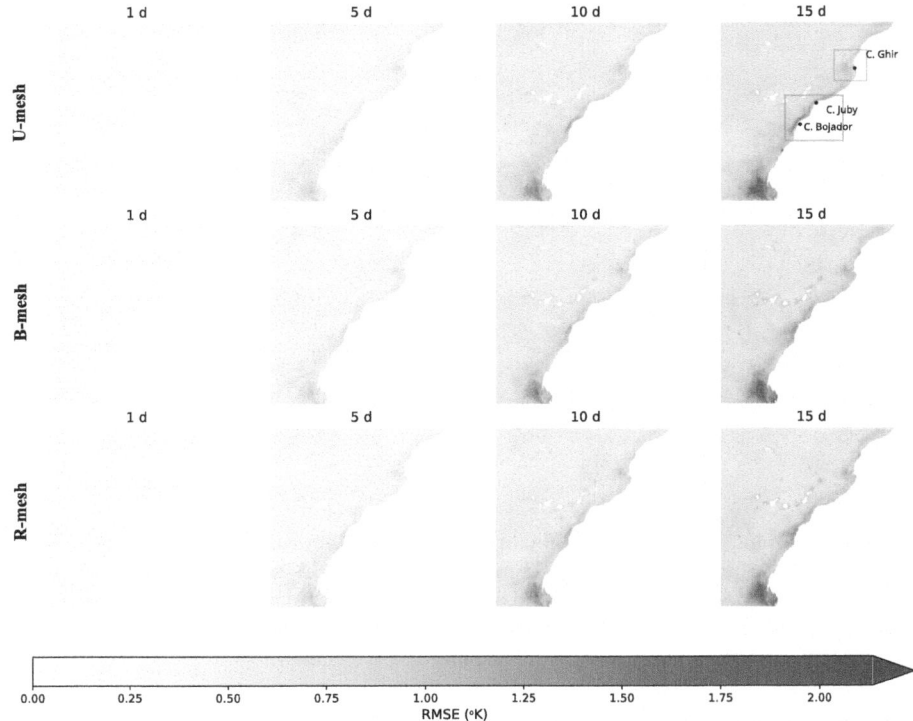

Fig. 3. Spatial RMSE maps in SST prediction for days 1, 5, 10, and 15, using the three mesh configurations. The U-mesh presents artifacts with square patterns, especially visible from day 10 onwards. In all cases, the error is mainly concentrated in coastal zones, especially near the capes. However, the B-mesh shows a qualitative improvement by reducing these localized errors and adapting better to the oceanic topography. Squares in the top-right map highlight regions where the differences are more significant (see Fig. 4 for a close-up).

showed no direct correlation with depth, resulting in an inefficient redistribution of computational resources. Therefore, although a quality improvement of the predictions was achieved, there was no significant improvement in the model's accuracy.

One important finding was that reducing the number of nodes did not compromise accuracy in regions of low variability. In the open ocean, where conditions approach a quasi-linear regime, the decrease in nodes allowed for maintaining predictive performance while reducing computational cost. This result challenges the common assumption that increasing the density improves the accuracy. Instead, it suggests that an intelligent distribution, based on the underlying dynamics, is more effective than a dense mesh.

These findings align with previous research that demonstrates higher precisions by strategically positioning nodes in areas of high nonlinearity [1, 10].

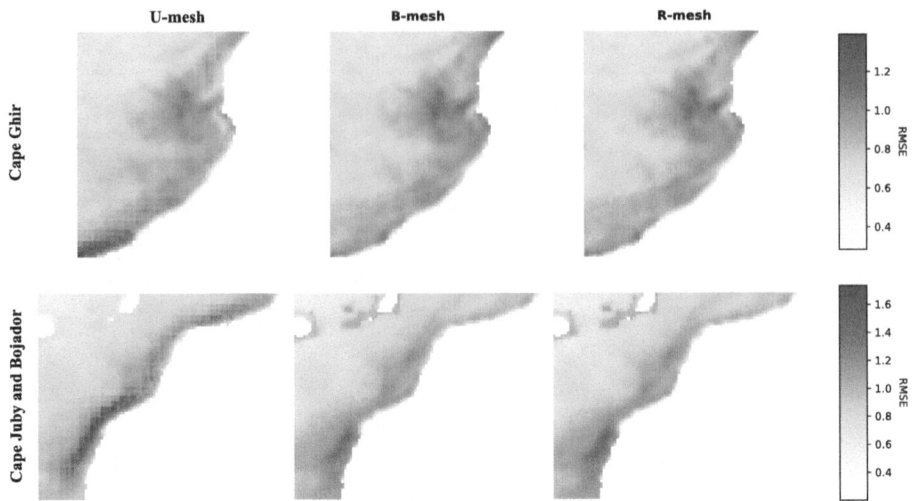

Fig. 4. RMSE in °K for the U-, B-, and R-mesh configurations across subdomains: Cape Ghir, Juby, and Bojador. The U-mesh exhibits square-shaped artifacts due to its grid configuration and higher RMSE near the coast. Some qualitative improvements are observed for the B- and R-meshes, especially in capes Juby and Bojador, as reflected in lower RMSE values. No artifacts are visible in these cases.

However, our approach emphasizes that it is not enough to guide densification by bathymetry but also by its physical dynamics.

4 Conclusion

This work presented a novel approach to SST forecasting by integrating physically informed adaptive meshes into a GNN framework. By leveraging bathymetric information to redistribute mesh node density—particularly in high-variability coastal zones—the proposed method addresses limitations inherent in uniform grid models, which often struggle to capture nonlinear ocean dynamics.

Three mesh configurations were evaluated: a uniform grid (U-mesh), a random distribution (R-mesh), and a bathymetry-based adaptive mesh (B-mesh). While global forecasting performance, as measured by RMSE, remained largely unchanged, the B-mesh demonstrated clear improvements in coastal regions and capes. It attains similar results with far fewer edges, reducing computational complexity. These localized accuracy gains underscore the value of incorporating physical priors into graph topology design, enhancing model performance where fine-scale resolution is needed.

The study revealed that reasonable node placement can match or surpass dense uniform meshes while reducing computational overhead. This finding suggests that strategic mesh adaptation offers a more efficient and scalable path for geophysical forecasting models. Qualitatively, B-mesh forecasts exhibited greater spatial coherence and finer representation of mesoscale structures.

Future work will explore multi-modal physical forcings, such as ocean currents, wind patterns, and temperature gradients, as drivers for mesh refinement. Integrating domain knowledge with geometric deep learning holds significant potential for advancing operational oceanography and climate modeling from remote sensing.

Disclosure of Interests. The authors have no competing interests to declare that are relevant to the content of this article.

References

1. Alet, F., Jeewajee, A.K., Bauza, M., Rodriguez, A., Lozano-Perez, T., Pack Kaelbling, L.: Graph element networks: adaptive, structured computation and memory. arXiv e-prints arXiv:1904.09019 (2019). https://doi.org/10.48550/arXiv.1904.09019

2. CMEMS: European north west Shelf/Iberia Biscay Irish seas - high resolution L4 sea surface temperature reprocessed (2024). https://doi.org/10.48670/moi-00153

3. Cuervo-Londoño, G.A., Sánchez, J., Rodríguez-Santana, A.: Deep learning weather models for subregional ocean forecasting: a case study on the canary current upwelling system. Technical report (2025). https://arxiv.org/abs/2505.24429

4. de Avila Belbute-Peres, F., Economon, T.D., Zico Kolter, J.: Combining differentiable PDE solvers and graph neural networks for fluid flow prediction. arXiv e-prints arXiv:2007.04439 (2020). https://doi.org/10.48550/arXiv.2007.04439

5. Hanocka, R., Hertz, A., Fish, N., Giryes, R., Fleishman, S., Cohen-Or, D.: MeshCNN: a network with an edge. arXiv e-prints arXiv:1809.05910 (2018). https://doi.org/10.48550/arXiv.1809.05910

6. Holmberg, D., Clementi, E., Roos, T.: Regional ocean forecasting with hierarchical graph neural networks. arXiv e-prints arXiv:2410.11807 (2024). https://doi.org/10.48550/arXiv.2410.11807

7. Keisler, R.: Forecasting global weather with graph neural networks. Technical report, Cornell University (2022). https://doi.org/10.48550/arXiv.2202.07575

8. Lam, R., et al.: Learning skillful medium-range global weather forecasting. Science **382**(6677), 1416–1421 (2023). https://doi.org/10.1126/science.adi2336

9. Li, X., Zhang, G., Huang, K., He, Z.: Towards spatio-temporal sea surface temperature forecasting via static and dynamic learnable personalized graph convolution network. Technical report, Cornell University (2023). https://doi.org/10.48550/arXiv.2304.09290

10. Lou, G., Zhang, J., Zhao, X., Zhou, X., Li, Q.: A non-uniform grid graph convolutional network for sea surface temperature prediction. Remote Sens. **16**(17) (2024). https://doi.org/10.3390/rs16173216

11. Narain, R., Samii, A., O'Brien, J.F.: Adaptive anisotropic remeshing for cloth simulation. ACM Trans. Graph. **31**(6), 147:1–10 (2012). https://doi.org/10.1145/2366145.2366171. Proceedings of ACM SIGGRAPH Asia 2012, Singapore

12. Nash, C., Ganin, Y., Eslami, S.M.A., Battaglia, P.W.: PolyGen: an autoregressive generative model of 3D meshes. arXiv e-prints arXiv:2002.10880 (2020). https://doi.org/10.48550/arXiv.2002.10880

13. Oskarsson, J., Landelius, T., Lindsten, F.: Graph-based neural weather prediction for limited area modeling. In: NeurIPS 2023 Workshop on Tackling Climate Change with Machine Learning (2023)
14. Perera, R., Agrawal, V.: Multiscale graph neural networks with adaptive mesh refinement for accelerating mesh-based simulations. Technical report, Cornell University (2024). https://doi.org/10.1016/j.cma.2024.117152
15. Sheng, L., Xu, L., Yu, J., Li, Z.L.: A graph multi-head self-attention neural network for the multi-point long-term prediction of sea surface temperature. Remote Sens. Lett. **14**(8), 786–796 (2023). https://doi.org/10.1080/2150704X.2023.2240506
16. Shi, N., et al.: GNN-surrogate: a hierarchical and adaptive graph neural network for parameter space exploration of unstructured-mesh ocean simulations. IEEE Trans. Visual Comput. Graphics **28**(6), 2301–2313 (2022). https://doi.org/10.1109/TVCG.2022.3165345
17. Wang, T., Li, Z., Geng, X., Jin, B., Xu, L.: Time series prediction of sea surface temperature based on an adaptive graph learning neural model. Future Internet **14**(6) (2022). https://doi.org/10.3390/fi14060171

Beyond Splines: Legendre and Shmaliy Polynomial-KANs for Robust Image Recognition

José Carlos Moreno-Tagle[1,2](✉)(iD), Jimena Olveres[2,3](iD),
and Boris Escalante-Ramírez[2,3](iD)

[1] Posgrado en Ciencia e Ingeniería de la Computación, Universidad Nacional
Autónoma de México, Mexico City, Mexico
jcmt.089@gmail.com

[2] Departamento de Procesamiento de Señales, Facultad de Ingeniería, Universidad
Nacional Autónoma de México, Mexico City, Mexico
jolveres@cecav.unam.mx, boris@unam.mx

[3] Centro de Estudios en Computación Avanzada, Universidad Nacional Autónoma de
México, Mexico City, Mexico

Abstract. This paper introduces orthogonal polynomials into the framework of Kolmogorov-Arnold Networks (KANs) by replacing spline parameterized edges with two variants: (1) discrete Shmaliy polynomials and (2) the classical Legendre polynomials, while keeping nodes as input summation points. We evaluate these polynomial-based networks against the original spline-based KAN across three benchmarks. On Fashion MNIST, Shmaliy reaches 87.6% accuracy while Legendre achieves 88.3% (KAN: 87.3%). For chest X-ray pneumonia detection, Shmaliy achieves 93.1% while Legendre matches KAN's 95.0%. For OrganAMNIST CT scan classification, Legendre attains 96.2%, outperforming both Shmaliy (89.3%) and KAN (94.5%). The results demonstrate that Legendre polynomials match or exceed KAN performance, while Shmaliy remains competitive. This establishes orthogonal polynomials as both viable alternatives to splines in KAN architectures and a promising direction for neural network design, particularly for medical imaging tasks where Legendre polynomials show significant advantages.

Keywords: Neural network architectures · orthogonal polynomials · medical image classification

1 Introduction

The design of neural network architectures has seen remarkable innovations in recent years, with approaches that challenge traditional paradigms of how neural networks process information. Among these innovations, Kolmogorov-Arnold Networks (KANs) [8] have emerged as a particularly interesting proposal. Introduced by Liu et al. in mid 2024, KANs represent a fundamental shift in neural

© The Author(s), under exclusive license to Springer Nature Switzerland AG 2026
M. Castrillón-Santana et al. (Eds.): CAIP 2025, LNCS 15622, pp. 373–384, 2026.
https://doi.org/10.1007/978-3-032-05060-1_32

network architectures by replacing both traditional learnable weights in the edges and fixed activations on the nodes with only learnable 1-D functions on the edges and nodes which serve as input summation point.

KANs draw their theoretical foundation from the Kolmogorov-Arnold representation theorem, which states that any multivariate continuous function can be expressed as a sum of continuous univariate functions. This powerful result suggests that complex functions can be constructed from simpler, learnable components, a principle that KANs implement directly in their architecture.

While KANs typically implement these learnable functions as splines, we propose an alternative approach using Legendre and Shmaliy polynomials. These polynomials offer unique advantages for function approximation and signal processing, making them particularly well-suited for learning complex patterns in image data. Our work explores how these polynomials can be integrated into a KAN-inspired architecture to create a flexible yet robust new approach to image classification.

2   Materials and Methods

2.1   Kolmogorov-Arnold Networks Background

KANs represent a revolutionary approach to neural network design. Unlike traditional networks that use learnable weights on the edges and fixed activation functions on the nodes, KANs place learnable functions on the edges while nodes simply sum their inputs. This design directly implements the Kolmogorov-Arnold representation theorem, which states that any multivariate function can be expressed as:

$$f(\mathbf{x}) = f(x_1, ..., x_n) = \sum_{q=1}^{2n+1} \Phi_q \left(\sum_{p=1}^{n} \phi_{q,p}(x_p) \right) \tag{1}$$

KANs utilize continuous univariate functions $\phi_{q,p}$ and Φ_q for transformations between layers. While the original Kolmogorov-Arnold representation (Eq. 1) suggests a two-layer structure, Liu et al.'s implementation [8] extends this to arbitrary depth. Each KAN layer transforms its inputs $\mathbf{x}_l$ through a matrix $\boldsymbol{\Phi}_l$ of learnable 1-D functions:

$$\mathbf{x}_{l+1} = \boldsymbol{\Phi}_l \mathbf{x}_l \tag{2}$$

where $\boldsymbol{\Phi}_l$ contains functions $\phi_{l,j,i}$ that connect neuron i in layer l $[(l, i)]$ to neuron j in layer $l+1$ $[(l+1, j)]$. More precisely, the learnable 1-D functions $\{\phi_{l,j,i}\}$ are organized into layer-specific transformation matrices $\boldsymbol{\Phi}_l$:

$$\boldsymbol{\Phi}_l = \begin{pmatrix} \phi_{l,1,1} & \cdots & \phi_{l,1,n_l} \\ \vdots & \ddots & \vdots \\ \phi_{l,n_{l+1},1} & \cdots & \phi_{l,n_{l+1},n_l} \end{pmatrix}, \quad \begin{matrix} i = 1, \ldots, n_l \\ j = 1, \ldots, n_{l+1} \end{matrix} \tag{3}$$

Each learnable function combines trainable splines with SiLU nonlinearities:

$$\phi(x) = w_s \cdot \text{spline}(x) + w_b \cdot \text{SiLU}(x) \tag{4}$$

where:

- $\text{SiLU}(x) = x/(1 + e^{-x})$ is the sigmoid linear unit
- $\text{spline}(x) = \sum_i c_i B_i(x)$ employs $B_i(x)$ basis splines
- w_b, w_s and c_i are trainable parameters

The complete KAN network stacks L such layers:

$$\text{KAN}(\mathbf{x}) = (\boldsymbol{\Phi}_{L-1} \circ \cdots \circ \boldsymbol{\Phi}_0)\mathbf{x} \tag{5}$$

with architecture specified by node counts $[n_0, \ldots, n_L]$. This formulation naturally adapts to supervised learning tasks where we approximate $y^{(i)} \approx f(\mathbf{x}^{(i)})$ from training pairs $\{\mathbf{x}^{(i)}, y^{(i)}\}$.

KANs have begun to be incorporated into other architectures, such as transformers, where they replace traditional multilayer perceptrons (MLPs) [13]. KANs' core idea has also been extended to 2D architectures for use in convolutional networks [1]. Notably, KANs have shown exceptional performance in time series forecasting tasks, outperforming other architectures like MLPs, transformers and ConvNets [6]. Beyond their original design, KANs have been recently proposed as a bridge between artificial intelligence (AI) and fundamental science, where modern AI relies on connectionism while pure sciences are based on symbolism [7].

2.2 Legendre Polynomials

The Legendre polynomials $P_n(x)$ (Fig. 1) form a family of orthogonal polynomials defined on the interval $[-1, 1]$, with wide applications in physics and numerical methods. They admit an explicit computation through the Rodrigues formula:

$$P_n(x) = \frac{1}{2^n n!} \frac{d^n}{dx^n} (x^2 - 1)^n, \quad n = 0, 1, 2, \ldots \tag{6}$$

These polynomials obey a three-term recurrence relation:

$$P_n(x) = \frac{(2n - 1)x P_{n-1}(x) - (n - 1)P_{n-2}(x)}{n}, \quad P_0(x) = 1, \; P_1(x) = x \tag{7}$$

and satisfy the orthogonality condition:

$$\int_{-1}^{1} P_m(x)P_n(x)dx = \frac{2}{2n + 1}\delta_{mn} \tag{8}$$

where δ_{mn} is the Kronecker delta, which equals 1 if $m = n$ and 0 if $m \neq n$. Their orthogonality enables key applications ranging from solving Laplace's equation

in physics, spectral methods [14] to practical implementations in signal processing (system identification and signal approximation) [2,5] and computer vision
(image registration) [3,10], alongside their classical use in Gaussian quadrature
integration.

2.3 Shmaliy Polynomials

The Shmaliy polynomials [9] (Fig. 2) are a family of discrete orthogonal polynomials (DOP) derived from unbiased finite impulse response (UFIR) functions.
Unlike classical DOP families such as Meixner, Charlier, Hahn, and Krawtchouk
polynomials, which depend on multiple parameters, Shmaliy polynomials rely
solely on a single parameter—the length of finite data [4].

The Shmaliy polynomials $h_n(x, N)$, where n is the polynomial degree ($n \geq 1$),
x the input variable, and N the number of discrete time points (e.g., $N = 100$),
are computed by the following recurrence relation:

$$h_n(x, N) = a_n h_{n-1}(x, N) - b_n h_{n-2}(x, N) \qquad (9)$$

with initial conditions $h_{-1}(x, N) = 0$ and $h_0(x, N) = \frac{1}{N}$. The coefficients a_n and
b_n in (9) satisfy (Fig. :

$$a_n = 2\frac{n^2(2N - 1) - x(4n^2 - 1)}{n(2n - 1)(N + n)} \qquad (10)$$

$$b_n = \frac{(2n + 1)(N - n)}{(2n - 1)(N + n)} \qquad (11)$$

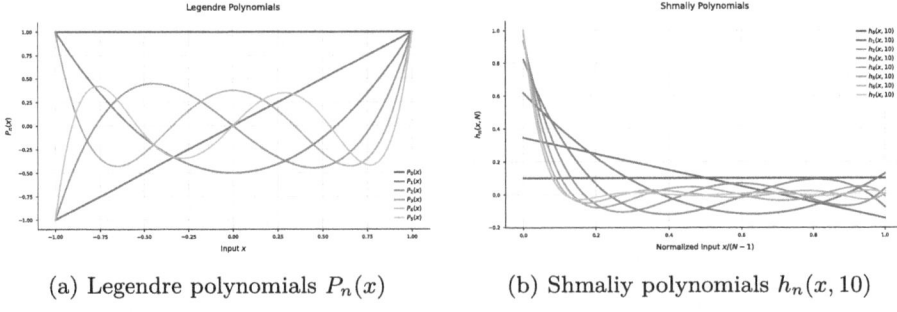

(a) Legendre polynomials $P_n(x)$ (b) Shmaliy polynomials $h_n(x, 10)$

Fig. 1. Orthogonal polynomial basis: (a) continuous Legendre polynomials and (b)
discrete Shmaliy polynomials. Both families serve as learnable transformations in our
network architecture.

Advantages and Applications. A key advantage of Shmaliy polynomials is their dependency on a single parameter, allowing for direct application without requiring extensive optimization when the data distribution is unknown. Moreover, as part of the UFIR framework, they naturally suppress noise without relying on explicit noise models. The recurrence relations defining these polynomials enable fast computation, reducing the computational burden in signal processing tasks.

These polynomials exhibit several fundamental properties that enhance their effectiveness in signal processing applications. First, they form a discrete orthogonal basis on finite intervals $[0, N - 1]$, enabling efficient representation of signals. Second, they provide unbiased estimation properties, which are crucial for accurate signal reconstruction. Third, they maintain numerical stability even for high-degree polynomials, ensuring reliable computations. Finally, they are fully specified by a single parameter N, simplifying their implementation.

The orthogonality condition for Shmaliy polynomials is expressed as:

$$\sum_{x=0}^{N-1} \rho(x, N) h_k(x, N) h_n(x, N) = d_n^2(N) \delta_{kn}, \tag{12}$$

where $\rho(x, N) = \frac{2x}{N(N-1)}$ is the weight function (a simple ramp function), and δ_{kn} is the Kronecker delta. The squared norm $d_n^2(N)$ is given by:

$$d_n^2(N) = \frac{(n + 1)(N - n - 1)_n}{N(N)_{n+1}}. \tag{13}$$

Here $(a)_0 = 1$, $(a)_k = a(a + 1) \ldots (a + k - 1)$, for $k = 1, 2, \ldots$ refers to the Pochhammer symbol, denoted as $(a)_k$.

These polynomials have proven particularly effective in applications requiring finite-horizon estimation, blind signal fitting, discrete data approximation, and real-time processing with minimal computational overhead. In our implementation, we set $N = 10$ for all experiments. This choice comes from two key considerations: First, N appears in the denominator of the recurrence relation, making smaller values preferable for numerical stability. Second, this value effectively prevents gradient vanishing during backpropagation while maintaining sufficient expressiveness. While larger N values could be explored, they risk causing vanishing gradients during network training.

2.4   Proposed Network Architecture

Our architecture builds upon the KAN framework by implementing the edge functions as Legendre or Shmaliy polynomial expressions. The network consists of five layers: one input layer, three hidden layers, and one output layer.

Both Legendre and Shmaliy networks process input data identically. Consider a Shmaliy Polynomial Neural Network trained on MNIST-like data (Fig. 2).

A 28×28 image is flattened into a 784-dimensional vector $(x_1, x_2, \ldots, x_{784})$ which serves as the input layer and then is fed into the network's first hidden

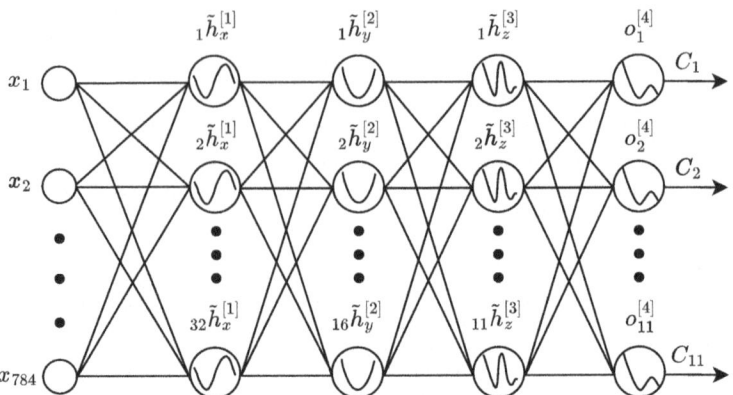

Fig. 2. Architecture of the Shmaliy neural network where $_m\tilde{h}_n^{[\ell]}$ denotes the m-th polynomial of degree n in layer ℓ. Flattened image pixels $(x_1, \ldots, x_{784})$ are transformed by 32 learnable Shmaliy polynomials in the first hidden layer $(_m\tilde{h}_n^{[1]}, m = 1, \ldots, 32)$, each producing 16 outputs for the subsequent layer. The network employs full connectivity between layers, with final outputs C_1 to C_{11} corresponding to the 11-class classification task.

layer. This layer contains 32 neurons, each implementing a learnable Shmaliy polynomial transformation.

Each neuron computes its output y_i by transforming all inputs through these polynomials:

$$y_i = \sum_{k=1}^{784} \tilde{h}_n(x_k), \tag{14}$$

where $\tilde{h}_n(x_j)$ represents the learnable Shmaliy polynomial transform:

$$\tilde{h}_n(x) = \sum_i c_i h_n(x). \tag{15}$$

Here $h_n(x)$ denotes the standard Shmaliy polynomial of degree n (initialized to $n = 4$), while c_i are trainable coefficients initialized via Xavier method and optimized with AdamW. Figure 3 illustrates this computation.

Both Legendre and Shmaliy neural networks use identical polynomial counts per layer across datasets. For Fashion MNIST, we used $[32, 16, 10, 10]$ polynomials from the first hidden layer to the output, while chest X-ray used $[32, 16, 8, 2]$ and OrganAMNIST used $[32, 16, 11, 11]$. The corresponding polynomial degrees per layer are specified in Table 1.

The KAN architectures mirror these layer dimensions exactly: $[32, 16, 10, 10]$ nodes/neurons for Fashion MNIST, $[32, 16, 8, 2]$ for Chest X-ray, and $[32, 16, 11, 11]$ for OrganAMNIST. KANs are built using learnable splines. Following the original KAN framework, we use cubic splines ($k = 3$) with a grid size of 5 throughout all network layers.

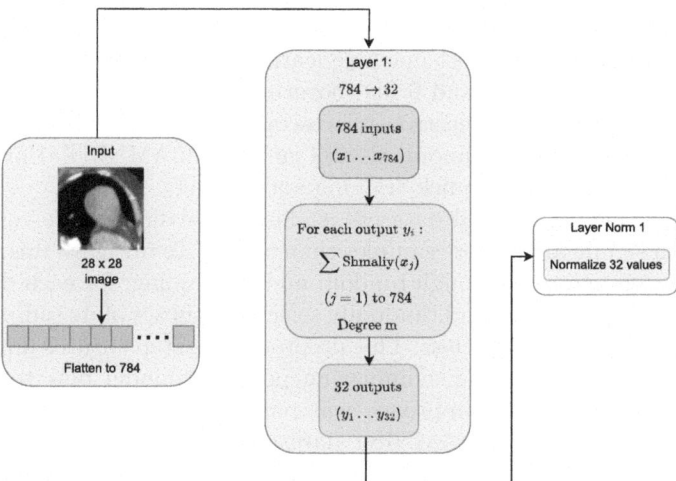

Fig. 3. Information processing in a Shmaliy polynomial neuron. The diagram illustrates the transformation of a flattened 28×28 input vector $(x_1, \ldots, x_{784})$ through the first hidden layer's Shmaliy polynomial neuron (here, $i = 32$ outputs). Each neuron computes $y_i = \sum_{j=1}^{784} \tilde{h}_n(x_j)$, where $\tilde{h}_n$ denotes the learnable Shmaliy polynomial basis.

Table 1. Architecture specifications showing hidden and output layer configurations (input layer omitted). Legendre and Shmaliy networks use orthogonal polynomials, while KANs employ cubic splines ($k = 3$) with grid size 5. All architectures contain 4 trainable layers.

Model	Specification	FashionMNIST	Chest X-ray	OrganAMNIST
Legendre	Layer config	4 layers	4 layers	4 layers
	Polynomial degrees	(4, 5, 3, 3)	(4, 5, 3, 3)	(6, 5, 4, 4)
	Polynomials per layer	32, 16, 10, 10	32, 16, 8, 2	32, 16, 11, 11
Shmaliy	Layer config	4 layers	4 layers	4 layers
	Polynomial degrees	(3, 5, 6, 3)	(3, 5, 6, 3)	(5, 5, 4, 3)
	Polynomials per layer	32, 16, 10, 10	32, 16, 8, 2	32, 16, 11, 11
KAN	Layer config	4 layers	4 layers	4 layers
	Spline parameters	$k = 3, g = 5$	$k = 3, g = 5$	$k = 3, g = 5$
	Nodes per layer	32, 16, 10, 10	32, 16, 8, 2	32, 16, 11, 11

2.5   Datasets and Experimental Setup

We evaluated our approach on three distinct datasets beginning with Fashion MNIST, a standard benchmark dataset comprising 70,000 grayscale images (60,000 training, 10,000 testing) across 10 clothing categories. All images have a resolution of 28×28 pixels. The original training/test split was used for training our neural networks.

The other two datasets come from MedMNIST [11, 12], a collection of biomedical images designed to help test machine learning models. MedMNIST includes 18 different datasets (12 2D and 6 3D) covering various medical imaging types like ultrasound, X-Ray, and electron microscope images. From MedMNIST, we specifically worked with PneumoniaMNIST and OrganAMNIST. Both datasets use 28×28 pixel images for quick training and testing.

A key challenge in medical datasets is class imbalance, where certain categories contain significantly more samples than others. To mitigate this, we implemented balanced training through random under-sampling: for each dataset, we identified the majority class and randomly selected a subset of its samples matching the count of the minority class. This preprocessing step ensures uniform class representation during training, eliminating potential model bias toward over-represented categories. This approach was particularly necessary for both the PneumoniaMNIST and OrganAMNIST datasets.

The PneumoniaMNIST dataset contains chest X-ray images for binary pneumonia detection, originally exhibiting significant class imbalance with 1,448 normal cases versus 3,884 pneumonia cases. After balancing, we obtained 1,448 images per class (2,896 total training samples).

The OrganAMNIST CT scan dataset presents an even more complex challenge with its 11-class organ classification task (including bladder, heart, liver, and other organs). Initial imbalance ranged from 2,150 to 9,449 samples per class. Our balancing procedure standardized this to 2,150 images per class (23,650 total), ensuring equal representation of all anatomical structures.

2.6 Implementation Details

Our experimental evaluation followed distinct approaches for general image recognition versus medical datasets. For Fashion MNIST, we preserved the standard 60,000/10,000 training/test split. All networks were trained for 30 epochs with a consistent batch size of 64, though we report results from the epoch achieving both peak accuracy and minimum loss. Learning rates were individually optimized for each architecture.

Medical imaging evaluation required special handling to both balance classes and maximize training data. Because 5-fold cross-validation requires repartitioning the full dataset, we combined original training and test sets before applying random undersampling, essentially matching all classes to the smallest category size. The balanced dataset was then subjected to 5-fold cross-validation, ensuring each fold maintained equal class representation while providing statistically robust performance metrics. Each fold was trained for 50 epochs with a fixed batch size of 128, while learning rates were individually optimized per network architecture. All experiments used Xavier-Glorot initialization and the AdamW optimizer.

3   Results

3.1   Fashion MNIST Classification

For Fashion MNIST classification, the Legendre polynomial network achieved 88.3% accuracy at epoch 23, using polynomial degrees $(4, 5, 3, 3)$ and $[32, 16, 10, 10]$ polynomials per layer from the first hidden layer to output. The Shmaliy network attained 87.6% accuracy at epoch 27 with degrees $(3, 5, 6, 3)$ and identical polynomial distribution across layers.

The original spline-based KAN reached 87.3% accuracy by epoch 6 (using cubic splines with $k = 3$ and grid $= 5$, node counts $[32, 16, 10, 10]$), after which test loss began increasing—indicating onset of overfitting. The complete results are presented in Table 2 and Fig. 4.

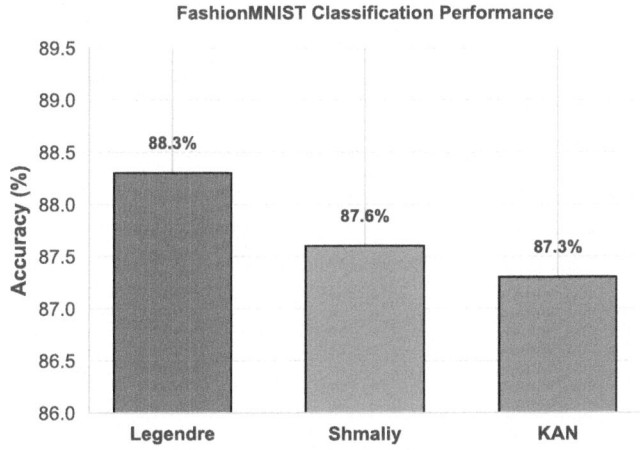

Fig. 4. Image classification performance on FashionMNIST. Results show peak accuracy achieved during 30-epoch training. Legendre and Shmaliy neural networks outperform the original spline-based KAN architecture.

Table 2. Performance comparison on FashionMNIST between Legendre, Shmaliy, and KAN models. All architectures contain 4 trainable layers, trained with Adam optimizer using Xavier-Glorot initialization.

Model	Accuracy (%)	Epochs	Learning Rate	Degrees/Parameters
Legendre	**88.3**	**23**	$\mathbf{3.0{\times}10^{-4}}$	**(4, 5, 3, 3)**
Shmaliy	87.6	27	1.0×10^{-3}	(3, 5, 6, 3), N = 10
KAN	87.3	6	$1.0 \times \mathbf{10^{-3}}$	Cubic splines (k = 3, grid = 5)

3.2   Medical Image Classification

Our polynomial-based neural networks demonstrated strong performance across all medical imaging tasks. All reported accuracies represent the mean of the highest validation accuracies achieved in each of the 5 cross-validation folds, with standard deviations calculated from the peak test accuracies achieved in each fold. Medical image classification results are summarized in Table 3 and Fig. 5.

Pneumonia Detection in Chest X-Rays. The PneumoniaMNIST dataset addresses binary pneumonia classification from chest X-ray images. Our Legendre polynomial network achieved 95.0% accuracy (±0.5% std) using polynomial degrees (4, 5, 3,3) and polynomial counts [32, 16, 8, 2] per layer. The Shmaliy variant attained 93.1% (±0.4% std) with degrees (3, 5, 6, 3) and identical layer-wise polynomial distribution. Notably, the original spline-based KAN matched Legendre's 95.0% accuracy (±0.2% std) while employing cubic splines (k = 3, grid = 5), demonstrating polynomial networks can equal state-of-the-art performance on binary medical tasks.

Multi-organ CT Scan Classification. For the more complex 11-class OrganAMNIST challenge, the Legendre network achieved superior performance at 96.2% (±0.1% std) using polynomial degrees (6, 5, 4, 4) and [32, 16, 11, 11] polynomials per layer. The Shmaliy network reached 89.3% (±0.5% std) with degrees (5, 5, 4, 3) and identical polynomial counts. The original KAN attained 94.5% (±0.3% std), placing second behind the Legendre variant. These results confirm that polynomial networks not only compete with but surpass spline-based KANs as task complexity increases, with Legendre showing a 1.7% accuracy advantage in multi-organ classification.

Table 3. Performance comparison of Legendre and Shmaliy networks versus KANs on medical image datasets (5-fold cross-validation). All model architectures consist of four trainable layers, optimized with AdamW using Xavier-Glorot initialization.

Dataset	Model	Accuracy (%) ± std. dev.	Degrees/Parameters
Pneumonia	**Legendre**	**95.0 ± 0.5**	**(4, 5, 3, 3)**
	Shmaliy	93.1 ± 0.4	(3, 5, 6, 3)
	KAN	95.0 ± 0.2	Cubic splines (k = 3, grid = 5)
OrganA	**Legendre**	**96.2 ± 0.1**	**(6, 5, 4, 4)**
	Shmaliy	89.3 ± 0.5	(5, 5, 4, 3)
	KAN	94.5 ± 0.3	Cubic splines (k = 3, grid = 5)

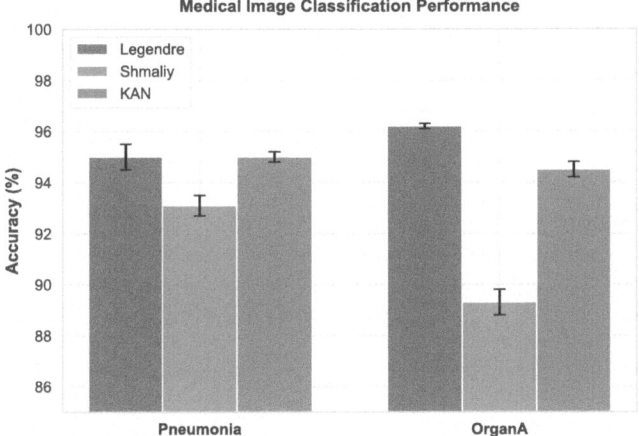

Fig. 5. Medical image classification performance comparison. Results show average accuracy ± standard deviation across 5-fold cross-validation for different models and medical datasets. Legendre neural networks match or exceed KAN performance, with error bars representing standard deviation of the peak accuracy across folds.

4   Discussion and Conclusions

Our results demonstrate that Legendre and Schmaliy polynomials can effectively serve as learnable 1D functions within a KAN-inspired architecture, achieving consistent accuracy across diverse domains. The Legendre variant attained 88.3% on FashionMNIST and 96.2% ($\pm$0.1%) on OrganAMNIST, surpassing the original KAN's 87.3% and 94.5% ($\pm$0.3%) respectively, while Shmaliy polynomials maintained competitive robustness with 93.1% ($\pm$0.4%) on PneumoniaMNIST versus KAN's 95.0%. Notably, polynomial degrees provided tunable performance: Legendre's (6,5,4,4) architecture outperformed KANs by 1.7% on OrganAMNIST, demonstrating how this approach combines architectural flexibility with outstanding performance. The strong performance across several classification tasks, particularly in medical imaging, where data quality and class imbalance often present significant challenges, suggests these polynomial based networks are particularly suited for clinical applications where accuracy gains carry practical significance. In summary, this approach demonstrates several advantages:

- These architectures maintain high accuracy across different domains, from general image classification to specialized medical imaging tasks.
- The approach shows particular robustness in handling class-balanced medical datasets.
- The polynomial degrees serve as an additional hyperparameter that can be tuned for optimal performance.

Future work will focus on three key directions:

– Developing automated methods for optimal polynomial degree selection.
– Extending the approach to diverse medical imaging modalities and clinical tasks.
– Designing hybrid architectures that combine complementary polynomial bases.

Acknowledgments. Authors wish to thank UNAM PAPIIT grants IT101624 and IN108624 for sponsoring this research. Jose Carlos would like to thank SECIHTI (Secretaría de Ciencia, Humanidades, Tecnología e Innovación), formerly CONAHCYT, for the financial support (CVU: 857565) during his PhD studies.

Disclosure of Interests. The authors have no competing interests to declare.

References

1. Bodner, A.D., Tepsich, A.S., Spolski, J.N., Pourteau, S.: Convolutional Kolmogorov-Arnold networks. arXiv preprint arXiv:2406.13155 (2024)
2. Carini, A., Cecchi, S., Romoli, L., Sicuranza, G.L.: Legendre nonlinear filters. Signal Process. **109**, 84–94 (2015)
3. Flusser, J., Zitova, B., Suk, T.: Moments and Moment Invariants in Pattern Recognition. Wiley, Hoboken (2009)
4. González, G., Nava, R., Escalante-Ramírez, B.: A comparative study on discrete shmaliy moments and their texture-based applications. Math. Probl. Eng. **2018**(1), 1673283 (2018)
5. Li, G., Wen, C.: Legendre polynomials in signal reconstruction and compression. In: 2010 5th IEEE Conference on Industrial Electronics and Applications, pp. 1636–1640. IEEE (2010)
6. Liang, Z., An, R., Fan, W., Rao, Y., Liang, Y.: iTFKAN: interpretable time series forecasting with Kolmogorov-Arnold network. arXiv preprint arXiv:2504.16432 (2025)
7. Liu, Z., Ma, P., Wang, Y., Matusik, W., Tegmark, M.: KAN 2.0: Kolmogorov-Arnold networks meet science (2024). https://arxiv.org/abs/2408.10205
8. Liu, Z., et al.: KAN: Kolmogorov-Arnold networks. arXiv preprint arXiv:2404.19756 (2024)
9. Morales-Mendoza, L.J., Gamboa-Rosales, H., Shmaliy, Y.S.: A new class of discrete orthogonal polynomials for blind fitting of finite data. Signal Process. **93**(7), 1785–1793 (2013)
10. Mukundan, R., Ramakrishnan, K.: Moment Functions in Image Analysis: Theory and Applications. World Scientific (1998)
11. Yang, J., Shi, R., Ni, B.: MedMNIST classification decathlon: a lightweight AutoML benchmark for medical image analysis. In: IEEE 18th International Symposium on Biomedical Imaging (ISBI), pp. 191–195 (2021)
12. Yang, J., et al.: MedMNIST v2-a large-scale lightweight benchmark for 2D and 3D biomedical image classification. Sci. Data **10**(1), 41 (2023)
13. Yang, X., Wang, X.: Kolmogorov-Arnold transformer. arXiv preprint arXiv:2409.10594 (2024)
14. Zheng, J., An, J.: Legendre-Fourier spectral approximation and error analysis for nonlinear eigenvalue problems in complex domains. J. Appl. Math. Comput. 1–28 (2025)

Author Index

The manufacturer's authorised representative in the EU is Springer
Nature Customer Service Centre GmbH, Europaplatz 3, 69115 Heidelberg,
Germany. If you have any concerns regarding our products, please
contact ProductSafety@springernature.com

Printed and bound by CPI Group (UK) Ltd, Croydon, CR0 4YY
29/04/2026
02099461-0015